Evolution of Networks

Evolution of Networks

From Biological Nets to the Internet and WWW

S. N. Dorogovtsev
Ioffe Institute, St Petersburg

J. F. F. Mendes
University of Porto and University of Aveiro

OXFORD
UNIVERSITY PRESS

Great Clarendon Street, Oxford OX2 6DP

Oxford University Press is a department of the University of Oxford.
It furthers the University's objective of excellence in research, scholarship,
and education by publishing worldwide in

Oxford New York

Auckland Bangkok Buenos Aires Cape Town Chennai
Dar es Salaam Delhi Hong Kong Istanbul Karachi Kolkata
Kuala Lumpur Madrid Melbourne Mexico City Mumbai Nairobi
São Paulo Shanghai Taipei Tokyo Toronto

Oxford is a registered trade mark of Oxford University Press
in the UK and in certain other countries

Published in the United States
by Oxford University Press Inc., New York

First published 2003

British Library Cataloguing in Publication Data

Data available

Library of Congress Cataloging in Publication Data

ISBN 0 19 851590 1

10 9 8 7 6 5 4 3 2 1

Typeset by the authors using LaTeX

Printed in Great Britain
on acid-free paper by
Biddles Ltd, Guildford & King's Lynn

PREFACE

This book is about the growth and structure of random networks. The book is written by physicists and presents the point of view of a physicist, but is addressed to all researchers involved in this subject and students.

Where was physics 50 years ago, and where is it now? At first sight, the role of physics is decreasing; other natural sciences are developing more rapidly. However, physics has penetrated into all sciences. A natural step for a physicist is to jump from the traditional topics of physics to new intriguing problems. Actually, our book describes a flight from physics to the new interdisciplinary field of networks. This escape is, however, still dependent on physics.

For many years the term 'random graphs' usually meant to mathematicians static, 'equilibrium' networks with a Poisson-type distribution of connections. Mathematicians have made truly great advances in the description of such networks.

Only recently have we realized that we reside in a world of networks. The Internet and World Wide Web (WWW) are changing our lives. Our physical existence is based on various biological networks. The extent of the development of communications networks is a good indicator of the level of development in a country. 'Network' turns out to be a central notion in our time, and the explosion of interest in networks is already a social and cultural phenomenon.

Graph theory has made great progress. However, the most important natural and artificial networks have a specific architecture based on a fat-tailed distribution of the number of connections of vertices that differs crucially from the 'classical random graphs' studied by mathematicians. As a rule, these networks are not static but evolving objects. Their state is far from equilibrium and their structure cannot be understood without insight into the principles of their evolution. Only in the last few years have physicists started extensive empirical and theoretical research into networks organized in such a way. Earlier, physicists' interest was rather in neural and Boolean networks where the arrangement of connections was secondary.

We think that the physics approach is the most advantageous for understanding the evolution of networks. Actually, what we physicists are now doing on this active topic is a direct generalization of the usual physics of growth, percolation phenomena, diffusion, self-organized criticality, mesoscopic systems, etc.

Our aim is to understand networks: that is, to understand the basic principles of their structural organization and evolution. We believe that this understanding is necessary to find the best solutions to the problems of real networks.

We decided to present a concise informative book which could be used even by students without a deep knowledge of mathematics and statistical physics and which would be a good source of reference material. Therefore we have tried to introduce the main ideas and concepts in as simple a manner as possible, with minimal mathematics. Special attention is given to real networks, both natural and artificial. We discuss in detail the collected empirical data and numerous real applications of existing theories. The urgent problems of communication networks are highlighted and discussed.

For a description of network evolution, we prefer to use a simpler continuum approach. We feel that it is more important to be understood than to be perfectly rigorous. Also, we follow the hierarchy of values in Western science: an experiment and empirical data are more valuable than an estimate; an estimate is more valuable than an approximate calculation; an approximate calculation is more valuable than a rigorous result. More cumbersome calculations and supplementary materials are placed in appendices. We hope that all of the results and statements that we discuss can be easily found in the text and understood without undertaking detailed calculations. Therefore, we ask our brave readers to skim over difficult pages without hesitation and not to pay any attention to footnotes. However, as this is a monograph written by theoretical physicists, we try to keep a 'physical level' of strictness in our explanations and definitions. Although, we try to avoid superfluous words, we are not afraid to repeat important statements at a different level. We hope that the book will also be useful to mathematicians, as a source of interesting new objects and ideas.

We thank our friends and colleagues for their help. Foremost among these are our collaborators in this field: Alexander V. Goltsev and Alexander N. Samukhin from the Ioffe Institute in St Petersburg. We did not reprint figures with empirical data from original papers but made sketches of data. We are grateful to Albert-László Barabási, Stefan Bornholdt, Jonathan Doye, Jennifer Dunne, Lee Giles, Ramesh Govindan, Byungnam Kahng, Ravi Kumar, Neo Martinez, Sergei Maslov, Mark Newman, Sidney Redner, Ricard Solé, Alessandro Vespignani, and their coauthors for permission to use data from their original figures for derivative reproduction. We are much indebted to John Bulger, Ester Richards, Chris Fowler, David Duckitt, Goutam Tripathy, and Neville Hankins, the copy editor at Oxford University Press for correcting the English of our book. Our computers did not crash only thanks to Miguel Dias Costa and João Viana Lopes. When this book was written, one of us (SND) was on leave from his native Ioffe Institute, and he acknowledges the Centre of Physics of Porto for their support and hospitality.

Porto
May 2002

S.N.D.
J.F.F.M.

CONTENTS

0

MODERN ARCHITECTURE OF RANDOM GRAPHS

The first natural questions to ask about a network are:

- What does it look like?
- What is its structure and its topology?
- Is it large or small?
- Why does it have the features it has?
- How did it emerge and develop?
- What can we do with it?

In fact, this book is devoted to a discussion of just these issues. Note that we are not very interested in the internal states of the vertices or edges of networks which are of primary importance in neural and Boolean nets. We restrict ourselves to problems related to the topological structure of random networks, to the evolution of this structure, and to the direct consequences of the particular structural organization of nets.

For many years, the structure of networks with random connections was an object of immense interest for researchers in various sciences, namely mathematics (graph theory), computer science, communications, biology, sociology, economics, etc. Physics may be practically omitted from this incomplete list if we forget neural networks. These sciences provided separate views of distinct networks. A general insight was absent.

In the late 1990s the study of the evolution and structure of networks became a new field of physics. Now we can speak about the statistical physics of networks. What happened?

Here we must explain the strange sounding title of this introduction. How can graphs be modern or traditional? The point is that a few years ago the common interest moved from graphs with a rapidly decreasing distribution of connections (a Poisson degree distribution) to those with a fat-tailed degree distribution; that is, those with many highly connected vertices. The difference between these two architectures is so striking that the transition to the study of the latter actually leads to a revolution in network science.

This transition has been induced by empirical observations of power-law distributions of connections in many real networks, above all, the WWW (Albert, Jeong, and Barabási 1999, Huberman and Adamic 1999) and the Internet (Faloutsos, Faloutsos, and Faloutsos 1999, Govindan and Tangmunarunkit 2000). Mathematical graph theory turned out to be rather a long way from real needs since it focused mainly on 'too simple' static random graphs with Poisson distributions of connections (Erdős and Rényi 1959, 1960), where hubs are not essential. In these random graphs of graph theory, edges are distributed at random between a fixed number of vertices. In our book, we call such simple nets *classical random graphs*. Moreover, mathematicians did not really study evolving random networks.

By the middle of the 1990s, the impact of large, growing communications nets with complex architectures, the Internet and the WWW, on our civilization became incredible. However, understanding of their global organization and functioning was absent.

On the other hand, knowledge of the principles of the evolution and structuration of networks was of vital practical importance. For example, the effective working of search engines is hardly possible without this knowledge. We should note that the first concepts of functioning and practical organization of large communication networks were elaborated by one of the 'parents' of the Internet, Paul Baran (1964). Actually, many present studies develop his initial outstanding ideas and use his terminology. What is the optimal design of communication networks? How can one afford their stability and safety? These and many other vital problems were first studied by Baran at a practical level.

Communication networks are well documented. The data on their structure can be obtained using special programs—*robots*. The most difficult aspect of these empirical studies is that really very large nets are necessary for good statistics. Indeed, the results of such observations are usually various statistical distributions. Finite-size effects cut off their tails and narrow the field of observation. One should note that a number of effects in networks cannot be explained without accounting for their finite sizes. In this sense, most real networks are *mesoscopic objects*. However, for the observation of the fat tails of degree distributions, these finite sizes are only a complicating factor.

Several years ago large artificial nets had approached sizes that allowed statistically reliable data to be obtained. At present, the largest of them, the WWW, contains about 10^9 vertices (the documents of the WWW, that is pages) connected by about 10^{10} edges (hyperlinks). The first empirical study of the WWW (Albert, Jeong, and Barabási 1999) showed that (1) *it is a surprisingly compact network*: the average length of the shortest directed path between two of its randomly chosen pages is only about 19 steps ('clicks'), that is of the order of the logarithm of the WWW size and (2) *the distribution of the numbers of connections of its vertices has an unusual fat-tailed form.*

At present, the first result seems quite obvious. Indeed, this value of the average shortest-path length is typical of networks with random connections. This smallness of the average shortest-path length is usually referred to as the *small-world effect.* Even the introduction of a single shortcut between widely separated sites in a finite lattice essentially reduces the average shortest-path length. A low concentration of shortcuts between randomly chosen lattice sites produces the shortest-path lengths typical of classical random graphs (Watts and Strogatz 1998). In random networks where all edges are actually such shortcuts, the average shortest-path length must be small. Therefore, an effect produced by the smallness of the WWW (and of the modern world) is of a rather psychological nature. The smallness of another important net, a network of acquaintances, was spectacularly demonstrated in the experiment of Milgram long ago (1967). Several hundred randomly chosen individuals in Omaha received letters with a simple instruction: each one was asked to send a letter with this instruction to his or her acquaintance whom he or she supposed to be closer connected to a target person X in Boston. One-fifth of the letters from Omaha successfully approached X, and the average length of the observed acquaintance chains between the individuals in Omaha and X in Boston turned out to be only six: 'six degrees of separation'.

As for the second observation of Albert, Jeong, and Barabási (1999), that is the fat-tailed distribution of connections, it was in sharp contrast to what one might expect, namely to the rapidly decreasing Poisson distributions of 'classical random graphs'. Moreover, the empirical distribution has a form close to a power law. It has no natural scale, unlike, for instance, the Poisson or exponential distributions, and, hence, such networks are also referred to as 'scale-free' nets. In fact, this important observation indicates that the WWW has a 'scale-free' architecture, which is based on a wide spectrum of vertices in terms of the numbers of connections they have.

Extensive studies have shown that many other large natural and artificial nets also have such fat-tailed and scale-free structures (Amaral, Scala, Barthélémy, and Stanley 2000). For example, to this class of nets belong the Internet (Faloutsos, Faloutsos, and Faloutsos 1999, Govindan and Tangmunarunkit 2000), nets of citations in the scientific literature (Redner 1998, networks of metabolic reactions (Jeong, Tombor, Albert, Oltvai, and Barabási 2000), nets of protein–protein interactions (Jeong, Mason, Barabási, and Oltvai 2001), etc., so the WWW is by no means an exception; such nets are everywhere, even in language (Ferrer i Cancho and Solé 2001a).

Note that the particular forms of the distributions of connections in some of these nets have not been clearly identified and are still being discussed, so, here, as a rough guide, we simply point out that *numerous real networks have an architecture based on a fat-tailed distribution of connections.*[1] Such distributions and the factor of growth determine the spectrum of unusual properties of these networks. Perhaps, their most impressive property is *their unique stability against failures and random damage.* To destroy such (infinitive) networks, that is to decay to a set of small unconnected clusters, one often has to remove at random almost all their vertices or edges (Albert, Jeong, and Barabási 2000, Cohen, Erez, ben-Avraham, and Havlin 2000). The resilience to failures is obviously necessary for biological and communication networks. This partly explains why scale-free nets are so widespread in nature.

[1]It is not so easy to find empirically whether the distribution is of a power-law form or not. At this point, we only indicate that, in the large-net limit, the observed distributions have divergent higher moments.

The superstability is a direct consequence of this specific architecture, but the same origin has a contrasting phenomenon, *the absence of an epidemic threshold* in such networks (Pastor-Satorras and Vespignani 2000). Diseases may easily spread within them, and this is a weak point of such networks.

Thus, the architecture that is based on fat-tailed degree distributions, with the key role of strongly connected vertices (hubs), is very important. Where does it come from? Is it a result of the imposition of some external will, a lucky product of special design, etc.?[2] Does somebody intentionally create such an architecture? The answer turns out to be: 'No, while growing, the networks *organize themselves* into the "fat-tailed" and scale-free structures' (Barabási and Albert 1999, Barabási, Albert, and Jeong 1999). These structures are the direct result of the *self-organization* of networks. Hence, the evolution of networks turns out to be among numerous growth processes which have been studied by physicists for many years. One can say, by definition, that scale-free networks are in a critical state.[3] So, the problems of the network growth are directly related to *self-organized criticality*.

Here we must emphasize a principal difference between two distinct kinds of networks. Without going into detail, random networks that were studied in graph theory for many years were *equilibrium* graphs with a fixed total number of vertices and random connections between them. Edges connected randomly chosen vertices. By contrast, in the typical picture of the real network growth, new vertices are added to a network all the time, new edges emerge between the growing number of vertices, and the network is in a *non-equilibrium state*. Equilibrium and non-equilibrium networks have very different structures and properties even if their degree distributions are similar (Krapivsky and Redner 2001).

The next natural questions are: what are the general principles of this self-organization process that produces such an architecture of the networks? Why do various nets in quite different areas grow up into very similar structures? A key idea and explanation, namely the *preferential linking* or *preferential attachment* of edges, is very simple. While a network grows, its new edges become *preferentially* attached to vertices with a high number of connections (Barabási and Albert 1999, Barabási, Albert, and Jeong 1999). Then these vertices get on even better chance to attract the next new edges, and so on—the rich get richer. As a result, the distribution of the numbers of attached edges for vertices has a chance to be fat-tailed. Such a preference may take various forms, and each one produces the particular structure of the network. However, the most natural (linear) kind of preference results in scale-free networks. Simple models of networks growing under this mechanism can be solved exactly, and one may check that the

[2] In fact, there have been recent attempts to explain the appearance of general power-law distributions as a result of special design or, more explicitly, optimization (see Carlson and Doyle 1999, 2000, Doyle and Carlson 2000, Valverde, Ferrer i Cancho, and Solé 2001b).

[3] A critical state is characterized by power-law correlations and power-law distributions.

mechanism is efficient (Dorogovtsev, Mendes, and Samukhin 2000a, Krapivsky, Redner, and Leyvraz 2000).

One should note that the concept of preferential distribution is not completely new, although previously it was not applied to networks. More than 45 years ago, Herbert Simon proposed a model which produces power-law distributions in a similar way (Simon 1955, 1957).[4] The Simon model was used widely in the field of economics and for the explanation of the frequency of occurrence of words in language. In fact, numerous processes in various areas are based on the same natural general principle *popularity is attractive.*[5] In the WWW, more popular pages preferentially get new links; in the scientific literature, more popular papers preferentially get new citations, etc. All these are so-called 'multiplicative stochastic processes'.

Networks may consist of separate parts, and the notion of topology for networks includes the image of a network as a set of such separate subgraphs (connected components). The statistics of the connected components are related to percolation problems. All the standard problems of percolation theory can be easily formulated for nets. One can ask, for example: what is the size of the percolating cluster (*the giant connected component* as it is called in graph theory), and how are the sizes of connected components distributed? The relative size of the giant connected component indicates the stability of a network. The network is certainly not a unit organism if the percolation cluster (the giant connected component) is absent. This is why the percolation properties of networks are an issue of intense interest.

Percolation theory for equilibrium networks is more developed (Molloy and Reed 1995, 1998, Newman, Strogatz, and Watts 2001). However, it has been found that the percolation properties and statistics of connected components for growing networks (Callaway, Hopcroft, Kleinberg, Newman, and Strogatz 2001, Dorogovtsev, Mendes, and Samukhin 2001b) show a dramatic difference from the equilibrium situation. In this respect, growing networks provide truly exciting problems.

Furthermore, networks with directed connections (for example, the WWW) have an especially rich topology (Broder, Kumar, Maghoul, Raghavan, Rajagopalan, Stata, Tomkins, and Wiener 2000). Such networks contain a complex set of interpenetrating giant components.

In this book, we concentrate on the physics principles of networks. Chapter 3 is entirely devoted to the discussion of results of observations and empirical studies but we are still very far from a complete understanding of real networks. Network theory, in its present state, proposes stunning ideas and minimal models demonstrating various impressive effects which are certainly related to reality. The field is open, and there are future challenges ahead.

[4]Also, we must mention the Flory–Stockmayer theory of polymer growth (Flory 1941, Stockmayer 1943/1944, Flory 1971.

[5]While thinking of the title of our paper (Dorogovtsev, Mendes, and Samukhin 2000b), we did not expect that this slogan would become popular.

1

WHAT ARE NETWORKS?

1.1 Basic notions

From a formal point of view, a network (or a graph) is a set of vertices connected via edges[1] (see Fig. 1.1). We do not discuss networks with 'weighted' edges (see Newman 2001e, Yook, Jeong, Barabási, and Tu 2001), and usually our networks do not include multiple edges and unit loops (an edge with both ends attached to the same vertex). Networks with undirected edges are called *undirected networks*, networks with directed edges are *directed networks*.

The total number of connections of a vertex is called its *degree* k (see Fig. 1.2).[2] In a directed network, the number of the incoming edges of a vertex is called its *in-degree* k_i, the number of the outgoing edges its *out-degree* k_o. Thus, $k = k_i + k_o$.

An important particular case of graphs is *trees*, that is graphs without loops (of any length). If a tree has no separate parts, it is called a *connected tree*. The

[1] Physicists often use the terms 'nodes' and 'sites', 'links' and 'bonds'. We prefer to use the standard terms of graph theory and computer science. However, note that mathematicians never say 'networks' but only 'graphs'.

[2] In modern statistical physics, 'degree' is often called 'connectivity'.

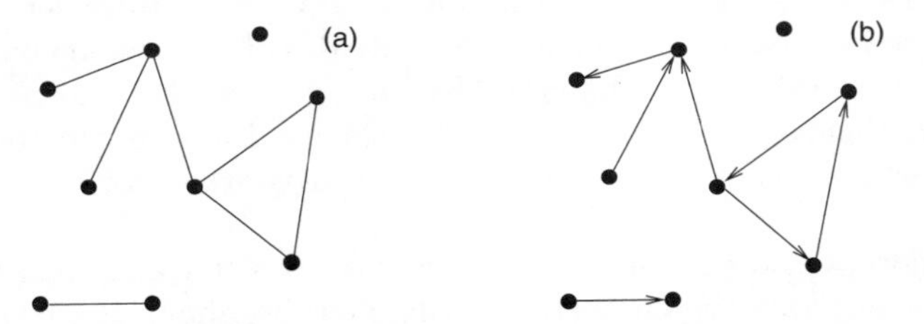

FIG. 1.1. The samples of networks. (a) Undirected graph. Vertices are connected via undirected edges. (b) Directed graph. Edges are directed. (A directed graph can also include undirected edges.) Both these graphs contain several connected components. Note that here we do not draw multiple edges and unit loops ('melons' and 'tadpoles').

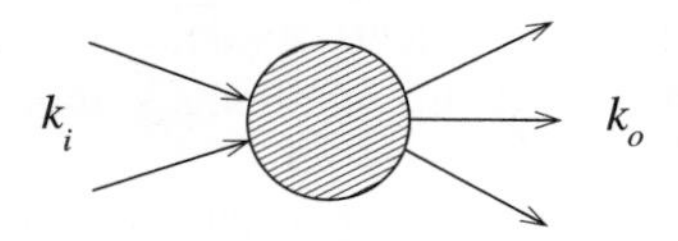

FIG. 1.2. In- and out-degrees, k_i and k_o, of a vertex. Degree $k = k_i + k_o$ is the total number of connections of a vertex.

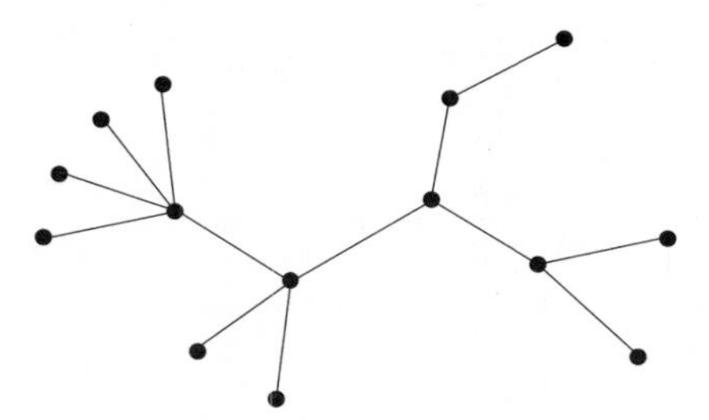

FIG. 1.3. A connected tree. Check that the total number of vertices, N, is greater by one than the total number L of edges: $N = L + 1$.

total numbers of vertices, N, and loops, L, in a connected tree are related to one another via $L = N - 1$ (see Fig. 1.3). In general, the number I of loops in an arbitrary undirected connected graph is related to the number of its edges and vertices (see Fig. 1.4 where loops are explained):

$$I = L + 1 - N\,. \tag{1.1}$$

This is one of the basic formulae in graph theory.

In general terms, random networks are networks with a disordered arrangement of edges. Note that, usually, in graph theory the meaning of a random graph is much more narrow (see discussion below). While speaking about random networks we have to keep in mind that a particular network we observe is only one member of a *statistical ensemble* of all possible realizations. Therefore, when we speak about random networks, we actually speak about statistical ensembles. Let us repeat: *'a random network' means 'an ensemble of nets'*. The complete statistical description of a random network suggests the description of the corresponding statistical ensemble. In principle, random networks may contain vertices of fixed degree. Usually, however, the degrees of vertices are statistically distributed.

From a physical point of view, random networks may be *equilibrium* and *non-equilibrium*. Let us introduce these notions using the simplest examples.

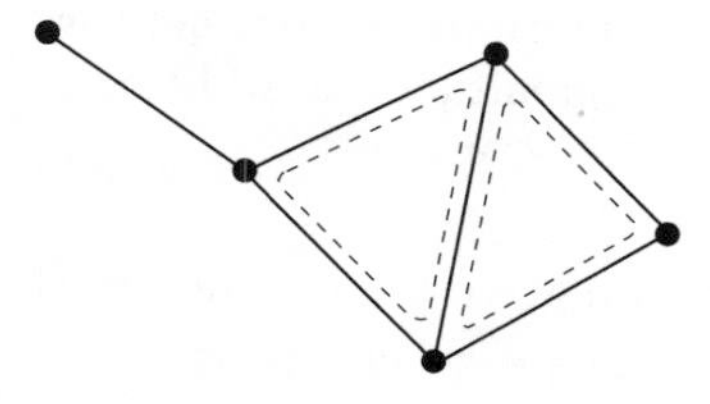

FIG. 1.4. A connected undirected graph. Note that it has only two and not three loops. Indeed, two of the loops cannot be reduced to a combination of smaller ones. Check that the number of loops, two, is equal to the number of edges of the graph, six, plus one minus the number of vertices, five.

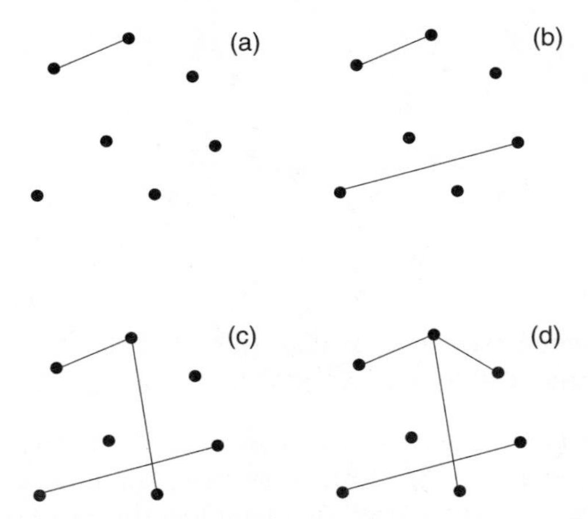

FIG. 1.5. Example of an equilibrium network, a classical random graph (the Erdős–Rényi model). (a), (b), (c), and (d) demonstrate the first steps of the construction procedure. Pairs of randomly chosen vertices are connected by edges. The total number of vertices is fixed.

(1) The example of an equilibrium random network, *a classical random graph*. This graph is defined by the following simple rules:[3]

(a) The total number of vertices is fixed.[4]

(b) Randomly chosen pairs of vertices are connected via undirected edges.

One may also specially forbid multiple links by definition but this is not essential for large graphs where such links are almost absent. One can say that the vertices of the classical random graph are statistically independent and equivalent. This simple model was proposed by Erdős and Rényi (1959, 1960). The construction procedure of such a graph can be thought of as the subsequent addition of new edges between pairs of vertices chosen at random (see Fig. 1.5). When the total number of vertices is fixed, this procedure obviously produces equilibrium configurations.

(2) The model for a non-equilibrium random network; this is a simple random graph growing through the simultaneous addition of vertices and edges (see Fig. 1.6). The definition of this graph is:

(a) At each time step, a new vertex is added to the graph.

(b) Simultaneously, a pair (or several pairs) of randomly chosen vertices is connected by an edge.

[3]This 'trivial' graph has been one of the main objects of graph theory for more than 40 years!

[4]Here we consider the simplest situation.

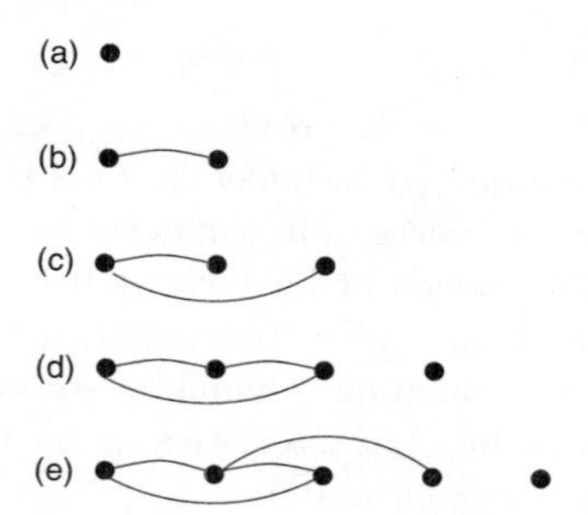

FIG. 1.6. An example of a non-equilibrium network. At each time step, a new vertex is added to the graph. Simultaneously, two randomly chosen vertices of the graph are connected by an edge.

One sees that the system is non-equilibrium. Edges are distributed inhomogeneously over the graph. The 'oldest' vertices are the most connected (in the statistical sense), and degrees of new nodes are the smallest. If, at some moment, we stop increasing the number of vertices but proceed with the random addition of edges, then the network will tend to an 'equilibrium state' but will never achieve it. Indeed, the edges of the network do not disappear, so the inhomogeneity survives. An 'equilibrium state' can be approached only if, in addition, we allow old edges to disappear from time to time.

(2′) Another example of a non-equilibrium random network—*a citation graph*. In the simplest version, this graph is defined in the following way (see Fig. 1.7):

(a) At each time step, a new vertex is added to the graph.

(b) It is connected with some old node via an undirected edge.

For example, initially, there may be one vertex without any connections. Different rules for the choice of an old vertex for the connection yield distinct citation graphs. Here, for us, it is only important that they are all non-equilibrium ones.

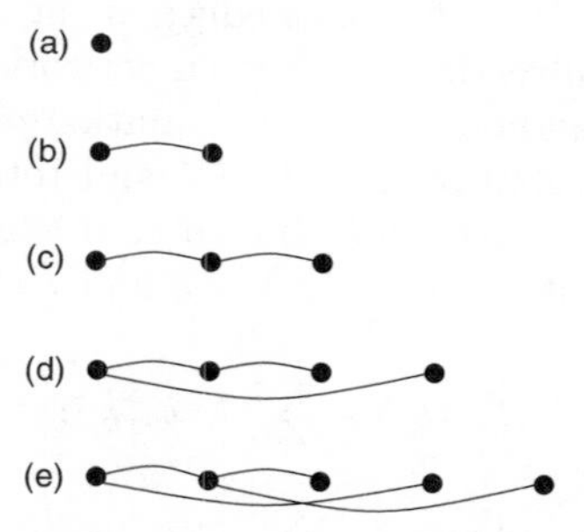

FIG. 1.7. Another example of a non-equilibrium network—a citation graph. At each time step, a new vertex is added to the graph. It connects to some existing vertex via an edge. Compare with Fig. 1.6.

1.2 Adjacency matrix

The *adjacency matrix* of a network provides its complete image. It indicates which of the vertices are connected (adjacent). This is a square $N \times N$ matrix, where N is the total number of vertices in the network. For directed networks, its element a_{ij} is equal to 1 if there is an edge coming from the vertex i to the vertex j.[5] The adjacency matrix of an undirected network is symmetrical: $a_{ij} = a_{ji}$. If unit loops are absent, the diagonal elements are equal to zero: $a_{ii} = 0$. In Appendix A, we show how the characteristics of an undirected graph may be expressed in terms of its adjacency matrix.

For random networks, an adjacency matrix completely describes only one particular realization. For the complete description of some random network, we have to describe the entire statistical ensemble of its realizations, that is the ensemble of adjacency matrices.

1.3 Degree distribution

As explained above, generally the degrees of vertices in random networks are statistically distributed. Let us start with undirected networks. If vertices can be distinguished, which is the standard situation in growing networks, for each vertex one introduces degree distribution $p(k, s, N)$. This is the probability that the vertex s in the network of size N has k connections (k nearest neighbours). Knowing the degree distributions of each vertices in a network, it is easy to find the *total degree distribution*

$$P(k, N) = \frac{1}{N} \sum_{s=1}^{N} p(k, s, N) \,. \tag{1.2}$$

If all the vertices of a random network are statistically equivalent, as in classical random graphs, each of them has the same degree distribution $P(k, N)$.

The first moment of the distribution, that is the mean degree of a network, is $\overline{k} = \sum_k kP(k)$. The total number L of edges in the network is equal to $\overline{k}N/2$.

Similarly, for individual vertices in directed networks one introduces in-degree distribution $p^{(i)}(k_i, s, N)$ and out-degree distribution $p^{(o)}(k_o, s, N)$. Also, one can define total in-degree distribution $P^{(i)}(k_i, N)$ and total out-degree distribution $P^{(o)}(k_o, N)$. Nevertheless, much more informative characteristics are the joint in- and out-degree distributions $p(k_i, k_o, s, N)$ and $P(k_i, k_o, N)$. Obviously,

$$P^{(i)}(k_i) = \sum_{k_o} P(k_i, k_o) \,, \tag{1.3}$$

$$P^{(o)}(k_o) = \sum_{k_i} P(k_i, k_o) \,, \tag{1.4}$$

[5] Here we assume that multiple and weighted edges are absent.

$$P(k) = \sum_{k_i} P(k_i, k - k_i) = \sum_{k_o} P(k - k_o, k_o)\,, \tag{1.5}$$

where the network size N is not explicitly indicated. Similar relations can be written for distributions of individual nodes. If the network has no connections to the exterior, then the average in- and out-degrees are equal:

$$\overline{k_i} = \sum_{k_i} k_i P(k_i, k_o) = \overline{k_o} = \sum_{k_o} k_o P(k_i, k_o) = \overline{k}/2\,. \tag{1.6}$$

Degree distribution is the simplest statistical characteristic of a network. This is the distribution of a 'one-vertex' quantity. So, the degree distribution characterizes only the local properties of a network but, often, even this poor information about the network structure is sufficient to determine its basic properties. Most of the empirical data are degree (or in-degree, or out-degree) distributions for various networks and are easy to measure (however, see Appendix B for a discussion about the problems in processing empirical degree distributions). Indeed, the main problem in the empirical studies of networks is poor statistics. Usually, we have at hand only one realization of a random network to obtain data from. Thus, the size of a network sets a strong restriction. When we plot degree distribution $P(k)$ using empirical data, we actually indicate the number of vertices of a degree k and then make the appropriate normalization. For a joint in-, out-degree distribution $P(k_i, k_o)$ we have not one but two variables, k_i, and k_o, and much more distinct $\{k_i, k_o\}$ points for the same total number of vertices of a network. This produces much more pronounced fluctuations.

When the vertices of a graph are statistically independent (equivalent), that is connections are random, the degree distribution completely determines the statistical properties of a network. Assume that degree distribution $P(k)$ is preset and the total number N of vertices is fixed. How can one construct a graph with random connections having this degree distribution (Bekessy, Bekessy, and Komlos 1972, Bender and Canfield 1978, Bollobás 1980, Wormald 1981a, 1981b)?

(1) Label N vertices.

(2) To the vertices $\{j\}$ of the graph ascribe degrees $\{k_j\}$ taken from the distribution $P(k)$. Now the graph looks like a family of hedgehogs: each vertex has k_j quills sticking out (see Fig. 1.8(a)).

(3) Connect at random ends of pairs of distinct quills belonging to distinct vertices (see Fig. 1.8(b)).

This is only a particular realization of the graph but, in such a way, it is easy to construct the entire statistical ensemble. Such simple equilibrium 'uncorrelated' random graphs are a natural generalization of classical random graphs.

The following examples demonstrate typical degree distributions for networks.

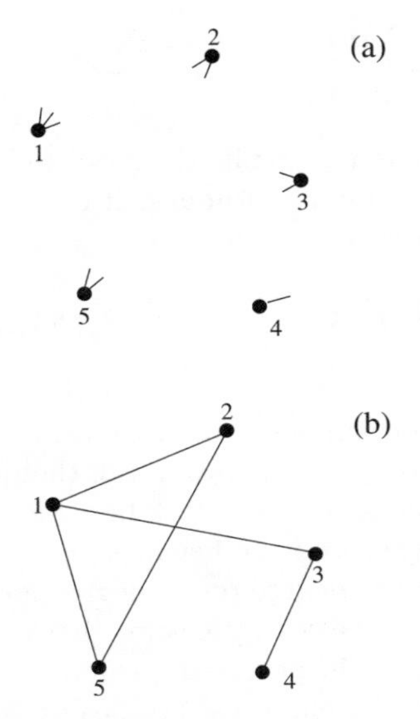

FIG. 1.8. The construction procedure for an equilibrium random graph with a given degree distribution $P(k)$. (a) Degrees k_j taken from the distribution are ascribed to the labelled vertices. (b) Pairs of random ends sticking out of different vertices are connected.

(1) *The Poisson distribution* (see Fig. 1.9(a)).

$$P(k) = \frac{e^{-\overline{k}}\,\overline{k}^k}{k!}\,. \tag{1.7}$$

Here, $\overline{k}$ is the average degree, $\overline{k} = \sum_{k=0}^{\infty} kP(k)$. A classical random graph asymptotically has just this degree distribution, if its number of vertices approaches infinity under the constraint that the mean degree is fixed.

(2) *Exponential distribution* $P(k) \propto e^{-k/\overline{k}}$ (see Fig. 1.9(b)). For instance, this is the degree distribution of the growing random graph described in Section 1.1, example (2).

Even for infinite networks, all the moments of distributions (1) and (2) are finite, $M_m \equiv \sum_{k=0}^{\infty} k^m P(k) < \infty$. Both of them have a natural scale of the order of the average degree.

(3) *Power-law distribution* $P(k) \propto k^{-\gamma}, k \neq 0$ (see Fig. 1.9(c)). Here, γ is the exponent of the distribution.

The power-law distribution contrasts with the Poisson and exponential distributions. It has no natural scale, and hence may be called scale-free. Networks with such distributions are called *scale-free*.

In infinite networks, all higher moments of order $m \geq \gamma - 1$ of the power-law distribution diverge. From this, we can see what values of the γ exponent

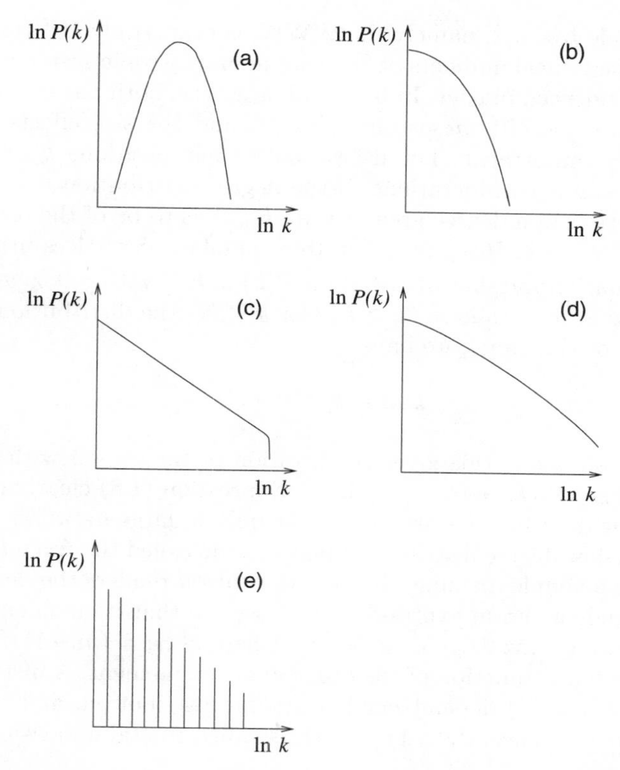

FIG. 1.9. Examples of degree distributions. (a) The Poisson distribution. The characteristic scale is the average degree $\overline{k}$. (b) The exponential distribution. The characteristic scale is the average degree. (c) The power-law distribution. The cut-off is the result of the finite-size effect. (d) The multifractal degree distribution. (e) A discrete spectrum of degrees in deterministically growing networks.

are possible for scale-free networks (of course, γ must be greater than 1 for normalizability):

(a) If the average degree of the scale-free network (that is the first moment of the degree distribution) is finite, than its γ exponent should be greater than 2, $2 < \gamma < \infty$.

(b) One can easily imagine a situation in which the average degree diverges. In growing networks, a total number of edges may grow faster than a linear function of a total number of vertices (Dorogovtsev and Mendes 2001a). This is true for many real networks. For this type of growth, if the degree distribution is stationary, $1 < \gamma \leq 2$.

In finite-size networks, fat-tailed degree distributions have natural cut-offs. At this point we must emphasize that all the real networks have restricted sizes.

The largest scale-free net, namely the WWW, contains only 10^9 vertices, that is far less than the typical numbers of particles in macroscopic systems in statistical physics and condensed matter. In fact, real networks, both natural and artificial, are not macroscopic but *mesoscopic* objects, and the size effects and cut-offs are of primary importance. Let us estimate their positions k_{cut} by applying the following simple considerations. If the degree distribution is stationary, the number of vertices of a degree greater than k_{cut} has to be of the order of 1, that is $N \int_{k_{cut}}^{\infty} dk\, P(k) \sim 1$. Here, N is the total number of vertices in the network. For the stationary power-law distribution $P(k) \propto k^{-\gamma}$ with $\gamma > 2$, assuming that this power-law form is valid for $k \geq k_0$ (for $k < k_0$, the distribution may be, for example, zero or constant), we have

$$k_{cut} \sim k_0 N^{1/(\gamma-1)} \,. \tag{1.8}$$

On a logarithmic scale, this gives the position of the cut-off with a reasonable accuracy, $\ln k_{cut} \approx \ln k_0 + (\gamma-1)^{-1} \ln N$. Expression (1.8) clearly demonstrates that power-law distributions are observable only in large networks.

The power-law degree distribution may also be called the *fractal* distribution. This term has a simple meaning. Indeed, the cut-off renders the degree distribution size dependent. From expression (1.8), we see that its moments depend on N in the following way: $M_m(N) \propto N^{\tau(m)}$, where $\tau(m) = (m-1)/(\gamma-1) - (\gamma-2)/(\gamma-1)$ is a *linear* function of the order m of the moment. A distribution with such a kind of $M_m(N)$ dependence is a fractal distribution, and the coefficient of the linear dependence, $d_f = 1/(\gamma-1)$, is called fractal dimension.

(4) *Multifractal distribution* (see Fig. 1.9(d)) (Mandelbrot 1983). This has richer properties. The exponent $\tau(m)$ of its size dependence is a *non-linear* function of the order m of the moment. Multifractals are actually statistical mixtures of fractals of different dimensions. For us it is important that multifractal distributions have no characteristic scales[6] and certainly cannot be characterized by a fixed exponent. Growing networks with multifractal degree distributions will be considered in Section 5.17.

(5) *Discrete distribution* (see Fig. 1.9(e)). A discrete spectrum of degrees is typical of deterministically growing networks, which we will discuss in Section 5.14.

1.4 Clustering

The *clustering coefficient* characterizes the 'density' of connections in the environment close to a vertex. Suppose that a network is undirected,[7] and one of its vertices has z nearest neighbours (see Fig. 1.10). The maximal linkage in this cluster (the maximal 'clustering') is approached when all $z(z+1)/2$ possible

[6]Hence, in principle, they also may be called scale-free. However, the term 'scale-free' is usually used for power-law distributions.

[7]Notice that the notion of clustering is only well defined for undirected networks.

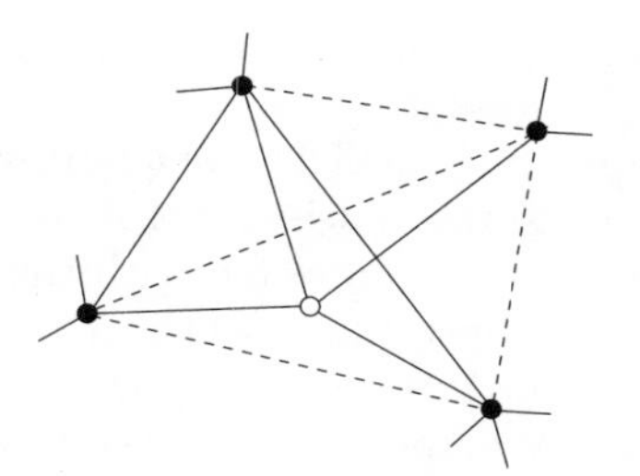

FIG. 1.10. A schematic illustration for the definition of the clustering coefficient. Only part of the connections between the nearest neighbours of the open vertex is present. To obtain the clustering coefficient, one has to divide the number of these edges by the total number of all possible edges between the nearest neighbours including absent (dashed) edges. In this configuration, the open vertex has four nearest neighbours. There are two edges between them. Hence the clustering coefficient is equal to $2/6 = 1/3$.

edges in the cluster are present. This also means that all $z(z-1)/2$ possible edges between the nearest neighbours exist. Conventionally (Watts and Strogatz 1998, Watts 1999, Strogatz 2001), the clustering coefficient C of a vertex is the ratio between the total number y of the edges connecting its nearest neighbours and the total number of all possible edges between all these nearest neighbours,

$$C = \frac{2y}{z(z-1)} \,. \tag{1.9}$$

Therefore, it reflects 'cliquishness' of the closest environment of a vertex, that is the extent of the mutual 'acquaintance' of its closest 'friends'.[8]

One can introduce the distribution of C in a network but usually only the average value $\overline{C}$ is considered, which is the clustering coefficient of a network. It characterizes the average 'cliquishness' of the closest environment of a network vertex.[9] $\overline{C}$ actually demonstrates the probability that there is an edge between two nearest neighbours of a randomly chosen vertex (see Fig. 1.10). One may also say that the clustering $\overline{C}$ shows 'the density' of small loops of length 3 in the network. So, the notion of clustering is directly related to loops. The clustering coefficient of a tree is zero by definition. The structure of highly clustered networks is very far from tree-like. The clustering (or, more generally, the presence of loops) is a specific form of correlations in networks. If the clustering coefficient

[8] Also, we can define the local 'density of linkage' D (Dorogovtsev and Mendes 2002a) around this vertex by dividing the total number of edges in the cluster containing the vertex itself and its nearest neighbours, $z+y$, by the total number of all possible edges between vertices of this cluster, $z(z+1)/2$. Therefore, D and C are simply related.

[9] On the other hand, $\overline{C}$ is the probability that if a triple of vertices of a network is connected together by at least two edges, then the third edge is also present. One can easily check that $\overline{C}$ is equal to the ratio of one-third of the number of triples of vertices connected together by three edges and the total number of connected vertex triples in a network.

of an 'infinite' net does not approach zero, then the correlations between vertices of the network are certainly present.

One can see that the clustering coefficient of a fully connected network is equal to 1. In the classical random graph consisting of N vertices randomly connected by L edges, each pair of vertices is connected with the same probability $\cong z/N$. Here $z = \overline{k} = 2L/N$. So, the clustering coefficient of a classical random graph is $\overline{C} = z/N$. The smallness of this value for large graphs is natural: classical random graphs have very few loops. We shall see that this is not the case for many real nets.

1.5 Small worlds

It is not so surprising that networks are compact. However, the extent of this compactness is truly astonishing. The terms 'small worlds' and 'small-world effect' are standard in network science (Milgram 1967). The compactness, here, means the smallness of the 'linear size' of a net. There is no question what the linear size of a lattice is. Distances between pairs of sites on a lattice are Euclidean, and so, for the measurement of a distance, one may simply use a ruler. In networks, the situation with distances is more complex. Thus, what is the linear size of a network?

At first, suppose that a network is undirected. Let all the edges of a network be of unit length. Thus, the distance between two vertices of a network is the length of the shortest path (the geodesic) between them. The distances ℓ between pairs of vertices of a random network are distributed with some distribution function $\mathcal{P}(\ell)$. This is the probability that the length of the shortest path between two randomly chosen vertices is equal to ℓ. $\mathcal{P}(\ell)$ is one of the main structural characteristics of the network. It is the distribution $\mathcal{P}(\ell)$ that allows us to evaluate the 'linear size' of a network. For rapidly decreasing distributions, the characteristic distance is of the order of the average length of the shortest path $\overline{\ell} \equiv \sum_\ell \ell \mathcal{P}(\ell)$. Another natural characteristic distance is the length of the longest shortest path existing in a network, ℓ_{max}.[10] Of course, both $\overline{\ell}$ and ℓ_{max} depend on the total number N of vertices in the network.

Look at the problem from another point of view. The natural generalization of the notion of degree k of a vertex, that is of the number of the nearest neighbours, $z \equiv z_1$, is the number of second nearest numbers, z_2, third nearest numbers, z_3, and so on. In general, there are z_m vertices in the m-th sphere centred at the vertex, $z_0 = 1$. The distance between a vertex and any of its m-th nearest neighbours is equal to m, and this is an equivalent way to introduce the notion of distance for a network. One can introduce distribution of the numbers of the m-th nearest neighbours, $P_m(z)$. This is the probability that a randomly chosen

[10] In distinct papers, both the average shortest-path length and the length of the longest shortest path are called by the same term, 'diameter' of the network (for example, compare Albert, Jeong, and Barabási 1999 and Broder, Kumar, Maghoul, Raghavan, Rajagopalan, Stata, Tomkins, and Wiener 2000). This is why we do not use this term at all.

vertex of the network has z m-th nearest neighbours. Obviously, the distributions $\mathcal{P}(\ell)$ and $P_m(z)$ are related quantities. The relations between them are

$$\mathcal{P}(\ell) = \frac{\sum_z zP_\ell(z)}{\sum_{\ell,z} zP_\ell(z)} = \frac{\overline{z}_\ell}{\sum_\ell \overline{z}_\ell} = \frac{1}{N}\sum_z zP_\ell(z)\,, \tag{1.10}$$

where $\overline{z}_\ell$ is the average number of the ℓ-th nearest neighbours of a vertex in a network, $\overline{z} = \overline{z}_1$, and

$$P_\ell(z) = \binom{N-1}{z}\mathcal{P}^z(\ell)[1-\mathcal{P}(\ell)]^{N-1-z}\,. \tag{1.11}$$

Here, $\binom{N-1}{z}$ is the binomial coefficient.

Equation (1.10) is obvious: $\mathcal{P}(\ell)$ should be proportional to the average number of the ℓ-th nearest neighbours. The denominator $\sum_\ell \overline{z}_\ell = N$ is introduced for the proper normalization. Equation (1.11) is obtained in the following way. In principle, a vertex can be connected to the remaining $N-1$ vertices of the network. The probability that any of these $N-1$ vertices is its ℓ-th nearest neighbour is equal to $\mathcal{P}(\ell)$. Then, simple combinatorics immediately leads to eqn (1.11). One can check that substituting eqn (1.11) into eqn (1.10) yields an identity. Thus, we have demonstrated that not only $P_m(z)$ but also the distribution of distances $\mathcal{P}(\ell)$ are natural generalizations of the degree distribution.

We postpone further discussion of the distributions $\mathcal{P}(\ell)$ and $P_m(z)$ until later. Now, we demonstrate how to estimate the average shortest-path length for an uncorrelated network, that is its linear size. If a random network is large, then it usually has a *tree-like local structure.* This means that loops (closed paths on a network) are almost absent in the neighbourhood of a vertex. This environment may be sufficiently large. The term 'almost' means that the probability of such an event is negligible for large networks. The tree-like local structure of networks is one of the key points of graph theory. Here we do not discuss conditions for its realization (see Chapter 6) but only claim that usually it is sufficient to assume a fast enough decrease of the degree distribution. Indeed, in a contrasting situation, when a finite fraction of the total number of edges is attached to the finite number of vertices, the network is certainly not tree-like.[11]

In any case, without going into details, let us make the simplest estimation for a tree-like network. In this case, about $\overline{z}^\ell$ vertices are at a distance ℓ or closer from a vertex. We can then estimate the typical distance, that is the average length $\overline{\ell}$ of the shortest path, using the relation $N \sim \overline{z}^{\overline{\ell}}$. From this, we obtain the well-known expression for $\overline{\ell}$,

$$\overline{\ell} \approx \frac{\ln N}{\ln \overline{z}}\,, \tag{1.12}$$

[11]Note that the structure of the giant connected component (see Section 1.6) of a random network deviates noticeably from the tree-like one. Also note that networks with large clustering coefficients are not tree-like.

so the average shortest-path length is small even for very large networks. This striking smallness is usually referred to as the 'small-world effect' (Watts and Strogatz 1998, Watts 1999). Compare the estimation (1.12) with the linear size of a lattice in d-dimensional space containing N vertices, $\overline{\ell} \sim N^{1/d}$, and with the value for a fully connected network, $\overline{\ell} = 1$. We note that practically all the networks in this book display the 'small-world effect'.[12]

For a directed network, the situation is more complex. On account of the directness of edges, there may be two different shortest paths between the pair of points. The shortest path from one point to another does not coincide with the shortest path in the opposite direction. Their lengths do not necessarily coincide. One can ignore the directness of edges and introduce the average length $\overline{\ell}$ of the shortest undirected path for a directed network. The average length of the directed path is no less than $\overline{\ell}$ and may essentially exceed it.

Betweenness

The shortest path is one of central notions in network science. Now we briefly discuss other standard structural characteristics of networks which are based on this notion.

In general terms, ***betweenness*** $\sigma(m)$ of a vertex m is the total number of shortest paths between all possible pairs of vertices that pass through this vertex. If the number of shortest paths between a pair of vertices exceeds one, then the path passing through the vertex m contributes to the betweenness with a corresponding reduced weight. Note that all pairs with the vertex m are also taken into account. This quantity was introduced in sociology (see Freeman 1977) to characterize the 'social' role of a vertex. Vertices with a larger betweenness are more 'influenced'. This quantity is also called the ***load*** (Goh, Kahng, and Kim 2001a, 2001b) or the ***betweenness centrality*** of a vertex.[13] In fact, it indicates whether or not a vertex is important in ***traffic*** on a network. If all pairs of vertices of a network communicate with the same rate, and the traffic goes by the shortest paths, then the traffic through a vertex coincides with the betweenness of this vertex. One may introduce the distribution of the betweenness of vertices, the average betweenness of a vertex in a network, and so on.

If the total number of the shortest paths between vertices i and j is $B(i,j) > 0$, and $B(i,m,j)$ of them pass through a vertex m, than the ratio $B(i,m,j)/B(i,j)$ shows how crucial the role of the vertex m is in connections between i and j. In fact, it is supposed that only the shortest paths between vertices are essential. If, for example, none of the $B(i,j)$ shortest paths pass through m, then the role of m for i and j is zero. If, in contrast, all the $B(i,j)$ paths pass through m, then the role of m is dramatic: the vertex m completely controls the shortest connections. The ***betweenness*** $\sigma(m)$ of the vertex m is

[12] The only exception is the fractal, mentioned in Section 5.14.

[13] For brevity, we do not differentiate the betweenness, the load, and the betweenness centrality.

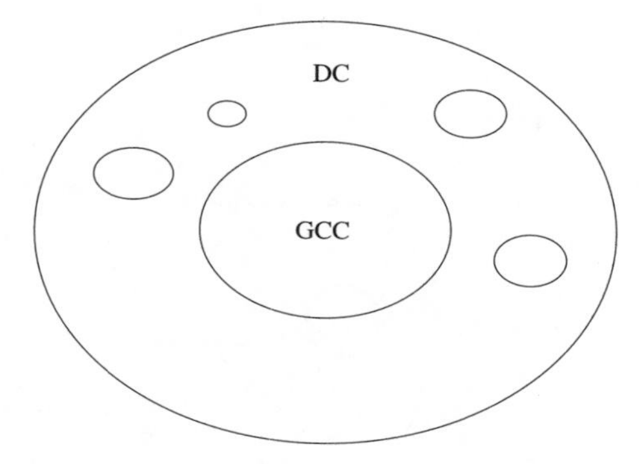

FIG. 1.11. The general structure of an undirected network with the giant connected component (GCC). The GCC plays the role of a percolating cluster. The rest of the network includes separate finite-size clusters: finite connected components. Usually, this part is referred to as 'disconnected components' (DC).

$$\sigma(m) \equiv \sum_{i \neq j} \frac{B(i,m,j)}{B(i,j)} , \tag{1.13}$$

where the sum is over all pairs of vertices for which at least one path exists, that is, with $B(i,j) > 0$. The vertices with high betweenness actually control a net. As is natural, one can suggest that the betweenness of a vertex strongly correlates with its degree.

Similarly, one may define betweenness of an edge. This characteristic may be used for detecting and indexing the hierarchical structure of nets in a natural way (Girvan and Newman 2002, a similar approach was applied by Holme, Huss, and Jeong 2002). A network is decomposed into a pair of separated subnets by successive deletion of the edges with maximum betweenness.[14] This procedure is repeated for each of the resulting subnets and so on.

1.6 Giant components

The characteristics of networks discussed so far do not allow us to imagine their global topology. To get a real image of a network, we have to know its percolation properties. As Fig. 1.1 demonstrates, a network may consist of a number of disjoint parts—connected components. The standard notions of percolating cluster and percolation threshold for networks are introduced in the following way.

To begin with, suppose that the network is undirected. A distinct connected component of a network is a set of mutually reachable vertices. The size of a connected component is the total number of vertices in it. When the relative sizes of all connected components of a network tend to zero as the number of

[14] Note that each deletion of an edge changes the betweennesses of other edges.

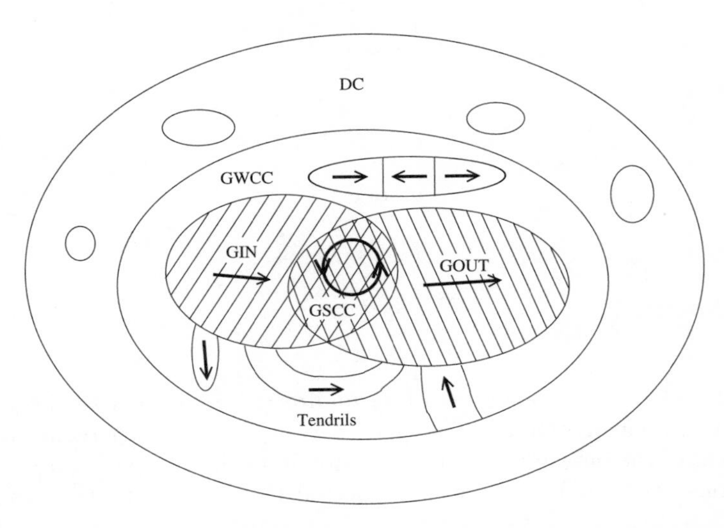

FIG. 1.12. The general structure of a directed network where the giant strongly connected component is present (Dorogovtsev, Mendes, and Samukhin 2001a), which is also the global structure of the WWW (Broder, Kumar, Maghoul, Raghavan, Rajagopalan, Stata, Tomkins, and Wiener 2000). Ignoring the directness of edges, the network consists of the *giant weakly connected component* (GWCC)—actually the usual percolating cluster and 'disconnected components' (DC). With regard to the directness of edges, the GWCC contains the following components: (a) the *giant strongly connected component* (GSCC), that is the set of vertices reachable from every its vertex by a directed path; (b) the *giant out-component* (GOUT), the set of vertices approachable from the GSCC by a directed path; (c) the *giant in-component* (GIN), which contains all the vertices from which the GSCC is approachable; (d) the *tendrils*, the rest of the GWCC, namely the vertices which have no access to the GSCC and are not reachable from it. They include 'tendrils' going out of the GIN to GOUT but there are also 'tubes' from the GIN to GOUT without passing through GSCC and numerous clusters which are only 'weakly' connected to other components of the GWCC.

vertices in the network approaches infinity, the network is below the percolation threshold. If the relative size of the largest connected component approaches a finite (non-zero) value in the limit of a large network, the network is above the percolation threshold. In such an event, the huge connected component plays the role of a percolating cluster. In graph theory, this is called the *giant connected component* (GCC). The general structure of an undirected graph, when the giant connected component is present, is shown in Fig. 1.11. The rest of the network consists of separate finite connected components. Traditionally (see Broder, Kumar, Maghoul, Raghavan, Rajagopalan, Stata, Tomkins, and Wiener 2000), these parts are together referred to as 'disconnected components' (DC).

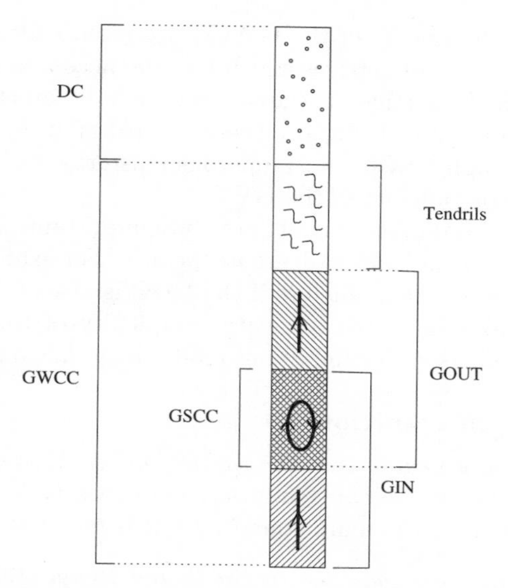

FIG. 1.13. Another schematic representation of the global structure of a directed network (compare with Fig. 1.12). The column corresponds to the entire network. Notice that the giant strongly connected component (GSCC) is the interception of the giant in-component (GIN) and the giant out-component (GOUT).

The general structure of a directed network may be much more complex. Generally, it includes several giant components (see Fig. 1.12 and Fig. 1.13). If we ignore the directness of edges, the network can be treated as undirected. The resulting connected components are called *weakly* connected components. The giant one, if it exists, is the *giant weakly connected component* (GWCC). In turn, with regard to the directness of the network, the GWCC has an intriguing structure.

First and foremost, it may include the *giant strongly connected component* (GSCC) that is the core, the most important part of a directed network. The GSCC is the set of vertices which are mutually reachable by a directed path. In other words, from one of the vertices of the GSCC, one can always approach another by moving either along or against the directions of edges. In this sense, every pair of the vertices of the GSCC are connected in both directions.

The *giant out-component* (GOUT) is the set of vertices approachable from the GSCC by a directed path. Therefore, it includes the GSCC. The *giant in-component* (GIN) contains all vertices from which the GSCC is approachable. One can see that it also includes the GSCC. Hence, the GSCC is the interception of the GIN and GOUT.[15]

[15]Note that in physical literature, other definitions of the GIN and GOUT are often used. For example, according to Broder, *et al.* (2000) and Newman, Strogatz, and Watts (2001), the

The remaining components of the GWCC are often called *tendrils* (Broder, *et al.* 2000). These consist of vertices which have no access to the GSCC and are not reachable from it. In particular, they indeed include something like 'tendrils' going out of the GIN to GOUT but also there are 'tubes' from the GIN to GOUT without passing through GSCC and numerous clusters which are only 'weakly' connected to other components of the GWCC.

The notions of the giant components are truly important. They characterize a network as a unit 'organism' and indicate its 'health'. For example, an undirected graph is only a set of separate clusters if the GCC is absent.

Figure 1.12 demonstrates the intricate complexity of the structure of networks. How does this arise? The impatient reader may take a look at Section 6.2.

1.7 List of basic constructions

Throughout the book we claim that *the basis of network science is construction procedures.* Indeed, before we can study a network we have to build it. Let us list in chronological order the main constructions from network theory.

(1) *Classical random graphs due to Erdős and Rényi* (1959, 1960) (see Sections 1.1 and 4.2). Roughly speaking, these are equilibrium graphs with connections between randomly selected pairs of vertices.

These graphs have a Poisson degree distribution. Their average shortest-path length is of the order of the logarithm of the total number of vertices. Large classical random graphs have tree-like 'local' structure. Loops are observable only at a large scale. Any clustering disappears as the size of such a graph approaches infinity. Correlations are absent.

(2) *Equilibrium random graphs with a given degree distribution* (Bekessy, Bekessy, and Komlos 1972, Bender and Canfield 1978, Bollobás 1980, Wormald 1981a, 1981b) (see Sections 1.3 and 4.3). Roughly speaking, these are equilibrium networks, which are maximally random under the constraint that their degree distribution is equal to a given one. In a specific case, when this degree distribution is binomial (or Poisson in the large-network limit), we obtain a classical random graph.

Their average shortest-path length is usually proportional to the logarithm of their size. Correlations and clustering in the large network limit are absent, and the network is locally tree-like.

(3) *'Small-world networks' due to Watts and Strogatz* (1998) (see Section 4.7). Roughly speaking, these are equilibrium networks, which are made from regular lattices by connecting pairs of vertices chosen at random. In fact, this is a superposition of a regular lattice and a classical random graph.

Even a very small number of such shortcuts, which do not change the local properties (a high clustering typical of a regular lattice), already produce the

GIN and GOUT do not include the GSCC. We use the more logical and convenient definition (see Chapter 6).

values of the average shortest-path length, which are typical of classical random graphs. Small-world networks have a Poisson-like degree distribution.

(4) *Networks growing under the mechanism of preferential linking due to Barabási and Albert* (1999) (see Sections 2.2 and 5.2). Roughly speaking, these are growing networks where new edges become attached to vertices preferentially. By definition, this means that more connected vertices have a better chance to get new connections.

The linear type of preferential attachment produces fat-tailed degree distributions. On the other hand, if any preference of linking is absent, and new connections become distributed at random, the degree distribution decreases rapidly. These networks are strongly correlated. As a rule, there are strong correlations between degrees of their vertices. The average shortest-path length is of the order of the logarithm of the size of such a network.

In Section 4.6 we introduce one more construction: *an equilibrium random graph with given degree–degree correlations.* Roughly speaking, these are equilibrium networks, which are maximally random under the constraint that the correlations between the degrees of the nearest-neighbour vertices are equal to a given correlation function. This means that the joint degree–degree distribution of the end vertices of edges is fixed.

A number of times, we say 'roughly' since each of the above construction procedures is actually a class of constructions with numerous versions and variations.

1.8 List of main characteristics

Here we list the main structural characteristics of networks. Some of them that have been introduced above are obvious, the others will be necessary in what follows. In the latter case we give definitions together with short explanations. Some standard notations are also presented, but be careful: sometimes we do not follow them too strictly in favour of brevity.

- *The total number of vertices*: N. Sometimes we call it the 'size' of a network.
- *The total number of edges*: L.
- *Degree of a vertex*: k. See Section 1.1.
- *In- and out-degrees of a vertex.* Usually we use $k_i = q$ and k_o.[16] See Section 1.1.
- *The total degree of a network*: K. This is the sum of degrees of vertices.
- *The mean degree of a network*: $\overline{k}$. This is the same as the average number of the nearest neighbours of a vertex in a network.
- *Degree distribution*: $P(k)$. See Section 1.3.
- *γ exponent of a degree distribution.* γ: $P(k) \sim k^{-\gamma}$. See Section 1.3.
- *In- and out-degree distributions*: obvious. See Section 1.3.
- *Exponents of in- and out-degree distributions*: γ_i and γ_o. See Section 1.3.

[16]Note that in safe situations k_i may be used to denote degree of a vertex i.

- *Joint in- and out-degree distribution of vertices*: $P(k_i, k_o)$. See Section 1.3.
- *Joint degree–degree distribution of the nearest neighbours*: $P(k, k')$. Roughly speaking, this is the probability that the end vertices of a randomly chosen edge have degrees k and k'. See Section 4.6.
- *Undirected-shortest-path length*: ℓ. See Section 1.5.
- *Directed-shortest-path length*. See Section 1.5.
- *Shortest-path-length distribution*: $\mathcal{P}(\ell)$. See Section 1.5.
- *The average shortest-path length*: $\overline{\ell}$. This is the average of the shortest-path length over all pairs of vertices between which a path exists. See Section 1.5.
- *The maximum shortest-path length*. This is the maximum shortest-path length for all pairs of vertices between which a path exists. See Section 1.5.
- *The number of m-th nearest neighbours of a vertex*: z_m. See Section 1.5.
- *The distribution of the number of m-th nearest neighbours*: $P_m(z)$. See Section 1.5.
- *Clustering coefficient of a vertex n*: $C(n)$. See Section 1.4.
- *Clustering coefficient of a network*. Usually this means the average clustering coefficient of the network, $\overline{C}$ or simply C. See Section 1.4.
- *Betweenness (betweenness centrality or load) of a vertex n*: $\sigma(n)$. See Section 1.5.
- *Sizes of giant connected components*. See Section 1.6.
- *Size distribution of finite connected components*. See Section 6.1.

2

POPULARITY IS ATTRACTIVE

Preferential linking (preferential attachment of edges) will be one of the key ideas in this story, so that we shall first introduce the matter at a basic level. For demonstration purposes, we will use simple undirected networks.

2.1 Attachment of edges without preference

Equilibrium network

Let us recall the equilibrium classical random graph (the Erdős–Rényi network). We could imagine the construction of the network in the following way. There is a given large number of edges, and we connect their ends to randomly chosen vertices (see Fig. 2.1(a)). This means that the attachment of edges occurs without preference: the probability that an edge becomes attached to a vertex is independent of the degree of this vertex. One edge is added per time step until the total number of edges reaches a given number. If this number is not too great, we need not worry about the multiple connections between the same pair of vertices. Moreover, in such a case we can allow both ends of an edge to become attached to the same vertex. Indeed, these events are not essential.

If we study a degree distribution, we can forget about edges themselves and consider only the attachment of independent edge ends. In this respect, the problem is equivalent to the distribution at random of balls among a fixed number of boxes. Suppose that the balls fall into boxes (see Fig. 2.1(b)). We do not want to go into details and hence use a continuum approach. Roughly speaking, this is the same as substituting balls for 'water': falling balls are now water that is 'rained' into the set of buckets beside each other. The impenetrable walls of the buckets mean that vertices can only attach edges and never lose them.

In this case, a random distribution simply means homogeneous rain (see Fig. 2.1(c)). This produces an equal level of water in the buckets, that is the δ-function degree distribution. In principle, the δ-function is a consequence of our 'continuum approach'. One can easily solve the problem exactly without resorting to naive 'rain analogies' and obtain a Poisson degree distribution (see Fig. 2.1 and Section 4.2). However, for now, what is important for us is that the degree distribution of an equilibrium network with linking without preference appears to be a rapidly decreasing function.

Growing network

Suppose that the attachment of edges in a growing network takes place randomly, that is without any preference. For example, consider the network (2)

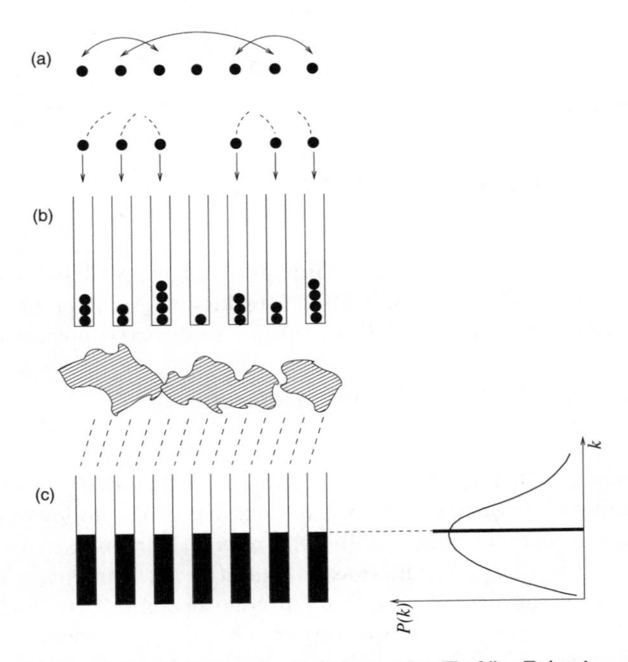

FIG. 2.1. Scheme of distribution of edges over the Erdős–Rényi equilibrium random graph. The total number of vertices is fixed. (a) Edges connect randomly chosen vertices. When we are interested only in the distribution function of a one-vertex characteristic, namely degree, we can consider the attachment of the independent edge ends. This is equivalent to the random distribution at random of balls among boxes (b). In the continuum approach, this is the same as rain homogeneously falling into buckets drawn together (c). This produces an equal level of water in the buckets, that is the δ-function degree distribution. An exact degree distribution has a Poisson form.

from Section 1.1: At each time step, one vertex is added and a pair of randomly chosen vertices is connected. In our 'continuum approach', this means homogeneous rain falling over an increasing number of buckets (see Fig. 2.2(a)). The overall amount of water falling into the buckets per unit time is constant, and so the rain becomes lighter with time.

The first buckets (vertices) are exposed to the rain longer than later ones, therefore the level of water in the first buckets is higher. Let us estimate the resulting distribution of water over the buckets (the degree distribution). We can label the buckets according to time s when each was added to the set, $0 \leq s \leq t$. Let the amount of water in the rain per time step equal two units; this accounts for the fact that an edge has two ends. During one time step at time t, the same amount $2/t$ of 'water' falls into each bucket. Let the bucket walls be vertical. If s is large, then the level $k(s,t)$ of water (degree) in a bucket (vertex) s at time t is

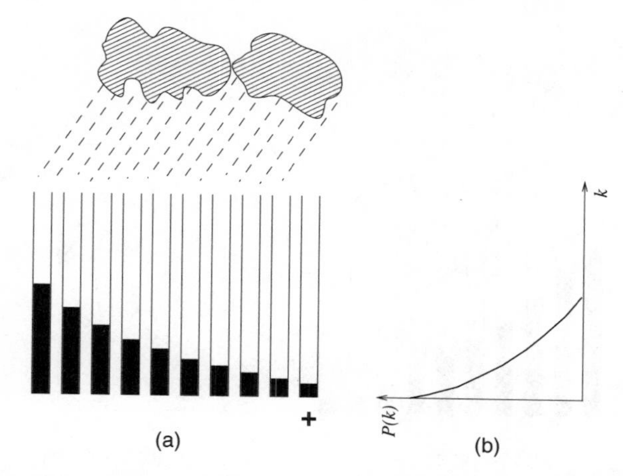

FIG. 2.2. Scheme of distribution of edges over the network that grows under the mechanism of random linking (a) (see the network (2) of Section 1.1). In our 'continuum' approach, vertices are buckets with impenetrable vertical walls. At each time step, a new bucket is added. In such an approach, the addition of edges is homogeneous rain falling over existing buckets (a). The overall amount of water that rains into the buckets is constant. This produces a slow rise in the level of water k (degree) in the direction of the first bucket and the exponential distribution of the level of water in the entire set of buckets (b).

$$\frac{\partial k(s,t)}{\partial t} = \frac{2}{t} \tag{2.1}$$

where the boundary condition is $k(s=t,t)=0$: a new bucket is empty (a new vertex without connections). This gives the general solution $k(s,t) = 2\ln t + C(s)$, where $C(s)$ can be obtained by using the boundary condition $k(s,s) = 0 = 2\ln s + C(s)$. Hence the solution is

$$k(s,t) = 2\ln(t/s)\,, \tag{2.2}$$

or one can write $s(k,t) = t\,e^{k/2}$. Obviously the level of water increases as we approach the first bucket but this increase is very slow (logarithmic function of s, that is the singularity at $s=0$ is weak).

It is easy to estimate the distribution of the level of water in the buckets, $P(k,t)$ (degree distribution). It is proportional to the number of buckets with a level k in the interval $(k-\delta, k+\delta)$, where δ is some small variation. This number is proportional to $[\partial s(k,t)/\partial k]2\delta$, so that

$$P(k,t) = -\frac{1}{t}\frac{\partial s(k,t)}{\partial k} = \frac{1}{2}\,e^{-k/2}\,. \tag{2.3}$$

The factor $1/t$ yields the proper normalization of the distribution.

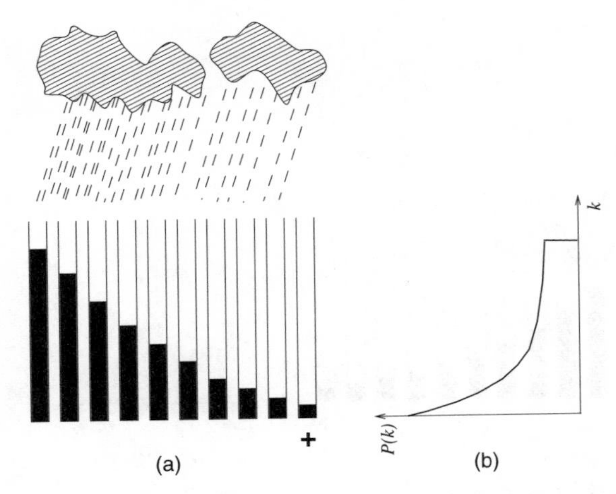

FIG. 2.3. Scheme of distribution of edges over the network that grows under the mechanism of preferential linking (a). In our 'rain approach', at each time step, a new bucket is added. 'Clouds' are attracted by fuller buckets (the first ones). The overall amount of water that rains into the buckets is constant. This produces a power-law rise in the level of water k (degree) in the direction of the first bucket and the power-law distribution of the level of water in the entire set of buckets (b).

Thus we have obtained the stationary exponential distribution. This very simple estimate actually coincides with the exact result for this network. We again have a rapidly decreasing distribution, although the decrease is not so rapid as that for a Poisson distribution. To get fat tails, we have to implement a stronger mechanism.

2.2 Preferential linking

Let us change the way the new edges become attached to vertices. We use the 'attractiveness of popularity'. Here we consider a linear type of preferential linking: the probability that the end of a new edge becomes attached to a vertex of degree k is proportional to $k+A$. Here the constant A is greater than 0, otherwise all the 'water' is collected in the first bucket. This form of preference generalizes that for the Barabási–Albert model (Barabási and Albert 1999). Note that we choose the linear linking not for simplicity but since this type produces the effect we want to demonstrate.

Let us again apply our 'rain approach'. The essential difference from the previous network is that the rain is no longer homogeneous (see Fig. 2.3(a)). The rain over the first (that is, fuller) buckets is heavier. Again let the 'intensity of rain' equal two units, which corresponds to two ends of one new edge per time step. The amount of water that falls down into a bucket s is $2[k(s,t) + A]/\sum_{u=0}^{t}[k(u,t) + A] = 2[k(s,t) + A]/[(2 + A)t]$. Indeed, $\sum_{u=0}^{t} k(u,t)$ is equal to the total amount of water in the buckets, that is $2t$ ($t \gg 1$). Then the level of water (degree) in buckets is determined by

$$\frac{\partial k(s,t)}{\partial t} = 2\frac{k(s,t)+A}{(2+A)t} \tag{2.4}$$

with the boundary condition $k(s=t,t)=0$ (recall that any new bucket is empty).

The solution of eqn (2.4) is

$$k(s,t)+A = \left(\frac{s}{t}\right)^{-1/(1+A/2)}. \tag{2.5}$$

$k(s)$ grows rapidly with decreasing s, and the singularity at the first bucket, $s=0$, is strong (a power law). Using the left equality of eqn (2.3) gives the power-law stationary distribution

$$P(k) \propto k^{-(2+A/2)} \equiv k^{-\gamma} \tag{2.6}$$

at large k. The γ exponent changes from 2 to ∞ (an exponential distribution) as A grows from 0 to ∞ (absence of preference). Thus we see that the combination of the growth and *linear* preferential linking naturally leads to power-law degree distributions.

Originally, the idea of preferential linking was demonstrated by a simple citation graph, in which each new vertex becomes attached to old ones with probabilities proportional to their degree, that is $A=0$ (Barabási and Albert 1999). If, for example, each new vertex has only one connection (in principle, this is not essential), the 'rain picture' for such a situation is very similar to Fig. 2.3 with only one exception: now each new bucket already contains a unit of water (see Fig. 2.4(a)). This means a single connection of a new vertex. The 'intensity of rain' now is unity: only one edge end becomes attached at each time step.

The amount of water that appears in a bucket s is $k(s,t)/\sum_{u=0}^{t} k(u,t) = k(s,t)/(2t)$. Indeed, the total amount of water in the buckets is $\sum_{u=0}^{t} k(u,t) = 2t$ ($t \gg 1$). Then the equation for $k(s,t)$ is

$$\frac{\partial k(s,t)}{\partial t} = \frac{k(s,t)}{2t} \tag{2.7}$$

where the boundary condition is $k(s=t,t)=1$. The solution of eqn (2.7) with this condition is of the form

$$k(s,t) = \left(\frac{s}{t}\right)^{-1/2}, \tag{2.8}$$

which gives

$$P(k) \propto k^{-3} \equiv k^{-\gamma} \tag{2.9}$$

(we have used the left equality of eqn (2.3)). $\gamma = 3$ is only one particular value of the exponent.

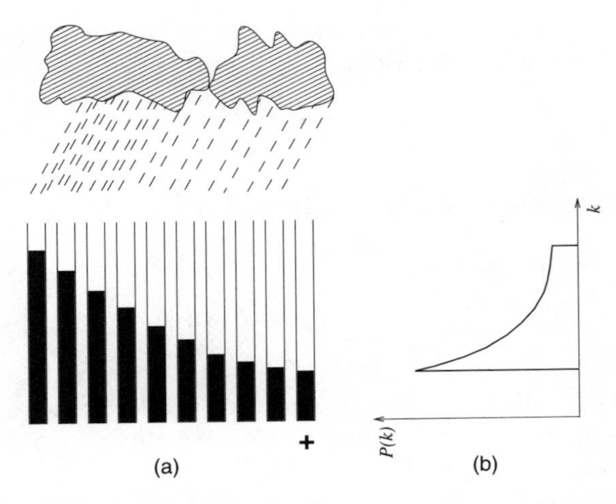

FIG. 2.4. Another example of the growth under the mechanism of preferential linking: the scheme of distribution of edges in the Barabási–Albert model (a). This is a citation graph where each new vertex becomes attached to an old one with probability proportional to the degree of the latter. In our 'rain approach', at each time step a new, partially filled bucket is added. The same total amount of water, as in a new bucket, rains into all the buckets with this proportional preference. This produces a power-law rise in the level of water k (degree) in the direction of the first bucket, $k(s) \propto s^{-1/2}$ and the power-law distribution of the level of water in the entire set of buckets: $P(k) \propto k^{-3}$ (b).

Note that this very naive continuum approach for equilibrium networks with preferential linking does not show power laws. Although preferential linking in equilibrium networks, in principle, may also result in power-law degree distributions, this is a rarer and more complex situation (see Chapter 4).

3

REAL NETWORKS

After having looked through the book, the reader will have noticed Fig. 3.32 where each point corresponds to a particular scale-free network. Several years ago this plot would have been empty: the fat tails in the distributions of connections of all these nets were not noticed. In this chapter we shall consider numerous real networks with an architecture based on fat-tailed distributions of connections.

3.1 Networks of citations of scientific papers

Let us begin with simple growing networks. It seems that scientists are overly concerned by the citation of their papers. Lists of the most cited scientists resemble tennis ratings. However, the statistics of citations were studied long before these 'scientific' ratings appeared (for example, see Lotka 1926 and Shockley 1957).[1] In fact, citations are connections between scientific papers and their statistics reflect a specific aspect of the evolution of science.

Citations in the scientific literature form citation graphs, which we introduced in Section 1.1 (see Fig. 1.6). The vertices of the networks of citations in scientific literature are papers, and their directed edges are citations. In Fig. 3.1 we show the general scheme of the growth of a citation graph. The main simplifying feature of these networks is that old papers cannot cite 'younger' ones. This is valid for books and publications in scientific journals. However, in electronic archives of scientific publications, files can be updated, and, in principle, new citations to 'younger' papers can be added. Therefore, here we consider only scientific pub-

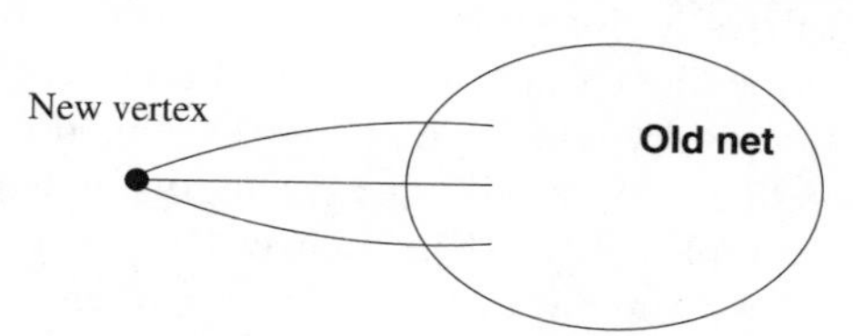

FIG. 3.1. Scheme of the growth of a citation graph. New connections can arise only between new and old vertices. New connections between already existing vertices are impossible. In general, the number of edges attached to a new vertex is not fixed. In networks of citations in scientific papers, all edges (citations) are directed to older vertices (papers).

[1]This is transistor inventor William Shockley, a Nobel Prize winner in physics (1956).

lications in paper form.[2] Naturally, all edges in these nets are directed to older papers.

The largest existing data on scientific citations are collected by the Institute for Scientific Information (ISI database), although other very large data sets on citations exist (take a look at the tremendously thick *Physical Review B* or *E* of the American Physical Society). Data in these databases are stored in such a form that it is much simpler to get the information on the present state of a growing network of scientific citations than to investigate the dynamics of the growth. So, usually the topic of empirical study is the statistics of the instant state of a network of scientific citations. Meanwhile the same is valid for most empirically studied networks.

In- and out-degree distributions

Redner (1998) obtained the empirical distributions of citations to scientific papers by using data from the ISI database for the 1981–June 1997 period and citations from *Physical Review D*, volumes **11–50** (1975–1994). In fact, for these data sets, Redner found the number of papers that have been cited k_i times.[3] This directly corresponds to an in-degree distribution.

The net of the ISI database consists of 783 339 papers and 6 716 198 citations (the average in-degree is $\overline{k}_i = 8.57$). The network of citations in *Physical Review D* contains 24 296 papers with 351 872 citations ($\overline{k}_i = 14.48$). The empirical in-degree distribution obtained by Redner (1998) for citations in the ISI database is represented in Fig. 3.2. The in-degree distribution of the citation network of *Physical Review D* has a similar form.

These were among the first networks reported as having a fat-tailed distribution of connections. Originally, the large-degree parts of the distributions were fitted by a power-law dependence $P(k_i) \propto k_i^{-\gamma_i}$ with exponent close to 3. The fitting by the dependence $(k_i + \text{const})^{-\gamma_i}$ was proposed by Tsallis and de Albuquerque (2000). The exponents were estimated as $\gamma = 2.9$ for the ISI net and $\gamma_i = 2.6$ for the *Physical Review D* citations. However, Fig. 3.2 demonstrates that even the ISI database is not sufficiently large to conclude solidly that the distribution of connections is actually fat-tailed. Later Krapivsky, Redner, and Leyvraz (2000) and Krapivsky and Redner (2001) poined out that these distributions can equally be fitted by rapidly decreasing dependences (see Section 5.9 and Appendix D for examples of the dependences that they used). Furthermore, Vázquez (2001b) described the data set of *Physical Review D* as a combination of two power-law dependences. Thus, various interpretations are possible, and we cannot appear to favour one over another.

The distribution of the number of references in a paper, that is the out-degree distribution, has also been studied (Vázquez 2001b). This distribution is

[2]In non-paper (electronic) scientific journals with refereeing, an update also is impossible.

[3]One can construct a network of citations to authors, but this will be not a citation graph by definition. The statistics of citations to authors are less studied (see Laherrère and Sornette 1998).

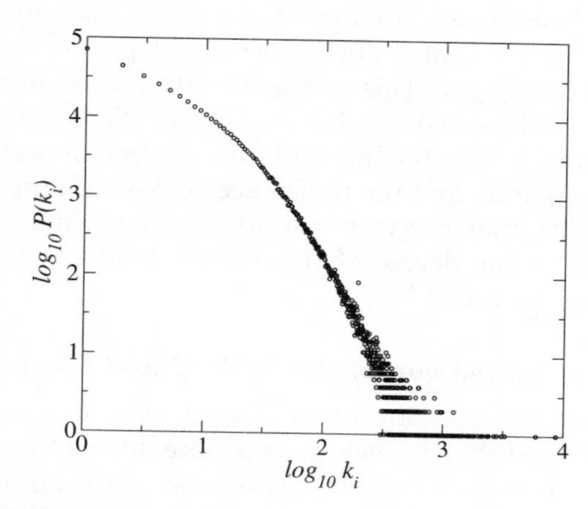

FIG. 3.2. Empirical in-degree distribution of the largest network of citations in scientific literature (after Redner 1998): the net of citations of the ISI database (783 339 papers in total). The distribution function is proportional to the number of papers which have been cited k_i times. Hereafter empirical degree distributions are shown without normalization.

a rapidly decreasing function, and is, as such, of little interest for us.

Evolution of the average number of references

We have noted that the empirical study of the dynamics of the growth of a network is a hard task. Nevertheless, something can be done. Is the average length of a citation list in scientific papers the same as, say, 20 years ago? The data of Vázquez (2001b) show that this is not the case. Vázquez studied the variation of the average number of references in a paper in various scientific journals with increasing time of publication. It transpires that we put more and more references in our papers. In all of the journals studied the average number of references per paper increases. This growth is not so astounding but it is quite noticeable. In various journals the number of references per paper grows by 5%–20% during the period, whilst the number of publications in a journal increases tenfold. As such, connections in networks of citations become denser.

Preferential linking

In the previous chapter we discussed, on a basic level, the nature of fat-tailed distributions of connections in networks, that is the preferential linking mechanism. These were purely 'theoretical' considerations. How can one check that this mechanism actually operates in real networks? Jeong, Néda, and Barabási (2001) empirically studied how papers are cited in a growing citation network or, in other words, the way in which vertices attach edges. They selected all 1736 papers published in *Physical Review Letters* in 1988 and analysed how these papers are cited during some period. Before the accumulation of data started, these 1736 papers had already been cited. So, it was easy to find out how the rate of the increase in the number of citations of a paper depends on the number of citations it has already got. This rate is directly related to the probability of attaching new edges (citations). In this way, Jeong, Néda, and Barabási (2001) arrived at the conclusion that the linking in the citation network of *Physical Review Letters* is preferential, and the preference is close to proportional. In other words, the probability that a vertex will attach a new edge is proportional or nearly proportional to the degree of this vertex. If this is true, the in-degree distribution must be fat-tailed.[4]

3.2 Communication networks: the WWW and the Internet

The WWW and the Internet are often treated as synonyms, but this is not correct. The WWW and the Internet are two quite different networks. In general terms, the Internet is a global net of computers, which are interconnected by wires. This network provides electronic exchange and transmission of information between computers. The vertices of the Internet are

(1) hosts that are the computers of users;
(2) servers (computers or programs providing a network service, which also can be hosts);
(3) routers that arrange traffic across the Internet.

Connections are undirected, and traffic (including its direction) changes constantly.

A scheme of the structure of the Internet, where routers are united in domains, is shown in Fig. 3.3 (Faloutsos, Faloutsos, and Faloutsos 1999). These domains (*autonomous systems*) are subnetworks in the Internet. Within autonomous systems, the routing of information is advertised by some internal rules and algorithms (*internal protocols*). In principle, the internal protocols of distinct autonomous systems should not coincide. To route information between autonomous systems, a special unitary algorithm is used (*the Border Gateway Protocol*).

In January 2001 the Internet contained about 100 million (10^8) hosts. However, it is not the hosts that determine the structure of the Internet but routers and 'domains'. So, one can consider the topology of the Internet at a router level or interdomain topology. In July 2000 there were roughly 150 000 routers in total (Govindan and Tangmunarunkit 2000). One year later the number rose to 228 265 (2×10^5) according to Yook, Jeong, and Barabási's data (2001). In 2001 the total number of domains (autonomous systems) was about 10^4.

[4]The effect of non-linear preferential linking is discussed is Section 5.9 and Appendix D. If the preference function, that is the dependence of the probability that a new edge become attached to a vertex of degree (or in-degree, or out-degree) k is a power law, then two situations are possible: (1) when the exponent of the power-law dependence is greater than 1, a single vertex (hub) attracts an essential fraction of edges; (2) if this exponent is less than 1, the distribution of connections is a rapidly decreasing function.

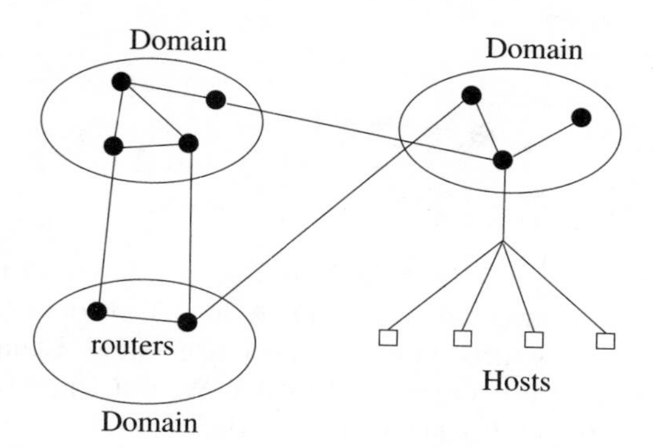

FIG. 3.3. Scheme of the structure of the Internet (Faloutsos, Faloutsos, and Faloutsos 1999). The global structure of the Internet is determined by routers (the router level) and domains (domain level).

The WWW is an array of documents (pages) connected by hyperlinks, which are mutual references in these documents. Hyperlinks are directed, although pairs of counter-links, in principle, may produce undirected connections. Web documents are accessible through the Internet (wires and hardware), and this determines the relation between the Internet and the WWW.

3.2.1 *Structure of the WWW*

In this section we mainly consider empirical facts on the growth and structure of the WWW. However, we start with a very schematic description of this growth process.

How new pages appear in the WWW

Here we describe two simple ways of adding a new document to the WWW (see Fig. 3.4).

(1) Suppose you create your own personal home page. At first you prepare it, put references to some pages of the WWW (usually, there are several such references but these references may be absent), etc. This is the first step. You then have to make your page accessible on the WWW; that is, launch it. You go to your system administrator, who puts a reference to your page (usually one reference) in the home page of your institution, and that is all: your page is on the Web.
(2) There is another way of adding new documents to the WWW. Suppose that you already have your personal home page and want to launch a new document. The process is even simpler than that described above. You simply insert at least one reference to the document into your page, and that is enough for the document to be included on the WWW.

Note that old documents can be updated, and so new hyperlinks between them can emerge. So, the growth of the WWW is a more complex process than

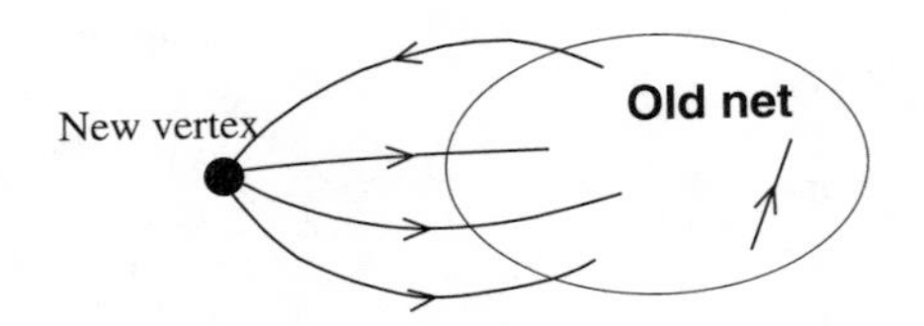

FIG. 3.4. Scheme of the growth of the WWW (compare with the scheme of the growth of a citation graph in Fig. 3.1). A new Web document (page) must have at least one incoming hyperlink to be accessible. Usually it has several references to existing documents of the WWW, but these references may, in principle, be absent. Old pages can be updated, and so new hyperlinks can appear between them.

that of citation networks.[5] What is the result of this process?

In- and out-degree distributions

The first significant results on the distribution of connections in the WWW were reported by Albert, Jeong, and Barabási (1999). Using the complete map of the nd.edu domain of the WWW (325 729 pages in total) they found that both the in-degree and out-degree distributions are of a power-law form and obtained their exponents. An extensive investigation of the structure of the entire WWW has been conducted by Broder, Kumar, Maghoul, Rahavan, Rajagopalan, Stata, Tomkins, and Wiener (2000). In this study the crawl from Altavista was used.

The appearance of the WWW from the point of view of Altavista is as follows:[6]

- In May 1999 the Web consisted of 203×10^6 vertices (URLs, that is pages) and 1466×10^6 hyperlinks. The average in- and out-degrees were $\overline{k}_i = \overline{k}_o = 7.22$.
- In October 1999 there were already 271×10^6 vertices and 2130×10^6 hyperlinks. The average in- and out-degrees were $\overline{k}_i = \overline{k}_o = 7.85$.

Notice that during this period, 68×10^6 pages and 664×10^6 hyperlinks were added, and 9.8 extra hyperlinks appeared per additional page. The average in- and out-degrees increased. Thus the number of hyperlinks grows faster than the number of vertices. In this sense, the growth of the WWW is a non-linear process.

The empirical in- and out-degree distributions of the WWW in October 1999 obtained by Broder *et al.* are represented in Fig. 3.5. The distributions were found to be of a power-law form with exponents $\gamma_i = 2.1$ and $\gamma_o = 2.7$, which confirms the earlier data of Albert, Jeong, and Barabási (1999) for the nd.edu domain of the WWW.

[5]In contrast to citation networks, many pairs of Web documents contain mutual references. In other words, a pair of Web pages may be interconnected by, e.g. two opposing hyperlinks. The fraction of these 'undirected' connections is high: 57%, as was observed in nd.edu domain by Newman, Forrest, and Balthrop (2002).

[6]Strictly speaking, this is not the entire WWW.

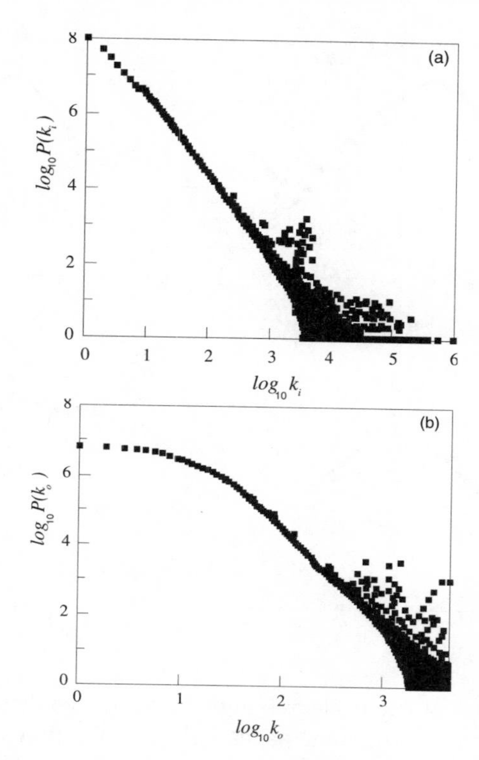

FIG. 3.5. The empirical in-degree (a) and out-degree (b) distributions of the WWW in October 1999 obtained by using the crawl from Altavista (after Broder, Kumar, Maghoul, Rahavan, Rajagopalan, Stata, Tomkins, and Wiener 2000, Copyright 2000 Elsevier Science). The in- and out-degree distributions were fitted by power laws with exponents $\gamma_i = 2.1$ and $\gamma_o = 2.7$.

Deviations from the power-law behaviour are naturally strong at sufficiently small in- and out-degrees. The deviation is more pronounced in the out-degree distribution. However, the fitting can be easily improved by using the functions $\propto (k_{i,o} + c_{i,o})^{-\gamma_{i,o}}$ (Newman, Strogatz, and Watts 2001).[7] The constants $c_{i,o}$ and exponents $\gamma_{i,o}$ are the parameters of the fitting:

- for the in-degree distribution: $\gamma_i = 2.10, \ c_i = 1.25$;
- for the out-degree distribution: $\gamma_o = 2.82, \ c_o = 6.94$.

Various categories of Web pages

Web pages are diverse: the incredibly popular URL http://www.yahoo.com has 630 000 incoming hyperlinks, and the home page of one of the authors of this

[7]This choice for the fitting function is natural. This form arises in the next to lowest approximation to degree distributions of networks growing under the mechanism of preferential linking (see Chapter 5).

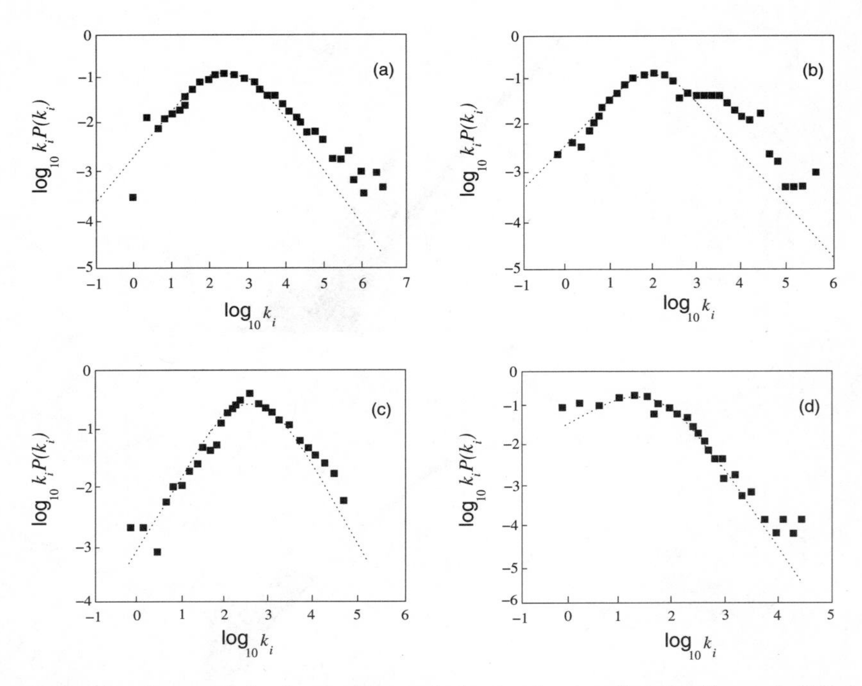

FIG. 3.6. Products of the in-degree and the in-degree distributions, $k_i P(k_i)$, of the four different sets of home pages: (a) a set of public company home pages, (b) a set of US newspapers' home pages, (c) a set of US university home pages, (d) a set of computer scientists' home pages (after Pennok, Flake, Lawrence, Glover, and Giles 2002, Copyright 2002 National Academy of Sciences, U.S.A.). Dotted curves show the fitting by the function $k_i(k_i + c_i)^{-\gamma_i}$ with the parameters from Table 3.1.

book has a single incoming link. Moreover, these two URLs certainly belong to two distinct categories of Web sites: a scientist's home page is nothing like Yahoo.

In this respect the WWW is an inhomogeneous net. Pages of distinct categories evolve differently. What are the results of this evolution? Pennok, Flake, Lawrence, Glover, and Giles (2002) studied the in-degree distributions of home pages of several different categories. Four sets of home pages were analysed: (a) a set of public company home pages, (b) a set of US newspapers' home pages, (c) a set of US university home pages, (d) a set of computer scientists' home pages. It turned out that the in-degree distributions of each of these sets are power-law-like, though with distinct exponents.

To distinguish the region of small in-degrees, Pennok *et al.* considered the products of the in-degree and the in-degree distributions and fitted them by the function $k_i(k_i + c_i)^{-\gamma_i}$. The results are represented in Fig. 3.6 and in Table 3.1. Notice that the fitting is reasonable even at $k_i = 1$, which corresponds to a single

Table 3.1 *The results of the fitting of the in-degree distributions for various categories of home pages in the WWW (Pennok, Flake, Lawrence, Glover, and Giles 2002). The empirical distributions are fitted by the function* $\propto (k_i + c_i)^{-\gamma_i}$, *where* c_i *is a constant. For comparison, in the two last lines the results for the entire WWW are represented: the fitting by the same dependence for the in-degree distribution, and the fitting of the out-degree distribution by the function* $\propto (k_o + c_o)^{-\gamma_o}$.

Category of home pages	γ_i	c_i
Companies	2.05	193
Newspapers	2.05	92
Universities	2.63	1370
Computer scientists	2.66	12
The WWW as a whole (in-)	2.10	0
The WWW as a whole (out-)	$\gamma_o = 2.72$	$c_o = 14$

incoming hyperlink. For comparison, in Table 3.1 the results of these authors, obtained for the in-degree and out-degree distributions of the entire WWW, are represented.[8] The sets of pages of companies and newspapers have in-degree distribution exponents γ_i close to that of the entire WWW. However, γ_i exponents of the university and computer scientist sets are noticeably larger. Thus, distinct sets of Web pages have very different degree distributions.

Components of the WWW

Now the reader has to recall the definitions of the components of a directed graph from Section 1.6: the giant weakly connected component (GWCC) and disconnected components (DC), the giant strongly connected component (GSCC), the giant in-component (GIN), the giant out-component (GOUT), and the tendrils. Figures 1.12 and 1.13 demonstrate the global structure of a general directed graph, which is also the global structure of the WWW. In Fig. 3.7 we represent the sizes of the distinct components of the WWW in May 1999, measured by Broder *et al.*

In Fig. 3.7, one can see the relative sizes of the components. Notice that the heart of the Web, that is, its GSCC, is not so large. It contains only 28% of the total number of pages in the WWW. On the other hand, Fig. 3.7 demonstrates that the WWW is far from either of the birth points of the giant components.[9]

[8]Note that the values c_i and c_o for the entire WWW from Table 3.1 moderately deviate from the values of these parameters of the same fitting, which were obtained by Newman, Strogatz, and Watts (2001).

[9]In Section 6.2 we will demonstrate that there may be two birth points in a directed graph: the birth point of the GWCC and the birth point of the GSCC, GIN, and GOUT. In other words, these are two percolation thresholds: that for the percolation without accounting for the directness of edges, and that for the percolation accounting for the fact that edges are directed.

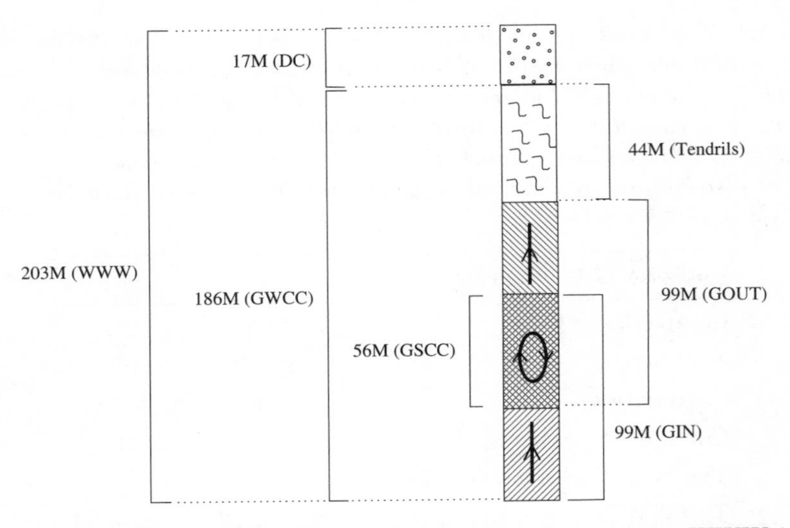

FIG. 3.7. The numbers of pages of distinct components of the WWW in May 1999. 1M represents 1 million pages. The giant weakly connected component (GWCC) and the giant strongly connected component (GSCC) contain 91% and 28% of the total number of pages in the WWW respectively.

Most of the Web pages (91%) belong to the GWCC. Furthermore, this component is extremely stable: Broder *et al.* found that even if all incoming hyperlinks to pages with three or more incoming hyperlinks are removed, the size of the remaining GWCC is still 15 million pages. Albert, Jeong and Barabási (2000) studied the effect of the random deletion of hyperlinks on the GWCC of the WWW (see Section 6.3). It turned out that it is practically impossible to eliminate the GWCC in this way.

Figure 3.8 schematically shows the structure of connections of individual vertices in the complements of the GSCC with respect to the GIN and the GOUT: the GIN – GSCC and GOUT – GSCC. Similarly to GIN and GOUT, one introduces the in- and out-components of a vertex in a directed network. By definition, the vertex can be reached from each of the vertices of its in-component. By definition, each of the vertices of the out-component of a vertex can be reached from this vertex.

Figure 3.8 demonstrates that the in- and out-components of vertices in the GIN – GSCC and GOUT – GSCC are extremely asymmetric.

- A vertex in the GIN – GSCC has an out-component of size close to the size of the GOUT, that is 99 million pages, and, according to the data obtained by Broder *et al.*, the size of its in-component is, on average, only 171.
- A vertex in the GOUT – GSCC has an in-component of size close to the size of the GIN, that is 99 million pages, and the size of its out-component is, on average, 3093.

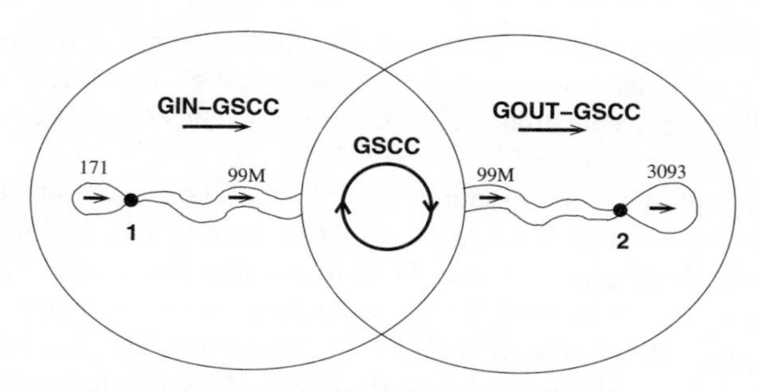

FIG. 3.8. Scheme of the structure of connections of individual vertices (1 and 2) in the complements of the giant strongly connected component with respect to the giant in-component and the giant out-component. $1 \in$ GIN − GSCC and $2 \in$ GOUT − GSCC. (Recall that the GSCC is the interception of the GIN and the GOUT.) The size of the out-component of vertex 1 is close to the size of the GOUT, 99 million pages. The size of the in-component of vertex 2 is close to the size of the GIN, 99 million pages. According to the data of Broder *et al.*, the size of the in-component of vertex 1 is, on average, 171. The size of the out-component of vertex 2 is, on average, 3093.

A noticeable fraction of Web pages do not belong to the giant components. Similarly to giant components, one can consider finite components: weakly connected, strongly connected, and finite in- and out- components. Broder *et al.* obtained the distributions of the number of pages in finite components of the WWW: namely, the size distribution of weakly connected components and the size distribution of strongly connected components. Unexpectedly, in both cases, the empirical size distributions are a power law with the same exponent, namely 2.5.

Why is this surprising? Both the theory of equilibrium networks (see Section 6.1) and the theory of growing nets (see Section 6.10) predict the exponential behaviour of the size distribution of finite components, providing that a giant component of a network is not small, which is just the case for the WWW. Nevertheless, the power-law behaviour was observed. Furthermore, the value of the exponent of the empirical size distribution of the growing WWW coincides with the standard exponent, $5/2$, of the size distribution of equilibrium networks at the birth point of a giant connected component.

Shortest paths in the Web

According to the data of Broder *et al.* for May 1999:

- for pairs of pages of the WWW between which directed paths exist, the *average shortest-directed-path length* is equal to 16;
- for pairs between which at least one undirected path exists (ignoring the directness of hyperlinks), the *average shortest-undirected-path length* is equal to 6.8.

The former value is especially important. In fact, this is the mean minimal number of steps ('clicks') necessary to reach one Web document starting from another by following hyperlinks. This number of clicks sets the characteristic linear scale of Web navigation.

However, it would be a mistake to think that this navigation is easy. We stress that between most Web pages, a directed path is absent. Suppose we have chosen at random Web pages A and B. What is the probability that a directed path exists from page A to page B? This path surely exists if page A belongs to the GIN, and page B belongs to the GOUT. Otherwise, it is hardly possible to get to B from A.[10] So, the probability that page B is reachable from page A is the product of the relative sizes of the GIN and the GOUT. Using the sizes of the giant components of the WWW from Fig. 3.7 gives only a 24% probability that a directed path from A to B exists.

The value of the average shortest-undirected-path length, 6.8, is less than one-half of the value of the average shortest-directed-path length. The former is not so informative in the directed Web. However, it is interesting to compare this value (6.8) with that for the classical random graph with the same number of vertices and the same average degree as the WWW. The standard formula for the average shortest-path length from Section 1.5 gives the same value 7 (more precisely, 7.2) as the one observed in the WWW.

The first data for the average shortest-directed-path length ('diameter') of the WWW were published by Albert, Jeong, and Barabási (1999). Using only data extracted from the nd.edu domain they estimated the average shortest-directed-path length of the entire WWW as 19. This value was obtained in a non-trivial way. In fact, it is not so easy to find the average shortest-path length for a large network. What did Albert *et al.* do to find the 'diameter' of the Web?

(1) The in-degree and out-degree distributions were measured in the nd.edu domain.
(2) A set of rather small model networks of different sizes (the numbers N of vertices) with these in- and out-degree distributions was constructed.
(3) For each of these networks, the average shortest-directed-path length $\overline{\ell}_d$ was found. Its size dependence was estimated as $\overline{\ell}_d(N) \approx 0.35 + 2.06 \log_{10} N$. For the nd.edu domain, this corresponds to $\overline{\ell}_d(325\,729) \approx 11.2$.
(4) $\overline{\ell}_d(N)$ was extrapolated to $N = 800\,000\,000$, that is the estimate of the size of the entire WWW in 1999.

The result, that is $\overline{\ell}_d(800\,000\,000) \approx 19$, is very close to the value $\overline{\ell}_d(200\,000\,000) \approx 16$ of Broder *et al.*, if we account for the difference in the sizes.

[10]For example, if A and B are in the region of 'disconnected components', it is necessary that both the pages be in the same finite connected component. The probability of this event is vanishing. Furthermore, it is evident that if both the pages belong to the complement of the GSCC with respect to the GIN, that is, $A, B \in \text{GIN} - \text{GSCC}$ or if $A, B \in \text{GOUT} - \text{GSCC}$, then the probability that there exists a directed path between A and B is also vanishingly small (see Fig. 3.8).

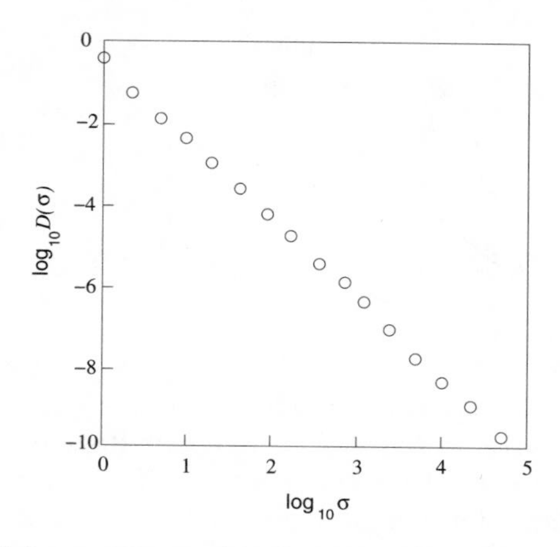

FIG. 3.9. Empirical betweenness distribution $\mathcal{D}(\sigma)$ obtained for the complete map of the nd.edu domain of the WWW (after Goh, Oh, Jeong, Kahng, and Kim 2002, Copyright 2002 National Academy of Sciences, U.S.A.). The distribution is fitted by the law $1/\sigma^2$.

We mainly discuss the average shortest-path length, but there is a wide spectrum of shortest-path lengths between pages of the WWW. As is natural, a shortest-path length between a pair of pages crucially depends on where these pages are situated (in which components). For example, Broder *et al.* estimated that the maximal shortest-directed-path length in the strongly connected component is at least 28 (or greater). On the other hand, according to their data, pairs of pages exist which are separated by a shortest directed path of length about 1 000 clicks long!

Betweenness

The distribution of the number of shortest paths through vertices of the WWW was studied by Goh, Oh, Jeong, Kahng, and Kim (2002). These authors analysed the betweenness distribution $\mathcal{D}(\sigma)$ (see Section 1.5) for the complete map of the nd.edu domain of the WWW. The resulting empirical distribution is presented in Fig. 3.9. The power-law behaviour of the betweenness distribution is evident. We should note that, in general, the betweenness distribution in scale-free networks is scale-free. Furthermore, the empirical distribution is fairly well described by the law $1/\sigma^2$ for scale-free networks. However, this $1/\sigma^2$ dependence is especially interesting. Firstly, it follows from this form that a finite fraction of shortest paths between vertices of the network passes through one or a few vertices. Secondly, the $1/\sigma^2$ behaviour seems to be universal. What are the limits of this universality? We do not know the answer yet (see Sections 3.11 and 6.11).

Cliques in the WWW

Cliques are specific bipartite subgraphs in directed graphs (see Fig. 3.10). A clique consists of two distinct kinds of vertices: *hubs* ('fans') and *authorities*

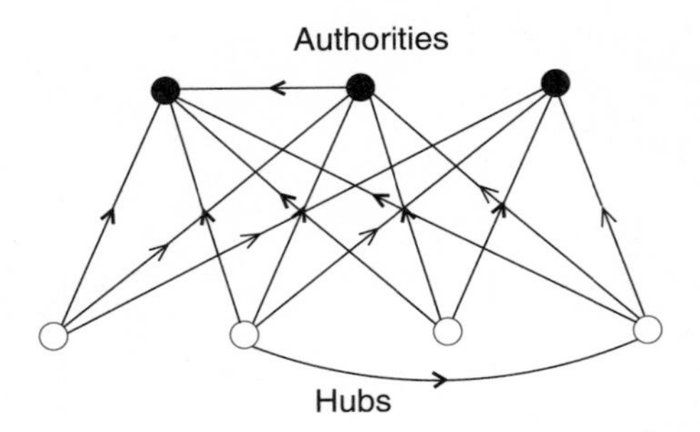

FIG. 3.10. A clique: a bipartite directed subgraph which is used for indexing cyber-communities in the WWW (Kumar, Raghavan, Rajagopalan, and Tomkins 1999a, 1999b, Kleinberg, Kumar, Raghavan, Rajagopalan, and Tomkins 1999). By definition, all possible edges between subsets of hubs and authorities are present. These edges are directed to authorities.

('idols' or centres). By definition, each hub connects to all the authorities of the clique through outgoing edges. If there are h hubs and a authorities in a clique, then the total number of edges connecting hubs and authorities is equal to ha. Connections between hubs, as well as connections between authorities, are arbitrary.

Note that the total number of cliques in a large growing network is large (see discussion in Section 5.4 and Appendix C). This is not the case in equilibrium uncorrelated graphs. In fact, cliques are specific correlations in a network. So, it is quite natural that at the thermodynamic limit of an uncorrelated graph, the relative number of cliques approaches zero.

We will explain below how the indexing of these specific substructures can be used to optimize a search in the WWW. The statistics of cliques in the WWW were studied by Kumar, Raghavan, Rajagopalan, and Tomkins (1999a, 1999b) and Kleinberg, Kumar, Raghavan, Rajagopalan, and Tomkins (1999). A very effective iterative algorithm for extracting cliques was proposed by Kleinberg (1998). By this method, a wide spectrum of cliques in the WWW was extracted. The statistics turned out to be sufficiently large, and the numbers of cliques with h hubs ($2 \le h \le 7$) and a authorities ($3 \le a \le 20$), $N_{\text{cliq}}(h, a)$, were analysed. Of course, it is hard to draw absolutely reliable conclusions by using so narrow a range of values. The empirical data were interpreted as follows:

- for a fixed number of authorities in a clique, $N_{\text{cliq}}(h, a = \text{fixed})$ decays as an exponential function of the number of hubs;
- for a fixed number of hubs in a clique, the decreasing dependence $N_{\text{cliq}}(h = \text{fixed}, a)$ can be fitted by a power law with exponent between 1.1 and 1.4.

The former result (the exponential form of $N_{\text{cliq}}(h, a = \text{fixed})$) is easily reproduced in simple models of growing networks (see Section 5.4 and Appendix C). However, as far as we know, the empirical power-law dependence $N_{\text{cliq}}(h = \text{fixed}, a)$ has not yet been explained.

3.2.2 *Search in the WWW*

The technology of searching in the Web is even more exciting than the Web itself. A hundred million users have a good chance to find and retrieve relevant information from over a hundred million Web documents in a few seconds. To a large extent, the possibility of this nearly unbelievable search is based on the rapid development of hardware, especially in storage systems. Huge arrays of hard drives store an appreciable portion of the Web documents in a cached form. Immensely powerful hardware rapidly processes queries. However, algorithmic ideas are no less important.

How do search engines operate? Our brief description will be very simplified (see, e.g. Lynch 1997). The basic elements of the standard technology of searching in the WWW are as follows:

- *Crawling.* Special programs (crawlers, robots, spiders) download and process documents from the WWW. These documents are usually stored in a transformed form on the hard drives of search engines. Of course, it is impossible to collect the entire Web, and so the crawlers pick up and process more 'interesting and important' documents. We do not discuss what 'interesting and important' means and how the crawlers make the selection. Naturally, this global inspection and downloading is a slow process, and it takes several months or even longer to scan the WWW. The retrieval of fresh information is a weak point of search engines.
- *Indexing.* Search engines examine collected pages and, for every term (word), maintain a list of all the pages with this word. The huge collection of these lists is an *index*. The queries usually arrive at a search engine in the form of a combination of words. Therefore, they can be answered by taking the intersection of the corresponding lists from the index.
- *Page ranking.* The resulting lists, which should be returned in response to queries, may be frighteningly long. Without additional processing such long lists of the page addresses would be useless. So, by using various criteria, more 'interesting and important' pages are placed at the head of a list which is returned to the user: pages are ranked in order of their importance to the user.

What information is indexed? For example, suppose we are searching for information about a scientific paper. If we use in our query the names of authors, words from the title of the paper, key words, etc., success is nearly guaranteed. Our chances are lower if we use terms from the abstract. There are few chances to find a trace of the paper if we send a request with a sentence taken from the body of the article.[11]

[11]Full-text indexing is realized only in some electronic databases. For example, see the NEC Research Institute ResearchIndex (http://www.researchindex.com/).

Knowledge of the structure of connections in the WWW is used in all three items of the list above. Pages with a higher number of incoming links are 'attended' by crawlers more frequently than 'unpopular' pages. The index is maintained by taking into account the connections of pages. The page ranking usually uses the number of incoming links of the page. Moreover, it is important from what pages these links are received. So, to rank a document, one has to know the structure of connections in the whole cluster of pages to which the document belongs.

The difficult task of search engines is to find and indicate pages which are most 'interesting and important' for a given query. One of the methods of location of these pages uses specific substructures of the WWW, which are called *cybercommunities* (Kleinberg 1999a). Roughly speaking, these are just the cliques which we discussed in the previous section (see Fig. 3.10). Cybercommunities can be effectively located, and it is their authorities which are chosen as pages with a top rank.

3.2.3 *Structure of the Internet*

We consider below the two levels of structural organization of the Internet: the interdomain level and the router level.

Interdomain level

The valuable data on the structure of the Internet on the 'interdomain' level are collected by the National Laboratory for Applied Network Research (NLANR). On its Web site, http://moat.nlanr.et/, one can find extensive Internet routing-related information collected since November 1997. For nearly each day of this period, NLANR has a complete map of connections of operating autonomous systems (AS). These maps are undirected connected networks,[12] and each of them represents the Internet at the interdomain level in the corresponding day.

Note that these graphs (including the total number of AS, which are observed by NLANR, and the total number of connections) fluctuate from day to day. For example, on 6 December 1999, 6301 AS numbers with 13 485 interconnections were observed, and so the mean degree of the graph was $\overline{k} = 4.28$, but two days later, on 8 December 1999, there were only 768 AS numbers and 1857 interconnections(!), so the mean degree was $\overline{k} = 4.84$. This dramatic drop is an exception; nevertheless the fluctuations are noticeable all the time.

The origin of these fluctuations is not only the emergence of new AS numbers and the mortality of AS (some Internet Service Providers have to go out of business) but also the permanent variation of interconnections. Connections between AS numbers may be rearranged and may flicker. If an AS number has few connections, there is a probability that all its connections become shut down from time to time. At these 'moments', this isolated AS number is 'invisible' from the outside, and so is absent on the corresponding map from the NLANR

[12]There graphs have no separate parts.

Table 3.2 *Basic parameters of the Internet graph (the interdomain level). The data are obtained from the analysis of the maps of operating autonomous systems. The parameters for the maps from November 1997 and April and December 1999 are taken from the paper of Faloutsos, Faloutsos, and Faloutsos (1999). The average values for 1997, 1998, and 1999 were obtained by Pastor-Satorras, Vázquez, and Vespignani (2001) and Vázquez, Pastor-Satorras, and Vespignani (2002). N is the total number of AS numbers, L is the total number of interconnections, $\overline{k}$ is the average degree, $\overline{\ell}$ is the average shortest-path length, C is the mean clustering coefficient, N_{new} is the number of new AS that emerged in a given year, N_{died} is the number of AS that disappeared during this year. The maps were collected since November 1997, and so the values of N_{new} and N_{died} for 1997 are actually related to the last two months of this year. Notice that the mean number of interconnections grows with time, and $\overline{\ell}$ is stable.*

	N	L	$\overline{k}$	$\overline{\ell}$	C	N_{new}	N_{died}
November 1997	3015	5156	3.42	3.76	—	—	—
Average 1997	3112	5450	3.5	3.8	0.18	309	129
April 1998	3530	6432	3.65	3.77	—	—	—
Average 1998	3834	6990	3.6	3.8	0.21	1990	887
December 1998	4389	8256	3.76	3.75	—	—	—
Average 1999	5287	10 100	3.8	3.7	0.24	3410	1713

collection. Highly connected AS numbers usually do not die and rarely lose their connections.

The maps collected by NLANR were analysed by Faloutsos, Faloutsos, and Faloutsos (1999) and later by Pastor-Satorras, Vázquez, and Vespignani (2001) and Vázquez, Pastor-Satorras, and Vespignani (2002). The basic parameters of these maps of the Internet are represented in Table 3.2.[13]

The table demonstrates that the Internet at the interdomain level is a rather small sparse net. However, the mean number of connections of its vertices, that is, the average degree, noticeably increases as the Internet grows. The number of AS which disappeared from the Internet is surprisingly large. In 1997, 1998, and 1999, for each disappeared AS number, there were only 2.4, 2.2, and 2.0 newborn AS numbers respectively. We conclude that competition between Internet Service Providers is increasing.

The data from Table 3.2 show that the Internet is a small world. The *average shortest-path length* $\overline{\ell}$ of the AS maps from NLANR is below 4. So, most of the vertices can be reached in four steps. The ratio of $\overline{\ell}$ and the average shortest-path length of a corresponding classical random graph is 0.6. Furthermore, Vázquez,

[13]To get rid of the fluctuations, the parameters of the NLANR maps are usually averaged over a period of time.

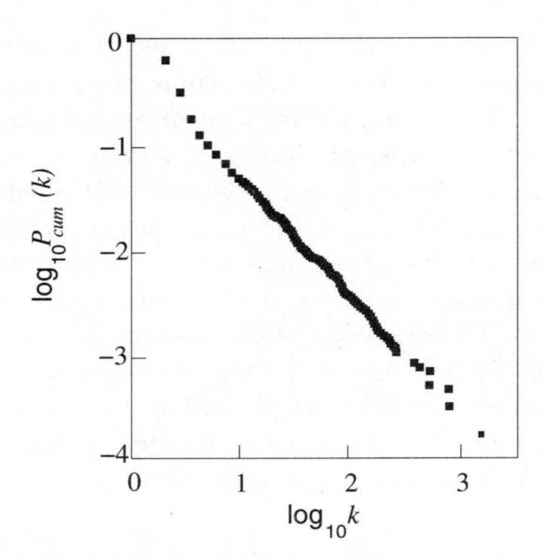

FIG. 3.11. Cumulative degree distribution of the map of autonomous systems in 1999 (after Pastor-Satorras, Vázquez, and Vespignani 2001). The empirical cumulative degree distribution was fitted by a power law with exponent 1.2, which corresponds to the exponent $\gamma = 2.2$.

Pastor-Satorras, and Vespignani (2002) obtained the degree distribution of the shortest-path length by using one of the maps from NLANR (see Fig. 5.19(b) from Section 5.14). From this plot, the reader can see that the maximum separation between a pair of vertices does not exceed 10 or 11, which also is not so large.

The *clustering coefficient* of the maps of AS is about 0.2. This is essentially greater than the clustering of the corresponding classical random graph (the relative difference is about 300). On the other hand, the fat-tailed degree distribution of the Internet graph is so distinct from a Poisson distribution that this difference is not surprising. In Section 6.1 we will see that networks with fat-tailed distributions of connections necessarily have much greater clustering than that of the classical random graphs with the same total numbers of vertices and connections.

Degree distributions are the central empirical results for these networks. We have mentioned that the degree distribution of the Internet at the interdomain level was originally obtained by Faloutsos, Faloutsos, and Faloutsos (1999). In Fig. 3.11 we reproduce the empirical cumulative degree distribution obtained by Pastor-Satorras, Vázquez, and Vespignani (2001) by using more recent data from NLANR. The degree distribution is described by a power law with exponent $\gamma = 2.2$.[14]

[14] One can estimate the clustering coefficient by inserting the first and the second moments of the empirical degree distribution into eqn (6.1) from Section 6.1 or make a simple analytical estimation with $\gamma = 2.2$. The result is much closer to the measured value than to the clustering coefficient of the corresponding classical random graph.

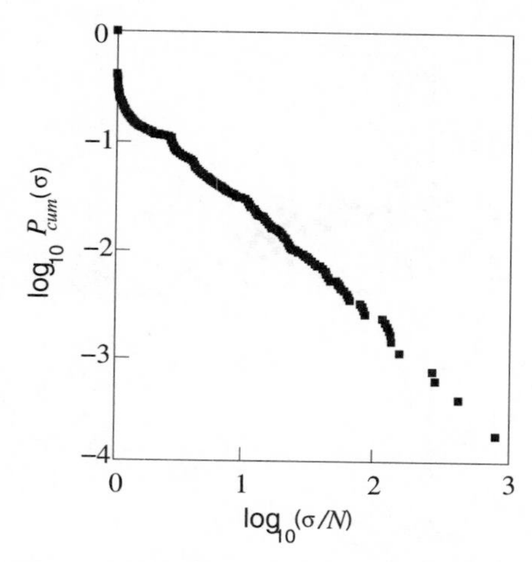

FIG. 3.12. Cumulative betweenness distribution of the map of autonomous systems in 1999 (according to Vázquez, Pastor-Satorras, and Vespignani 2002). σ is the betweenness, N is the total number of AS numbers in the network. The exponent of the corresponding betweenness distribution was evaluated as 2.1 ± 0.2.

The *betweenness distribution* which is determined by the distribution of the shortest paths over the Internet, also turned out to be of a power-law form. In Fig. 3.12 we show the cumulative betweenness distribution of the map of autonomous systems (Vázquez, Pastor-Satorras, and Vespignani 2002, see also Goh, Oh, Jeong, Kahng, and Kim 2002). The betweenness distribution can be described by a power law with the exponent close to 2.

It is natural that the number of the shortest paths that pass through a vertex should strongly correlate with the number of its connections. This also was observed empirically. By studying the AS maps, Vázquez, Pastor-Satorras, and Vespignani (2002) found that the dependence of the betweenness of a vertex (an AS number) on its degree (the number of its connections) is close to proportional.

Less simple kinds of correlations in networks are *pair correlations*. These are the correlations of the degrees of the nearest-neighbour vertices, which may be characterized by a joint distribution of degrees of the nearest neighbours, $P(k, k')$. The pair correlations are absent in classical random graphs and in many other equilibrium networks, but, in growing networks, are natural (see the discussion in Section 5.12). The joint degree–degree distribution $P(k, k')$ is inconvenient for empirical analysis owing to the necessarily poor statistics. Pastor-Satorras, Vázquez, and Vespignani (2001) found a way to study correlations without considering the joint degree–degree distribution. From the AS maps, they obtained the average degree of the nearest neighbours of a vertex as a function of the degree of this vertex, $\overline{k}_{nn}(k)$. In this dependence, only one variable (k) is present, and there are no problems with the statistics. The inde-

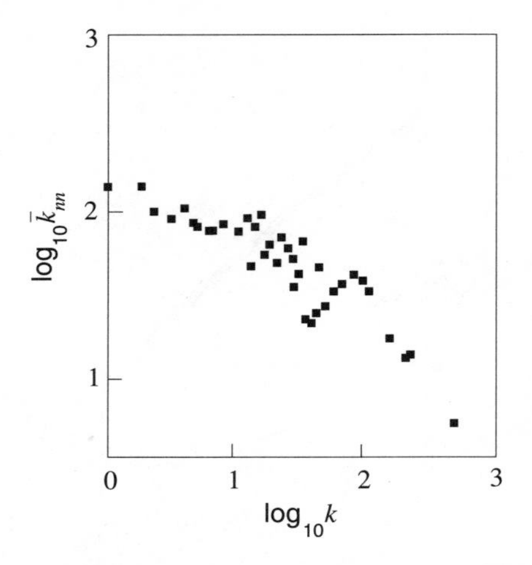

FIG. 3.13. Correlations of the degrees of nearest-neighbour vertices (autonomous systems) in the Internet at the interdomain level (after Pastor-Satorras, Vázquez, and Vespignani 2001). The empirical dependence of the average degree of the nearest neighbours of a vertex on the degree of this vertex is shown in a log–log scale. This empirical dependence was fitted by a power law with exponent approximately 0.5.

pendence of $\overline{k}_{nn}$ of k would indicate the absence of the pair correlations and vice versa.

The resulting empirical dependence $\overline{k}_{nn}(k)$ is presented in Fig. 3.13. The pair correlations are indeed strong. One sees that, as a rule, highly connected vertices have 'poor' nearest neighbours, that is vertices with small degrees. So, in this respect, the growing Internet is extremely far from a classical random graph. The authors of this paper fitted the empirical dependence by a power law with exponent of approximately 0.5. We will discuss the nature of the pair correlations in growing networks in Section 5.12.

The NLANR map collection provides a unique possibility to investigate the mechanism of *preferential linking* in the Internet. By comparing the sequences of the maps one can study the dynamics of the emergence and rearrangement of interconnections. In a similar way as for networks of scientific citations (see Section 3.1), one can find how the rate of attachment of new interconnections to a vertex (an AS number) depends on the degree of this vertex. This program was realized by Jeong, Néda, and Barabási (2001). The observed dynamics of the growth of the numbers of interconnections of AS indicate that the mechanism of preferential linking operates in the Internet. Furthermore, the preference turned out to be close to proportional.

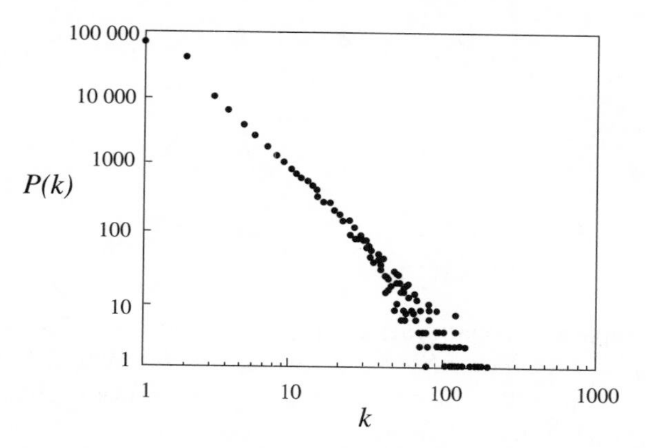

FIG. 3.14. The empirical degree distribution of the map of routers in the Internet in 2000 (after Govindan and Tangmunarunkit 2000, Copyright 2000 Institute of Electrical and Electronics Engineers, Inc.).

Router level

In 2000 the Internet contained about 150 000 routers connected by 200 000 links (Govindan and Tangmunarunkit 2000). In the next year 228 265 routers were observed. This may be compared with the total number (12 409) of AS approximately at the same time. One sees that the net of routers is much larger. However, the structure of the Internet at the router level is less studied.

According to relatively poor data from 1995 (Faloutsos, Faloutsos, and Faloutsos 1999), the map of the Internet at the router level consisted of 3888 vertices and 5012 edges, with a mean degree equal to 2.57. The empirical degree distribution of this network was fitted by a power-law dependence with exponent equal to 2.5. The degree distribution of the router graph, which was obtained by Govindan and Tangmunarunkit (2000) by using a more recent map of the Internet, is represented in Fig. 3.14. This plot 'lends some support to the conjecture that a power law governs the degree distribution of real networks'.[15] If this is true, then from Fig. 3.14, one can estimate that the γ exponent of the degree distribution is about 2.3.

From the data of Govindan and Tangmunarunkit (2000) one can also estimate the average shortest-path length or the Internet at the router level as $\bar{l} \approx 10$. This value is slightly less than that for the corresponding classical random graph—the Internet (at each of its levels) is a small world.

Geographic factor

We have not yet touched upon a factor which can seriously influence the structure of many networks: the geographic location of vertices. This factor is certainly important in many situations, where the probability of connection strongly depends

[15]Here we cite the original paper of Govindan and Tangmunarunkit (2000). Note that, as a rule, other empirical researchers are more fearless in their claims.

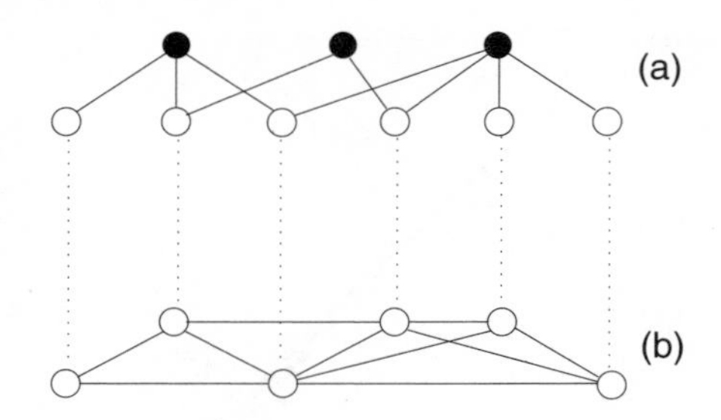

FIG. 3.15. A bipartite collaboration graph (a) and one of its one-mode projections (b) (according to Newman, Strogatz, and Watts 2001). Open and filled circles show collaborators and acts of collaboration respectively. In the one-mode projection, each two collaborating vertices are directly connected. Note that each act of collaboration with more than two collaborators leads to triangles of edges in the one-mode projection graph. Hence, as a rule, one-mode projection collaboration graphs have numerous loops of length 3 and, consequently, a high clustering coefficient.

on Euclidean distances between vertices. The interconnections in the Internet graph are wires, and so, by economic reasons, new connections emerge between geographically close routers with greater probability. Thus the geographic location of routers and AS is very essential. The role of the geographic factor in the Internet was empirically studied and modelled by Yook, Jeong, and Barabási (2001). In particular, the location of both the routers and the AS in North America turned out to be closely correlated with the population density.

3.3 Networks of collaborations

Collaborations necessarily imply the presence of two constituents: (1) collaborators and (2) acts of collaboration. So, the set of collaborations can be represented by the *bipartite graph* containing two kinds of vertices: collaborators and acts of collaboration (see Fig. 3.15). Collaborators are connected together through acts of collaboration, so in such bipartite graphs, direct connections between vertices of the same type are impossible. Edges are undirected. For example, in bipartite collaboration networks of scientists (Newman 2001a, 2001e), one kind of vertex corresponds to authors and the other one corresponds to scientific papers. In movie actor bipartite graphs, these two kinds of vertices are actors and films respectively.

Usually, instead of these bipartite graphs, their less informative *one-mode projections* are used. The projection procedure is shown in Fig. 3.15. In a one-mode projection graph, collaborating vertices are directly connected by edges. As a rule, just these projection graphs are referred to as collaboration networks.

Collaboration networks include a wide variety of nets: acquaintance networks, friendship networks, terrorist networks, various memberships, movie actor networks, networks of sexual contacts, networks of relations between enterprises,

Table 3.3 *Main characteristics of the networks of coauthorships of Medline and the Stanford Public Information Retrieval System (Newman 2001e). N is the total number of authors, $\overline{k}$ is the mean number of collaborators of an author (the average degree), C is the clustering coefficient, $\overline{\ell}$ is the average shortest-path length, and ℓ_{max} is the maximum length of the shortest path between a pair of authors in a network. Notice the high mean number of collaborators in experimental high-energy physics (the SPIRES database).*

	N	$\overline{k}$	C	$\overline{\ell}$	ℓ_{max}
Medline	1 520 251	18.1	0.066	4.6	24
SPIRES	56 627	173	0.73	4.0	19

and so on (Amaral, Scala, Barthélémy, and Stanley 2000). Moreover, in many cases, networks of distinct types show some features of collaboration nets. In principle, any interaction may be treated as a collaboration. In a wide sense, the relation between a predator and its prey is also an act of collaboration. There exist numerous variations of collaboration networks.

Small collaboration networks are a matter of permanent interest in social network analysis. For example, the structure of the network centred around the 19 terrorists of 11 September, 2001 was analysed, and all its basic characteristics were found (Krebs 2002). The clustering coefficient, the betweenness, the average shortest-path length, and so on were obtained for each dead terrorist. What can be concluded from this is a 'hard problem' of social analysis.

Physicists are mostly interested in the statistics of sufficiently large collaboration networks. As far as we know, all such nets display the small-world effect and have a large clustering. We present the parameters of two scientific collaboration networks in Table 3.3: the networks of coauthorships of the largest database Medline and of the Stanford Public Information Retrieval System, SPIRES (Newman 2001a, 2001e). In both these collaboration networks, a giant connected component contains about 90% vertices. So, few authors are in separated regions.

We should mention an interesting feature of the distribution of shortest paths over such collaboration networks, observed by Newman (2001e). As a rule, most of the shortest paths between a vertex and the rest of the network pass through only one of the nearest neighbours of the vertex. The rest of the nearest neighbours are insignificant in this respect. S. H. Strogatz has called this effect *funnelling*.

Many collaboration networks have a fat-tailed degree distribution, and in several cases it was interpreted as of a power-law form (Barabási and Albert 1999, Albert and Barabási 2000a, Barabási, Jeong, Néda, Ravasz, Schubert, and Vicsek 2002). In particular, an empirical degree distribution was obtained by using data for 2810 vertices of the web of human sexual contacts (Liljeros, Edling, Amaral, Stanley, and Åberg 2001). This distribution was interpreted as a power

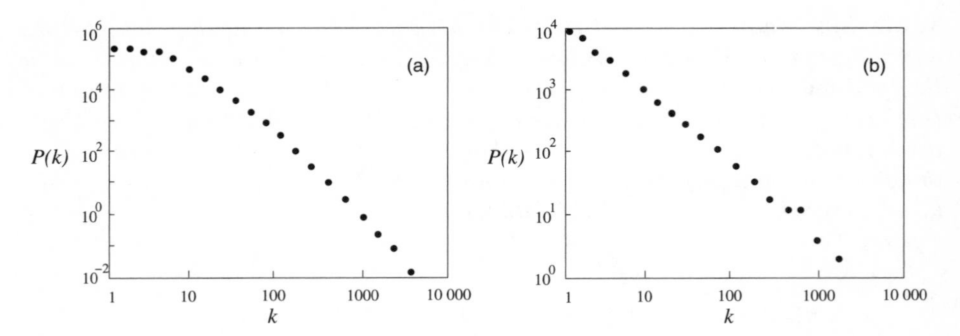

FIG. 3.16. Empirical degree distributions of networks of coauthorships in Medline (a) and the SPIRES database (b) (after Newman 2001e).

law with a γ exponent of about 3.4. We present the fat-tailed empirical degree distributions of the Medline and SPIRES collaboration nets in Fig. 3.16. Notice that the degree distribution of the Medline coothorship network cannot be fitted by a single power-law dependence.[16] The degree distribution of the SPIRES net may be described by a power law with an exponent approximately equal to 1.2. This shows that the exponent of a degree distribution, in principle, may be below 2.

How do collaborations evolve? Barabási, Jeong, Néda, Ravasz, Schubert, and Vicsek (2002) observed an important feature of the dynamics of networks of coauthorships which is typical of a wide circle of collaboration nets. *The mean number of connections of a vertex rapidly increases as a network grows* (see Fig. 3.17). The structure of connections in the network becomes more dense. The observed effect is much stronger than in the Internet and WWW. So, in the infinite network limit, the average degree, that is the first moment of a degree distribution, must diverge. Then, the degree distribution may have an exponent less than 2 (see Section 1.3).

Now, what about preferential linking in collaboration nets? In a similar way as was described in Section 3.1, Jeong, Néda, and Barabási (2001) studied the *dynamics of linking* in scientific collaboration networks and actor nets. It was found that the preferential linking mechanism operates but the preference deviates slightly from proportional.

3.4 Biological networks

3.4.1 *Neural networks*

The organization of neural networks is an increadibly wide field. Here we briefly discuss only a very narrow structural aspect of this topic, which is interesting for us.

The number of neurons in the human brain is of the order of 100 billion (10^{11}), and it is the largest network we mention in our book. The number of neurons

[16]The large-degree part of this degree distribution behaves as $\sim k^{-3.0}$.

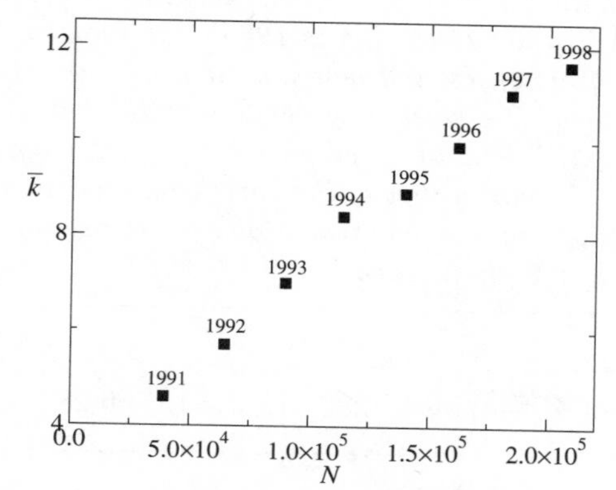

FIG. 3.17. Variation of the mean number of coauthorships (the average degree $\overline{k}$) of the network of coauthorships in neuroscience journals with increasing number of authors, N (according to Barabási, Jeong, Néda, Ravasz, Schubert, and Vicsek 2002).

in the octopus brain is slightly smaller: it is of the order of 100 million (10^8). The number of neurons in the neural network of a tiny organism *Caenorhabditis elegans* (*C. elegans*) is only about 300. Naturally, the small neural network of *C. elegans* is better studied.

Watts and Strogatz (1998) constructed a simple graph ignoring some subtleties of the real neural network and obtained its main structural characteristics (see Table 3.4). In fact, the structure of directed(!) connections between neural cells is rather complex. There exist various kinds of synaptic connections and also direct contacts between bodies of neurons. In the resulting graph, all these connections were treated as undirected edges. Moreover, multiple connections, which play an important role in neural networks, were replaced by single edges. The degree distribution of the resulting network is rapidly decreasing, and so one may reasonably compare the characteristics of this net with those of another network with a rapidly decreasing distribution of connections—that is, with a classical random graph.

The results of such a comparison are shown in Table 3.4. One can see that the neural network is a small world, and that there is some relative difference (5.6) between the value of its clustering coefficient and the clustering coefficient of the corresponding classical random graph. Alas, this relative difference is not great enough to draw the conclusion that neural networks crucially differ from classical random graphs and that the high clustering in neural networks has some special origin. On the other hand, the neural network of *C. elegans* is too small. As such, the clustering coefficient of the corresponding classical random graph is rather large ($C_r = 0.05$), and so the relative difference that we discuss cannot be great (many orders in magnitude).

It is quite possible that in larger neural networks, the clustering will indeed

Table 3.4 *Main characteristics of the neural network of a tiny organism, C. elegans (Watts and Strogatz 1998, Watts 1999): the number of neurons N (vertices), the number of synapses and gap junctions L (edges), the clustering coefficient C, the ratio of the clustering coefficient to that of the corresponding classical random graph, C/C_r, the average shortest-path length $\overline{\ell}$, the ratio of the average shortest-path length to that of the corresponding classical random graph, $\overline{\ell}/\overline{\ell}_r$. The directness of edges is ignored. Multiple connections between a pair of neurons are replaced by a single edge.*

	N	L	C	C/C_r	$\overline{\ell}$	$\overline{\ell}/\overline{\ell}_r$
Neural network of *C. elegans*	282	1.97×10^3	0.28	5.6	2.65	1.18

appear to be many times greater than in uncorrelated equilibrium networks. However, no statistical analysis of the structure of large neural networks is available yet.[17]

3.4.2 *Networks of metabolic reactions*

Life is a chemical process. A complex set of metabolic reactions in an organism provides all necessary energy for life. In general, chemical processes can be conveniently represented by chemical reaction graphs (Kauffman 1993). For systems of metabolic reactions, these directed graphs may be extremely complex. So, metabolic reactions are mostly documented for simple organisms, where the total number of molecular compounds involved in these reactions does not exceed 10^3.

The vertices of metabolic reaction networks are substrates, that is molecular compounds which play the role of educts or products in metabolic reactions. Note that the same vertex may participate in distinct reactions as an educt and product. Directed edges connect educts and products which participate in a metabolic reaction.[18] The incoming connection of a vertex means that this vertex is the product of the reaction. The outgoing edge means that the vertex is the educt of the reaction. In principle, undirected connections are also possible, which shows that a reaction can occur in both directions.

An extensive study of the structural organization of the networks of metabolic reactions of 43 different organisms was undertaken by Jeong, Tombor, Albert,

[17]The same difficulty is met while analysing food webs (see Section 3.4.4) and other small networks. It is rather easy to investigate infinite networks, but they do not exist. It is possible to draw definite conclusions for large networks; however, the reliable treatment of empirical data on small networks is hardly achievable.

[18]Each metabolic reaction connects several substrates. In principle, in addition to vertices–substrates, we can introduce a second kind of vertex for metabolic reactions. This leads to a directed bipartite network which resembles the bipartite collaboration graph shown in Fig. 3.15 from Section 3.3. Substrates play the role of collaborators and reactions play the role of acts of collaboration. In fact, the metabolic reaction network with vertices–substates, which is discussed here, is the one-mode projection of this bipartite reaction graph.

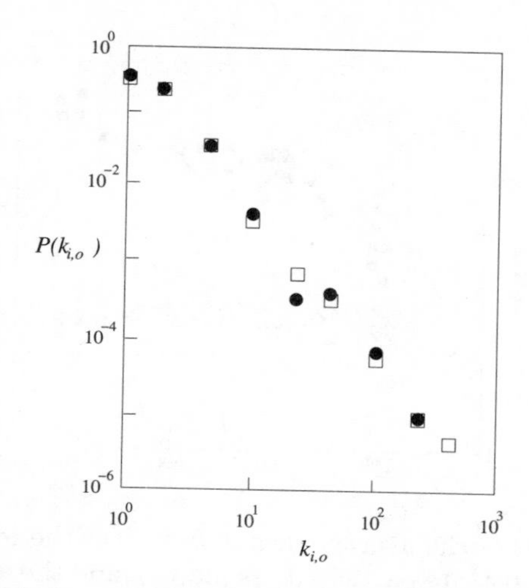

FIG. 3.18. Empirical in- and out-degree distributions of the network of metabolic reactions of the bacterium *E. coli* (after Jeong, Tombor, Albert, Oltvai, and Barabási 2000, Copyright 2000 Macmillan Publishers Ltd). The filled circles and open squares show the in-degree and out-degree distributions respectively. The out-degree distribution is, in fact, the probability that a substrate participates in k_o metabolic reactions as an educt. The in-degree is the probability that a substrate participates in k_i metabolic reactions as a product. The network has 778 vertices with average degree 7.4. Both the distributions were fitted by a power law with exponent equal to 2.2.

Oltvai, and Barabási (2000) (see also Wagner and Fell 2001). These organisms belong to all three domains of life: the Bacteria, the Eukaryota (animals, plants, and so on), and the Archaea. So, it is exciting that the networks of all the 43 organisms have the same scale-free structure.

The sizes of these networks vary from 200 to 800 vertices. In each of the networks, separated clusters contain less than 10% of the total number of substrates. As the size increases, the mean degree of a vertex grows from about 4 to 7.5. In Fig. 3.18 we represent the empirical in- and out-degree distributions of the network of metabolic reactions of the Bacterium *Escherichia coli*, which has a typical form for all the 43 organisms. The studied networks of metabolic reactions are very small, but the fat-tailed form of the degree distribution is clearly present. Both the in- and out-degree distributions can be fitted by a power law with exponent equal to 2.2 over a sufficiently wide range of degrees.[19]

[19]Networks of metabolic reactions are directed networks. Ignoring the directness of edges, one can find the clustering of a network. Wagner and Fell (2001) did this for a fraction of the network of metabolic reactions of *E. coli* (282 substrates with mean degree $\overline{k} = 7.35$). The resulting clustering coefficient is $C = 0.32$, which is 12 times greater than the clustering coefficient of the corresponding classical random graph.

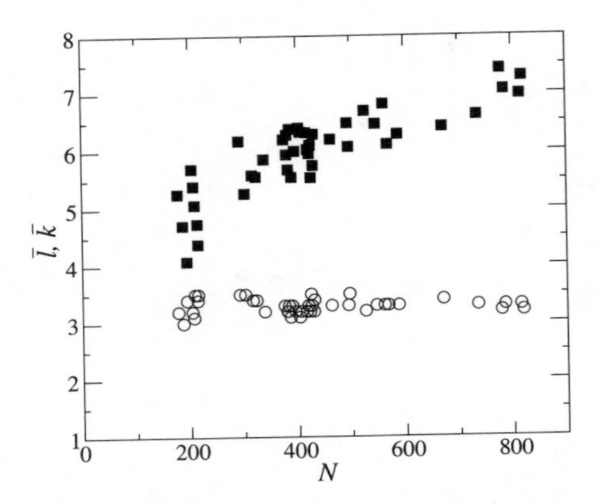

FIG. 3.19. Variations of the average degree $\overline{k}$, that is the mean number of reactions in which a substrate participate (squares), and the average shortest-path length $\overline{\ell}$ (circles) with the size of the networks of metabolic reactions of 43 organisms according to the data of Jeong, Tombor, Albert, Oltvai, and Barabási (2000). Errors of the empirical average shortest-path length are not shown. As the size of the nets increases, connections become more dense, and the average shortest-path length is practically constant.

Furthermore, this value (with non-essential variations) was obtained for all the 43 organisms.

The reader may be surprised. How can one obtain convincing results for such small networks? The reason for this quality of empirical degree distributions is the closeness of the exponent to 2 which takes the cut-off of the distribution away to the area of large degrees (see discussions in Sections 1.3 and 3.11). The observed structure is extremely robust against random defects, errors, mutations, and so on. For example, the random removal of 10%(!) of the total number of vertices in the metabolic network of *E. coli* does not produce any noticeable variation of the average shortest-path length. The measured value of the average directed shortest-path length is very close for all the 43 networks: $\overline{\ell} = 3.0$–3.5, which is a really small number (see Fig. 3.19). However, this is close to the average shortest-path length in the corresponding classical random graphs. One can suggest that this smallness is important: optimal biochemical pathways should not be long. The typical shortest path in the smallest of the studied metabolic reaction nets, roughly speaking, is the same as in the largest one.

This invariance of the average shortest-path length also is not surprising. We mentioned above that the average degree of these networks grows with their size (see Fig. 3.19). Let us estimate the average shortest-path length for the corresponding classical random graphs for the smallest and the largest nets from this set ignoring the directness of edges. The smallest is the net of *Mycoplasma pneumoniae*: 178 substrates, $\overline{k} = 5,26$, so $\overline{\ell} = \ln N/\ln \overline{k} = 3.12$. The largest net is the metabolic reaction network of *Salmonella typhi*: 819 substrates, $\overline{k} = 7.28$, and so $\overline{\ell} = 3.38$. One sees that the difference is subtle.[20]

Thus, nature has chosen the optimal construction of life. How this construction was selected is an open question.

3.4.3 *Genome and protein networks*

There are two kinds of protein networks:[21]

- *A network of physical protein–protein interactions.* The vertices of this network are proteins. The undirected edges are direct pairwise physical interactions between proteins.[22] These edges form biochemical and signalling pathways by which, in turn, the production and degradation of proteins is regulated.
- *A protein regulatory network* (genetic regulatory network of a cell). The vertices of this directed network are proteins. Two proteins are connected by a directed edge if the first protein directly regulates the production and degradation of the second. Oppositely directed pairs of edges are also possible, which corresponds to mutual regulations.

Genetic regulation actually includes a number of distinct types and levels of regulation, and so the complete genetic regulation network is extremely complex. To construct an interpretable subset of the complete genetic regulatory network, one has to select some specific type of regulation. Usually, this is important transcription regulation.

Jeong, Mason, Barabási, and Oltvai (2001) studied the network of physical protein interactions of the yeast *Saccharomyces cerevisiae* constructed by using the data of Uetz *et al.* (2000). The resulting net consists of 1870 vertices and 2240 edges. So the network is sparse. The average shortest-path length is equal to 6.8 which is 4/5 of the average shortest path of the corresponding classical random graph. A schematic view of a part of the network of physical protein interactions according to Maslov and Sneppen (2002a) is shown in Fig. 3.20.

The largest connected component of the network of the yeast contains 78% of the total number of vertices. Definitely, the distribution decreases slower than an exponent and is fat-tailed. It was fitted by the law $(k + k_0)^{-\gamma} e^{-k/k_{cut}}$, where γ is approximately equal to 2.4, $k_0 = 1$, and $k_{cut} = 20$. Wagner (2001a) studied a similar protein interaction network of the same yeast, and also interpreted the resulting empirical degree distribution as a power law with exponent approximately equal to 2.5. Unfortunately, this value of the exponent (that is, rather far from 2) necessarily leads to the cut-off at low degrees, which hampers the

[20] Since metabolism in all living cells is based on the generic set of substrates, biochemical pathways should be similar in distinct organisms. So, the invariance of $\overline{\ell}$ seems to be natural. Moreover, one may suggest that the constancy of $\overline{\ell}$ is among factors which determine the evolution of many networks (Puniyani and Lukose 2001).

[21] For a more serious introduction to protein networks, the reader should refer to the cell biology literature (for example, see Kauffman 1993).

[22] Roughly speaking, these interactions are direct contacts between proteins.

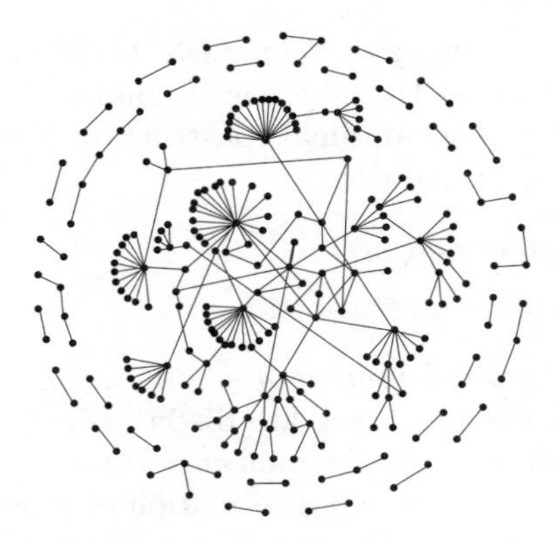

FIG. 3.20. Schematic view of a part of a typical network of physical protein–protein interactions according to Maslov and Sneppen (2002a).

interpretation (see Sections 1.3 and 3.11). The fat-tailed form of the distribution of connections in larger protein interaction networks should be more visible.

In Section 3.2.3 we discussed correlations between degrees of the nearest-neighbour vertices of the Internet. Such correlations are typical of growing (non-equilibrium) networks. Maslov and Sneppen (2002a) investigated the pair correlations in both the protein physical interaction network and the genetic transcription regulatory network of the yeast *S. cerevisiae*. The correlations of the total degrees of the nearest neighbours in these networks were studied. Moreover, it turned out to be possible to observe the correlations in the degree–degree distribution function for the nearest-neighbour vertices, $P(k, k')$. In Fig. 3.21 we present the empirical dependence of the average degree $\overline{k}_{nn}$ of the nearest neighbours of a vertex on the degree k of this vertex. This dependence clearly indicates the presence of the strong pair correlations in both the protein networks. The situation is very similar to the Internet (compare Fig. 3.21 with Fig. 3.13 from Section 3.2.3; also, notice that in Fig. 3.20 most of the highly connected vertices have nearest neighbours of low degrees). This similarity shows that, like the Internet, protein networks are non-equilibrium ones. Each of them may be treated as an intermediate state of a long evolutionary process.

The observed architecture of the protein networks of the yeast, which combines the fat-tailed distribution of connections and the small-world compactness, has two evident consequences: robustness against mutations and shortness of regulatory pathways. There are few doubts that this optimal structure is inherent for all protein networks.

3.4.4 *Ecological and food webs*

Food webs represent a very important 'gastronomic' aspect of relationships between species in ecosystems. In such a web, vertices are species in an ecosystem. Directed edges (trophic links) connect pairs: the predator and prey species. The direction is chosen from a predator to its food. In Fig. 3.24 we schematically show

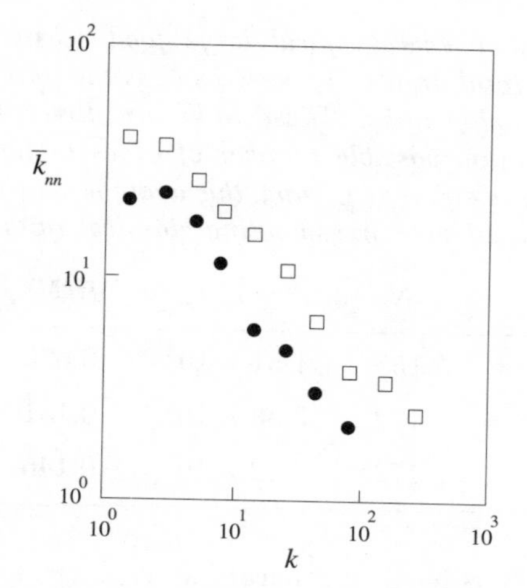

FIG. 3.21. Pair correlations in protein networks. The average degree $\overline{k}_{nn}$ of the nearest neighbours of a vertex versus degree of this vertex, k, for protein networks of the yeast *S. cerevisiae* (after Maslov and Sneppen 2002a). Squares and circles correspond to the physical interaction network and the regulatory network, respectively. The empirical dependence was fitted by the power law $\overline{k}_{nn} \propto k^{-0.6}$. The protein interaction network consists of 3278 proteins with 4549 physical interactions. The genetic transcription regulatory net contains 682 proteins connected by 1289 regulations.

the structure of a typical food web. One can see that every predator is eaten by something else. Some pairs of species eat each other. In addition, cannibalism is widespread (see a unit-length loop in Fig. 3.22). The structure of a food web actually reflects the complex ecological balance of an ecosystem.

The maximum possible number of trophic links in a food web with N trophic species is N^2. In this situation all the species eat each other, which gives $2 \times N(N-1)/2$ edges, and, in addition, they are all cannibals, and so there are N unit loops. So, in total, there can be precisely N^2 trophic links. Of course, not all the species are cannibals, so that the total number of edges L in food webs must be less than the maximum possible number. However, the values of the ratio L/N^2 may be rather large (see examples in Table 3.5).

There are two difficulties with food webs. Firstly, it is hard to construct

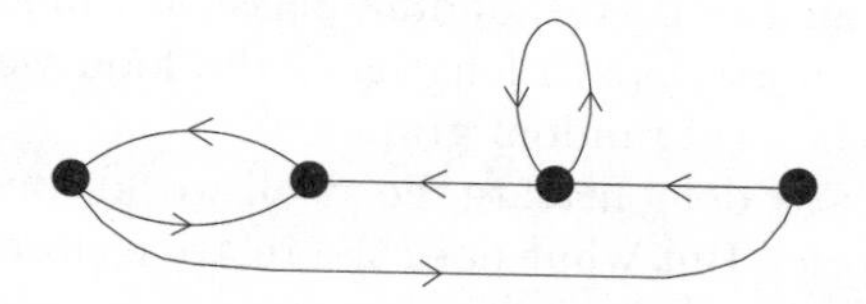

FIG. 3.22. Schematic structure of a typical food web: cannibalism and mutual eating are widespread.

Table 3.5 *Parameters of several typical 'large' food webs (Dunne, Williams, and Martinez 2002a). The total number of vertices N is the number of trophic species, L is the number of trophic links. These links are directed, and cannibalism is admitted, so the maximum possible number of edges including unit-length loops is N^2. The clustering coefficient C and the average shortest-path length $\overline{\ell}$ are relatively close to those of the corresponding classical random graphs.*

Food web	N	L	L/N^2	C	$\overline{\ell}$
El Verde Rainforest	155	1.51×10^3	0.026	0.12	2.20
Lake Tahoe	172	3.88×10^3	0.131	0.14	1.81
Mirrow Lake	172	4.32×10^3	0.146	0.14	1.76

a food web uniquely; it is hard to separate an ecological system perfectly; it is hard to fix the number of trophic species; and it is hard to fix some trophic links. Secondly, all known food webs are *very* small. In Table 3.5 we collect data on the largest food webs. The second circumstance strongly hampers the statistical analysis of these networks.

First empirical analyses of the distribution of connections in food webs suggested that these networks are scale-free (Solé and Montoya 2001, Montoya and Solé 2002). However, food webs are actually too small for such strong conclusions, and a more recent study by Dunne, Williams, and Martinez (2002a) demonstrated rapidly decreasing distributions.

Dunne, Williams, and Martinez (2002a) studied a rather large set of 16 food webs whose sizes varied from 33 to 172 trophic species. Ignoring the directness of trophic links, they found the main structural characteristics for these webs: the degree distributions, the clustering and average shortest-path lengths. In Fig. 3.23, the empirical cumulative degree distributions of the largest networks from this list are shown. These are just the food webs from Table 3.5. One can see that all the degree distributions are not fat-tailed but rapidly decreasing functions.

The empirical clustering coefficients of the resulting undirected webs are, at first sight, large (see Table 3.5). However, one can easily compare the clustering coefficients of the real food webs with those of the corresponding classical random graphs and find that the values are relatively close. Some real food webs are even less clustered than the corresponding classical random graphs. Similarly, the empirical average shortest-path lengths of the food webs are close to those of the corresponding classical random graphs.[23]

Thus food webs clearly demonstrate the small-world effect, which is quite natural for random networks. But what does the relative closeness of the clustering

[23] One can easily check this by substitution of the parameters of the networks from Table 3.5 into standard formulae for classical random graphs.

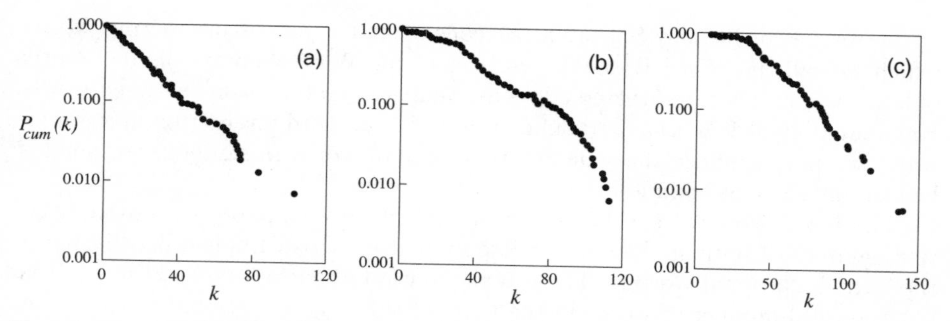

FIG. 3.23. Cumulative degree distributions of food webs from Table 3.5: of (a) El Verde Rainforest, (b) Lake Tahoe, and (c) Mirror Lake (after Dunne, Williams, and Martinez 2002a). All the plots are log–linear, so the distributions decrease rapidly.

of food webs to that of classical random graphs mean? Degree distributions both for food webs and for the classical random graphs are rapidly decreasing functions. So, if the vertices of a food web are connected at random, the clustering coefficient should be close to that of the corresponding classical random graph.

There is another simple reason: the minute size of food webs. The clustering coefficient of a classical random graph, $C_r = \overline{k}/N$, where $\overline{k}$ is its average degree, is relatively close to 1 (the maximum possible value) if the graph is small. The clustering coefficient of an arbitrary random network of the same size and with the same average degree, roughly speaking, may be in the range between 1 and C_r. If C_r is of the order, say, of 10^{-1}, a great relative difference is impossible in principle.

3.4.5 *Word Web of human language*

Briefly speaking, language is the result of the operation of grammatical rules which are defined on a set of words, that is, on a lexicon. In a close definition, language is a dictionary (lexicon) plus a set of rules of grammar imposed upon this dictionary:

$$\text{language} = \text{lexicon} + \text{grammar}.$$

Is it possible to understand the structure of language and the laws of its evolution without serious study of a particular language's grammar? The statistical analysis of texts yields a rather rough and restricted but useful global image of language, where subtleties of a concrete grammar turn out to be of secondary importance. In fact, texts are a very convenient object for statistical analysis, and studies on the statistics of language have a long history. However, these studies mainly focused on the frequency of *occurrence* of distinct words in texts (see Section 7.1). Little information about language can be extracted from this distribution. The approach of Ferrer i Cancho and Solé (2001a), which is based on the statistics of *co-occurrence and interaction* of distinct words in sentences, provides a more complete image of language.

Ferrer i Cancho and Solé (2001a) constructed a net of interacting words, which we call the *Word Web*. The vertices of the Word Web are distinct words of language, and the undirected edges are (pairwise) interactions between words. Unfortunately, it is not easy to define the notion of word interaction in a unique way. One may suppose that the co-occurrence of words in sentences or smaller lexical units means the interaction between words.

The Word Web was constructed in the following way. A representative set of text samples of both spoken and written modern British English is collected in the British National Corpus. These texts contain about 470 000 distinct words, but the total number of words in the texts of the British National Corpus is, of course, much greater, about 70 million words. These 470 000 distinct words are actually the complete dictionary (the lexicon) of modern British English. These distinct words are the vertices of the Word Web. So, the Word Web is really a large network.

If a pair of distinct words interact *at least in one sentence* from this collection of texts, they are connected by an undirected edge.[24] In a naive definition, two words interact if they are the nearest neighbours in a sentence. Of course, 'real' interactions may be more long-range. So, larger lexical units than those made up of two words were actually considered. This is just the point where a rigorous definition is impossible. Without going into detail (see Ferrer i Cancho and Solé 2001a) we note that two slightly different definitions of the pairwise word–word interaction result in networks with nearly the same degree distributions (see Fig. 3.24). In both the cases, the average number of connections of a vertex is $\overline{k} \approx 72$.

The reader can see that the empirical degree distribution is certainly fat-tailed. Furthermore, it has two regions with distinct power-law behaviour separated by a crossover point. The region below the crossover was fitted by a power law with exponent equal to 1.5. Fitting above the crossover is more difficult because of strong fluctuations. The exponent of the power-law dependence was estimated as 2.7. In Section 5.16 we will give a quantitative explanation of this complex form of the degree distribution. The theoretical degree distribution, which is obtained by using only the known parameters of the Word Web, is shown in Fig. 3.24 by a solid line. If impatient readers want to learn immediately about the evolution and self-organization of language, and why the distribution has such a complex form, they must skip to Section 5.16.[25]

[24] A more informative image of language can be obtained if one sets the number of connections between two distinct words to be equal to the number of sentences in which these words interact. This construction would represent not only the structure of interactions between words but also frequencies with which specific interactions occur.

[25] In fact, such degree distributions with two distinct regions of power-law behaviour are usual for various types of collaboration networks. The Word Web can also be considered as a collaboration net where words are collaborators in language. The parameters of the Word Web are so convenient that both these power-law regions are well resolved. In many other collaboration nets, the crossover may occur in the range of too large or too small degrees, the available range of degrees may not be sufficient, and so on. Then, such a clear resolution is

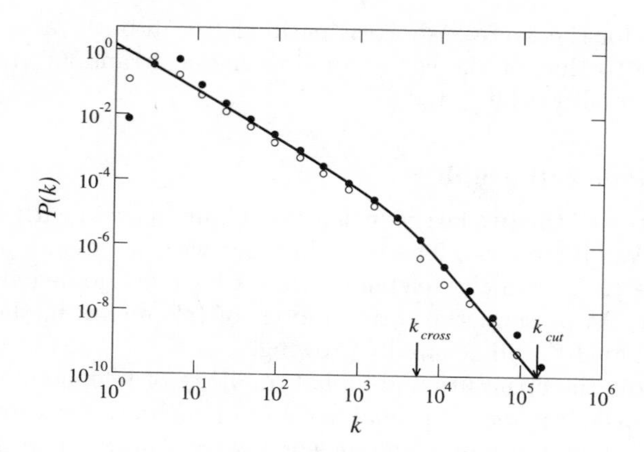

FIG. 3.24. The empirical distribution of the number of connections (the degree distribution) of words in the Word Web (Ferrer i Cancho and Solé 2001a). Open and filled circles show the distributions for the networks constructed with two distinct definitions of word–word interactions. The Word Web contains 470 000 distinct words (vertices); the average number of connections of a word is $\overline{k} \approx 72$. The solid line is the result of calculations that used only these two parameters (see the theory of the evolution of language in Section 5.16). The arrows indicate the theoretically obtained point of crossover, k_{cross}, between the regions with the exponents 3/2 and 3 and the cut-off k_{cut} of the power-law dependence due to size effect. For a better comparison, the theoretical curve is slightly displaced upwards.

Here we recall that if the value of the γ exponent is smaller than 2 for the main part of the degree distribution than the mean number of connections of a vertex in the limit of a large network becomes infinite. So, the total number of edges in the growing Word Web increases faster than the number of words.

Figure 3.24 shows that highly connected words have a different structure of connections from others. One can estimate the number of words above the crossover point. This number is only about 5000, which is much smaller then the total number of words in the Word Web, 470 000. These words play a key role in language.[26]

The Word Web is a compact, strongly clustered structure. For two different definitions of the word–word interactions, its empirical clustering coefficient is equal to 0.69 or 0.44 respectively. This is greater by 4.4×10^3 and 2.8×10^3 than the clustering coefficient of the corresponding classical random graph, re-

impossible. We can only guess that the origin of a complex form of degree distributions in collaboration nets is general. This origin is the specifics of the evolution of these networks where the total number of connections grows unproportionally to the total number of vertices (see more detailed discussion in Section 5.16).

[26]It is because of the smallness of this number, 5000, that the empirical value of exponent γ for the range above the crossover has to be rough.

spectively. As for the average shortest-path length, in both cases it is only 2.65, which is close to that of the corresponding classical random graph. Thus, our language is a small world!

3.5 Telephone call graph

We have mentioned the two largest networks: a human brain with its 10^{11} neurons and the WWW (10^9 pages). There exists a net with no less an impressive total number of vertices, namely telephone nets which are joined in a unit world telephone net. At present, the total number of telephones in the world is only slightly less than 10^9 and is rapidly growing.

Surprisingly, the structure and global topology of telephone nets practically have not yet been studied. A probable cause of such a poor knowledge of these networks is the reluctance of telephone companies to make their data on connections accessible to outsiders, for privacy reasons.

Here we briefly discuss one possible way to construct a graph that reflects the structure of telephone connections. Suppose telephone calls in a telephone net are registered over some period. This means that pairs of telephone numbers are registered: in each pair, the first number has made a call and the second has received this call. Then a telephone call graph is defined as follows: (a) Its vertices are the telephone numbers that have been registered. (b) Its directed edges are telephone calls between these numbers. The direction of an edge is determined by the number that made a call. Note that, in this construction, multiple edges between a pair of vertices are possible.

Aiello, Chung, and Lu (2000) studied the statistics of such a graph that was constructed by using long-distance telephone calls recorded in one day. The resulting directed graph has 47 million vertices and about 8×10^7 connections. Both the distributions of incoming and outgoing connections of this telephone call graph are fat-tailed. However, only the in-degree distribution can be fitted by a power law. Its exponent was roughly estimated as $\gamma_i \approx 2.1$.

3.6 Mail networks

Mail nets are very similar to telephone call graphs. The basic chain in all such networks is

$$\text{sender} \xrightarrow{\text{message}} \text{recipient} .$$

In principle, the actual meanings of the three elements of this chain are not so important for the structure of a net. Here we discuss e-mail networks, which are a more convenient object of study than ordinary mail nets.

In fact, electronic mail is not a less significant component of the information era than the Internet and WWW. We may guess that, for our readers, e-mail is more important than the postal service. Unfortunately, e-mail is also an ideal way for spreading computer viruses.

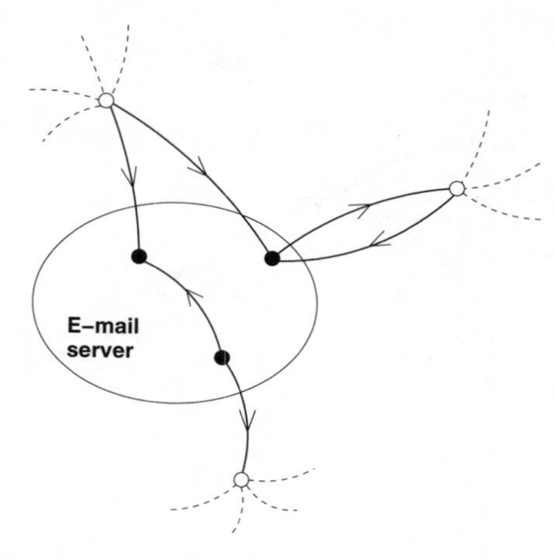

FIG. 3.25. Scheme of an e-mail network constructed by using a log-file taken from an e-mail server. The log-file contains complete lists of connections of e-mail addresses on this server. Nothing is known about connections between external addresses (the dashed lines).

The information on submission and receipt of messages is stored in log-files on e-mail servers. The difficulty is that a log-file contains all the necessary information about e-mail addresses on the host server but, naturally, very little is known about the corresponding addresses outside of this server. The only information about the latter is on the exchange of messages with the 'internal' addresses on the server. Figure 3.25 schematically shows the situation.

Ebel, Mielsch, and Bornholdt (2002) studied the statistics of a large e-mail net, based on students' accounts on the server of Kiel University. The source and destination addresses from the log-files of the server for all e-mail messages were analysed over a period. Using this information, Ebel *et al.* constructed a directed net (in total, 59 912 vertices), only a small part of which were student accounts (5165 vertices). The rest are 'external' addresses. The average degree of the 'complete network' accounting for 'external' addresses and their connections with 'internal' vertices is 2.88. The average degree of 'internal' vertices is much greater, $25.45 = 13.06 + 12.39$. For the internal addresses, we also show the average in- and out-degrees, $\overline{k} = \overline{k}_i + \overline{k}_o$.

In Fig. 3.26 we show the empirical degree and in- and out-degree distributions of the internal addresses. A fat-tailed form of all these distributions is clearly visible. The degree distribution (the $k < 10^2$ region) was fitted by a power law with the γ exponent equal to 1.3. The fitting of the in-degree distribution (the $k_i < 10^2$ region) by a power law gave exponent γ_i equal to 1.5. As for the out-degree distribution, it was impossible to fit it by a power-law dependence. Notice that all these empirical distributions have a complex form, which can hardly be described by a single power-law dependence. The reader may compare Fig. 3.26

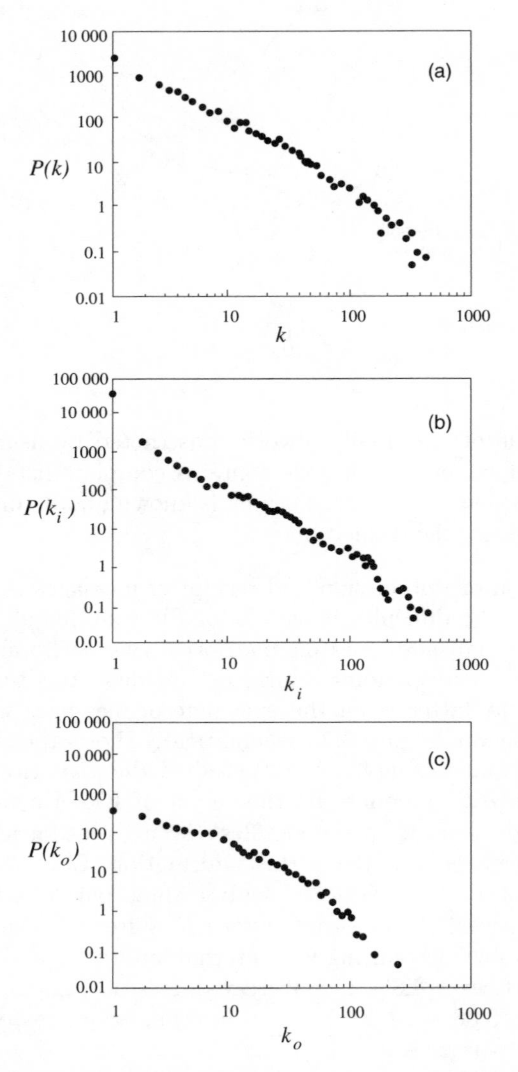

FIG. 3.26. Empirical degree (a), in-degree (b), and out-degree (c) distributions of the student accounts on the e-mail server of Kiel University (after Ebel, Mielsch, and Bornholdt 2002). The region $k < 10^2$ of the degree distribution was fitted by a power law with exponent γ equal to 1.3. The region $k_i < 10^2$ of the degree distribution was fitted by a power-law dependence with exponent γ_i equal to 1.5. A power-law fitting for the out-degree distribution was impossible.

with degree distributions for other kinds of collaboration networks, for example, Fig. 3.24 for the Word Web.

The clustering coefficient and the average shortest-path length were obtained for the network that includes external addresses. The directness of edges in this net was ignored. The resulting clustering coefficient (0.156) is more than 3000 times greater than that for the corresponding random graph. On the other hand, this empirical clustering coefficient is not as large as it seems at first sight. It is 'only' eight times greater than the clustering of the equilibrium uncorrelated network with the same degree distribution (see Section 3.11 for a more detailed discussion of the comparison with such networks, which can be used as a reference point).

The average shortest-path length of this undirected graph was found to be 4.95, which is even less than that for the corresponding classical random graph (two times less). On the other hand, it is 1.4 times greater than the average shortest-path length of the equilibrium uncorrelated network with the same degree distribution.

Mail graphs are evolving (growing) networks. Consequently, to understand the nature of a distribution of their connections one must know at least the relative dynamics of the total numbers of vertices and edges. Unfortunately, such empirical data are still absent.

3.7 Power grids and industrial networks

Our civilization is based on complex transportation systems (communications), which form a variety of large networks. The reader can easily make a long list of such networks. These nets are a basic component of our lives and industry. Further, the list can include not only railway and metro and automobile transport, but also the transportation of electric energy (power grids) and even the transportation of information (communication networks). Here, we briefly discuss the power grids.

Data on a large power grid were analysed by Watts and Strogatz (1998). They constructed an undirected graph, in which the vertices are transformers, substations, and generators of the Western States Power Grid, and the edges are high-voltage transmission lines of this grid. The degree distribution of this network is exponential-like. So, the distribution of connections is not of the fat-tailed form that we are mainly interested in. The empirical data on the clustering and the average shortest-path length are presented in Table 3.6.

The power grid is a sparse network. Its average degree is $\overline{k} = 2.67$, and so the average shortest-path length is not as small, $\overline{\ell} = 18.7$. This is rather close to the average shortest-path length of the corresponding classical random graph (actually, one and a half times greater than the latter). Also, it is worthwhile to compare $\overline{\ell}$ of the power grid with another typical value. Let us neglect long-range connections in the power grid. With this (certainly wrong) assumption, the network has the geometry of a two-dimensional lattice. A rough estimate of the average shortest-path length of the latter is $\sqrt{N}/2$. For $N = 4941$, this

Table 3.6 *Main characteristics of the Western States Power Grid (Watts and Strogatz 1998, Watts 1999): the total number of vertices N, the total number of connections L, the clustering coefficient C, the ratio of the clustering coefficient to that of the corresponding classical random graph, C/C_r, the average shortest-path length $\overline{\ell}$, and the ratio of the average shortest-path length to that of the corresponding classical random graph, $\overline{\ell}/\overline{\ell}_r$.*

	N	L	C	C/C_r	$\overline{\ell}$	$\overline{\ell}/\overline{\ell}_r$
Western States Power Grid	4941	6.61×10^3	0.08	1.5×10^2	18.7	1.5

gives an average shortest-path length ~ 35, which is only two times greater than the actual value for the power grid. In this situation, it is hard to claim that the power grid displays the small-world effect.[27] Essentially larger power grids must be studied to make reliable conclusions. However, are there power networks much greater than the Western States Power Grid?

The clustering of the Western States Power Grid is high: the empirical clustering coefficient is 150 times greater than that of the corresponding classical graph. This difference is essential since both networks have rapidly decreasing degree distributions. So, there must be some special origin for such a high clustering. Watts and Strogatz (1998) supposed that the reason is because power grids have a structure close to that of a lattice but spoiled by a fraction of short-range connections (see Section 4.7). The strong clustering of the resulting network is a direct consequence of the large clustering coefficient typical of many lattices.

3.8 Electronic circuits

Electronic circuits may quite naturally be considered as graphs. The vertices of these graphs are electronic components (resistors, diodes, capacitors, and so on in analogue circuits and logic gates in digital devices). Undirected edges are the wires. An old TV set contains only several hundred electronic components, but, of course, the sizes of modern large digital circuits are incomparably greater. These large sizes already allow a reliable statistical analysis of the structure of such graphs.

Ferrer i Cancho, Janssen, and Solé (2001) have studied the general structure of several large electronic circuits. They constructed large graphs with unspecified vertices, irrespective of the type of electronic component. The largest graphs for digital circuits contained over 20 000 components with the mean number of connections about 4. The degree distribution of the large digital circuits was found to be fat-tailed (see Fig. 3.27). Furthermore, the quality of the empirical distribution was sufficient to fit the distribution by a power-law dependence with exponent $\gamma \sim 3$.

[27] This difficulty is quite typical of small networks.

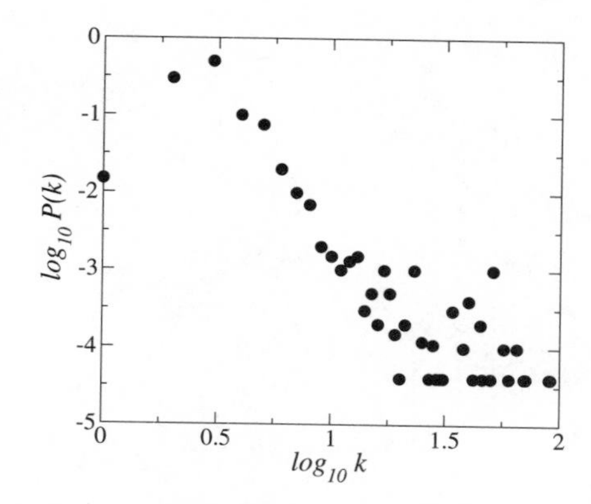

FIG. 3.27. Empirical degree distribution of a large digital circuit which contains over 20 000 components (after Ferrer i Cancho, Janssen, and Solé 2001).

The analysis implies that these networks are small worlds. The values of their average shortest-path lengths do not deviate significantly from those for the corresponding classical random graphs. The clustering coefficients of the largest nets are about 3×10^{-2}, which is much less than 1 but two orders higher than the clustering of the classical random graphs.

The 'compactness' of large electronic circuits is quite natural, as it is a necessary condition of a good design. In any case, circuits must contain 'long-range' connections between their components, which crucially decrease their 'linear sizes'. As for the nature of fat-tailed distributions of connections in these graphs, it is still unclear.

3.9 Nets of software components

Modern software is increasingly complex. Software engineers (developers) build programs (now called 'software products') which may contain thousands(!) of interacting software components (modules). This modular architecture makes software products cheap and flexible. The modular structure of software products is naturally represented by *class diagrams*. The class diagrams are, in fact, directed graphs whose vertices are the modules of a program (classes) and edges show (pair) interactions between modules.

The class diagram approach is the basis of the main trend in modern software engineering, which is object-oriented design. It is the main modelling technique. Leaving aside the technical details of the method of class diagrams,[28] what are the statistics of such graphs? What is their global structure?

Valverde, Ferrer i Cancho, and Solé (2002) analysed two large class diagrams: (1) the class diagram of the public Java Development Framework 1.2 (JDK 1.2)

[28]Class diagrams are not just used in software engineering; there are numerous versions and variations of this approach.

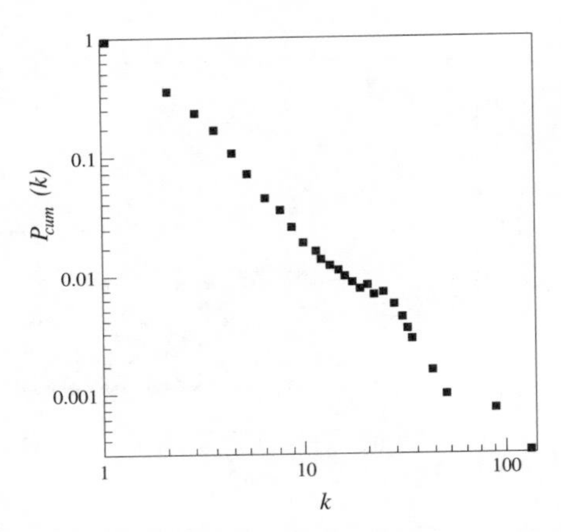

FIG. 3.28. Cumulative degree distribution of the largest connected component of the public Java Development Framework 1.2 (after Valverde, Ferrer i Cancho, and Solé 2002). This component consists of 1376 vertices and 2174 connections. The cumulative distribution was fitted by a power law with exponent equal to 1.5, so the γ exponent of the corresponding degree distribution is 2.5.

and (2) the class diagram of a large computer game. The former is a very large set of software modules for Java applications (9257 modules, in total), and hence this graph contains numerous separate connected components. Most of these 3115 connected components are single modules but two of them are really very large. The largest connected component contains 1376 vertices with 2174 connections, the second largest has 1364 vertices with 1930 connections. We stress that in this analysis, class diagrams were considered as *undirected* graphs. These two largest components were analysed.

In both cases, the degree distributions turned out to be fat-tailed and were fitted by power-law dependences with γ exponents equal to 2.5 and 2.65 (see Fig. 3.28).[29] The fitting by a power law is possible in a narrow range of degrees. However, a fat-tailed form of the distribution is clearly seen. Thus we have another example of a fat-tailed degree distribution.

Smaller components have an especially pronounced tree-like structure similar to that shown in Fig. 3.29. However, even the largest connected components have rather modest clustering. Their clustering coefficients are 0.06 and 0.08 for the largest connected component and the second largest, respectively. However, these values turn out to be 25 and 40 times greater than the clustering coefficients of the corresponding classical random graphs.

As expected these graphs show the small-world effect. The average shortest-path lengths of the first- and second-largest connected components are 6.39 and

[29]One might suggest that, since their sizes and structures are similar, these two connected components are two versions of the same software product.

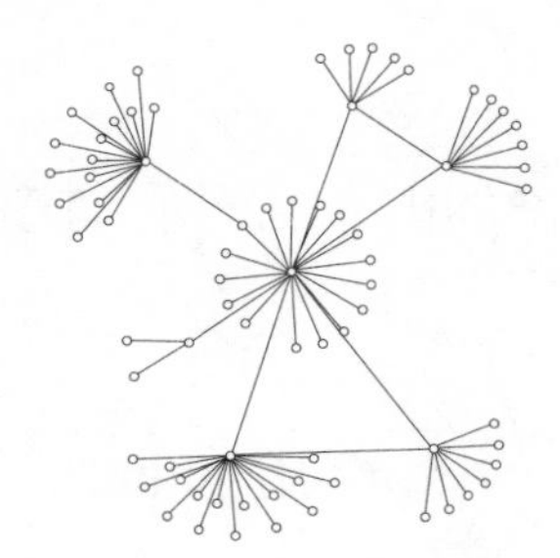

FIG. 3.29. The typical structure of a 'small' connected component of the class diagram of the Java Development Framework according to Valverde, Ferrer i Cancho, and Solé (2002). Note that it is very close to a tree: loops are nearly absent. Larger connected components resemble trees less.

6.91, respectively, which are similar to those in the corresponding classical random graphs.

The reader could guess that computer games have the largest class diagrams. The class diagram studied in Valverde *et al.*'s paper has 1989 vertices(!) and 4.78×10^3 connections. The clustering coefficient and the average shortest-path length of this net are close to the values for the Java Development Framework. $C = 0.08$, which is 35 times greater than for the corresponding classical random graph. The average shortest-path length is $\overline{\ell} = 6.2$ (1.28 times greater than that in the classical random graph).

The nature of fat-tailed distributions of connections in these nets remains unclear. Software products are, naturally, a product of design. What, however, is the role of optimal design in the resulting specific structure of the class diagrams? At the moment, we cannot answer this question. Instead, we note an important feature of these growing networks. As the class diagrams increase, their numbers of edges grow more rapidly than their numbers of vertices, and these nets resemble trees less and less. The mean number of connections becomes noticeably greater than 2, which is the value for a large tree. In this respect, the 'growth' of the class diagrams is essentially non-linear.

3.10 Energy landscape networks

Even small interacting systems may have extremely complex reliefs of potential energy with numerous mountains and depressions, valleys and passes. Potential energy landscapes determine the thermodynamics and dynamics of systems and their structure. Furthermore, such landscapes play a very important role in the processes of complex relaxation. Usually, a relaxing system has to pass through many local minima of potential energy before reaching the global minimum. This is why knowledge of these complex landscapes is crucial in the studies of glasses, biomolecules, protein folding, liquids, atomic clusters, etc.

How are these landscapes organized? What is their topography? We shall see that network science provides a useful insight into this problem.

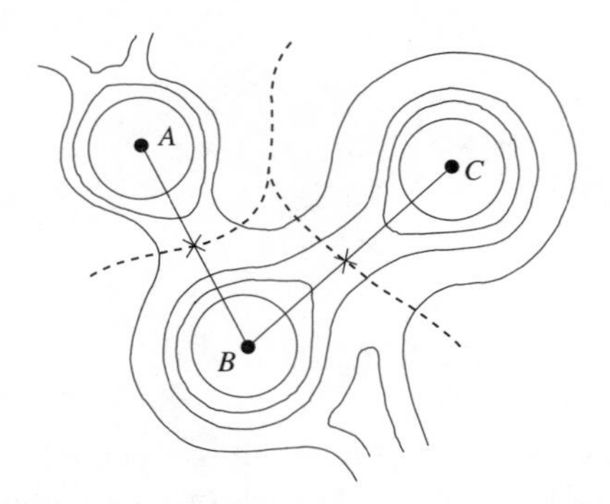

FIG. 3.30. The scheme of the construction of a potential energy landscape network according to Doye (2002). In fact, the potential energy landscape is defined on the multidimensional configuration space of a system, and this two-dimensional plot is oversimplified. The local minima of the potential energy landscape are vertices of the landscape network (see the points A, B, and C). Saddle (transition) points on the landscape indicate the pairs of the nearest-neighbour minima. Edges in the network connect such pairs. The dashed lines separate the basins of attraction of the minima.

For example, let us discuss the potential energy of atomic clusters. The atoms interact with each other, and the potential energy of a cluster, in principle, depends on the positions of all the atoms in that cluster. *The potential energy landscape* represents the potential energy of a cluster versus the relative coordinates of the atoms. One may say that this is the potential energy of a system, plotted in multi-dimensional configuration space.

The total number of minima in the potential energy landscape is huge: $\sim e^{cn}$, which is a standard estimate from condensed matter theory. Here, n is the number of atoms in the cluster, and c is an uninteresting constant, dependent on the system. Each local minimum has its basin of attraction, and the temporal evolution of the system can be treated as jumps from minimum to minimum of its potential energy landscape. Suppose that the system is in the basin of attraction of some minimum. Where can it appear next? In other words, what are the nearest-neighbour minima?

One may notice that, with overwhelming probability, the system jumps over potential barriers at saddle points of the potential relief.[30] So, *two potential minima are nearest neighbours if a saddle point exists between them* (see Fig. 3.30). One may also call it a transition point or states. Note that there can be only a single saddle point between a pair of minima. The total number of such transition (saddle) points is even greater than the number of minima. The number of saddles is $\sim ne^{cn}$.

[30]That is, if the transition takes place, then the easiest jump is over a saddle point. If a saddle point between a pair of minima is absent, then the system finds easier ways to evolve than to jump between these two minima.

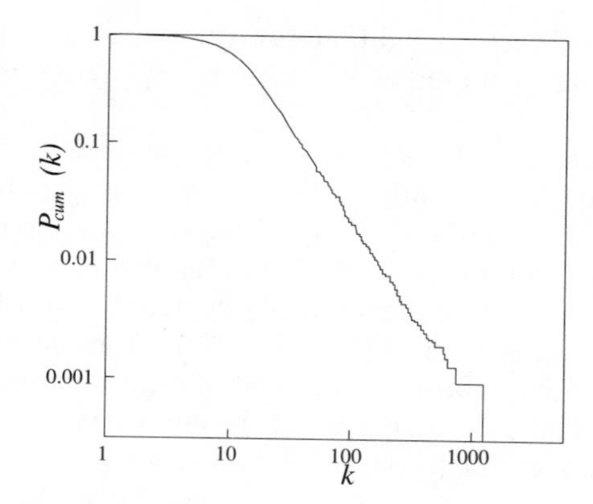

FIG. 3.31. The cumulative degree distribution of the potential energy landscape network for a 14-atom cluster (after Doye 2002). The γ exponent of the corresponding degree distribution is equal to 2.78.

Of course, if a basin of attraction has several or many transition points, the actual runaway from the basin depends on the relation between the barrier heights. The system chooses the easiest paths. In principle, to have a reasonably complete description of a potential energy landscape, one must know the potential energies of all its minima and all the transition probabilities between 'nearest-neighbour minima'.

Doye (2002) has found that an even less informative approach to potential energy landscapes can provide an important insight into the topography of these reliefs. Doye constructed a network whose vertices are local minima of the landscape and edges connect pairs of the nearest-neighbour vertices (minima). For a 14-atom cluster with a Lennard-Jones potential interaction between atoms, 4196 minima (vertices) and 87 219 transition states (edges) were located. This network was analysed in a standard manner. The reader already knows that this means the measurement of (1) the degree distribution, (2) the clustering coefficient, and (3) the average shortest-path length.

Surprisingly, the degree distribution of the network turns out to be scale-free. Its γ exponent is 2.78 (see Fig. 3.31). Why, however, is this surprising? The reason is that the configuration space of a system of atoms was studied earlier using similar ideas (Scala, Amaral, and Barthélémy 2001), but the constructed networks were found to have rapidly decreasing degree distributions. Scala *et al.* studied polymers on lattices, where interaction was modelled not by a Lennard-Jones potential but by a set of constraints for a polymer chain. In this case, all possible transitions in the configuration space are also well defined. However, one can see that the resulting networks are quite distinct.

The important point is that, in atomic clusters with the Lennard-Jones interatom interaction, the system can jump directly to the global minimum from most of the local minima. Doye (2002) found that 76% of the local minima for a 14-atom cluster are nearest neighbours of the global minimum, which is thus a hub in the energy landscape network. Roughly speaking, this landscape looks like a funnel!

The average shortest-path distance of this net, $\overline{\ell} = 2.32$, practically coincides with that of the corresponding classical random graph, $\overline{\ell}/\overline{\ell}_r = 1.04$. The clustering coefficient $C = 0.073$ is 'slightly' higher than in the classical random graphs, $C/C_r = 7.4$. This modest relative difference can be easily explained if we account for the fact that the form of the degree distribution is very far from Poisson. The reader can refer to the formula (6.1) from Section 6.1 for the clustering coefficient of an equilibrium uncorrelated random graph with an arbitrary degree distribution. Thus, the structure of the energy landscape network is close to that of the equilibrium uncorrelated random net with the same distribution of connections.

This approach can be applied to other interacting systems (Calvo, Doye, and Wales 2002), where in many cases fat-tailed degree distributions should arise. Such a structure of energy landscape nets (the significant role of hubs and the compactness) facilitates the approach to the global minimum and so strongly affects the dynamics and relaxation processes in such systems.

3.11 Overview

In this chapter a number of real networks where highly connected hubs play an important role have been discussed. Let us present a brief summary of the empirical results for these nets.

The reader has certainly noticed that the standard programme of the empirical research of a complex network usually looks as follows:

(1) find the (in-, out-)degree distribution;

(2) find the clustering coefficient; if the net is directed then find the clustering coefficient of the undirected projection of this network;

(3) find the average shortest-path length with and without accounting for the directness of edges.

As a rule, only these data on networks are available, and so we only consider these characteristics in the summary. Moreover, the data on the clustering and the shortest-path length are essentially poorer than on degree distributions.

We must stress a serious difficulty in the interpretation of empirical data. Most fat-tailed empirical degree distributions are fitted by a power-law dependence. However, several obstacles obstruct a convincing fitting:

(1) The size effect dramatically cuts off the tail of the degree distribution of even large networks (see eqn (1.8) from Section 1.3 and the discussion in Section 5.6).

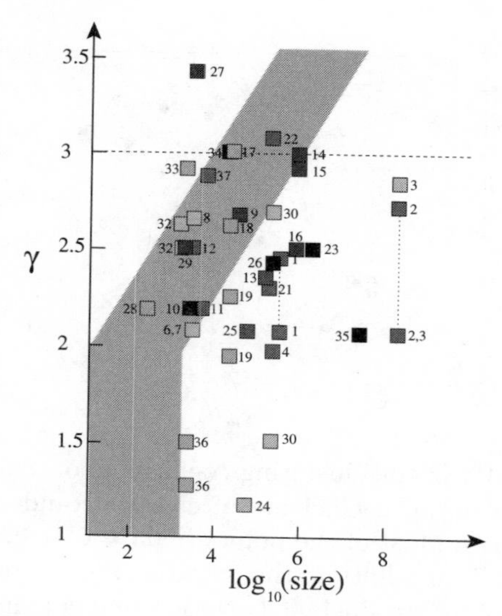

FIG. 3.32. Log–linear plot of γ (γ_i, γ_o) exponents of all the networks reported as having power-law (in-, out-)degree distributions, that is scale-free nets, versus their sizes (the total number of vertices, N). The grey region ($\gamma \sim 1 + \log_{10} t/2.5$ for $\gamma > 2$ and $N \sim 10^2$–10^3 for $\gamma < 2$) is the estimate of the finite-size boundary for the observation of the power-law degree distributions (see Section 5.6). Here 2.5 is the range of degrees (orders), which we believe necessary to observe a power law. In fact, these 2.5 orders include a region of small degrees, which, loosely speaking, cannot be used in the fitting. The dashed line, $\gamma = 3$, is the resilience boundary (see Chapter 6). Below this boundary, large networks are extremely stable against random failures and, simultaneously, extremely vulnerable to the spread of diseases. The points are plotted using the data from Table 3.7. Points for γ_o and γ_i from the same set of data are connected. The numbers indicate particular networks from Table 3.7. The precision of the right points is about ± 0.1(?) and is much worse for points in the grey region. A chance exists that some of these networks are, in fact, not in the class of scale-free nets.

(2) In the region of small degrees, the form of the distribution depends on subtleties and differs from the large-degree asymptote. One can suggest that, at least, degrees below 10^1 cannot be used in the fitting.[31]

(3) Very often, networks are inhomogeneous and contain distinct components with distinct properties. The resulting degree distributions are actually combinations of degree distributions of subcomponents.

Thus, the range of degrees, where power laws can, in principle, be observed,

[31] In fact, it is hard to estimate this range.

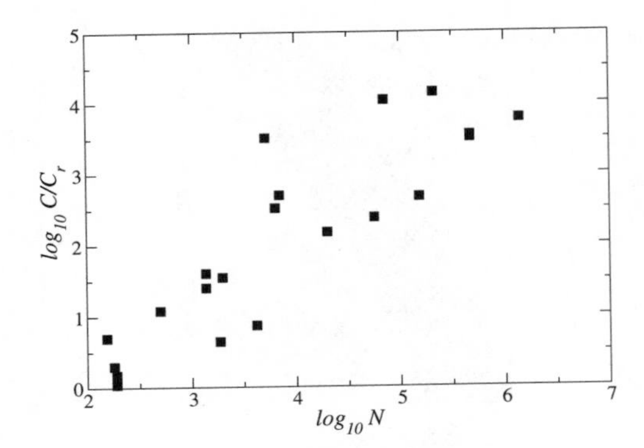

FIG. 3.33. Log–log plot of the clustering coefficients of scale-free networks with respect to the clustering coefficients of classical random graphs versus the sizes of the networks. Most of the points ($0.03 < C < 0.76$) are plotted using data from Table 3.7. In addition, the data on several non-scale-free networks (food webs and so on) are included. Notice that all the points are inside a rather restricted area (Dunne, Williams, and Martinez 2002a).

is restricted from both above and below. In Section 5.6 we will show that the cut-off moves to smaller degrees as exponent γ increases. Intentionally, we did not show the fitting lines on many plots with empirical degree distributions in this chapter. So, the reader can easily estimate the range of degrees where power-law dependences are valid and, therefore, the possible quality of the fitting.

Figures 3.32, 3.33, and 3.34 and Table 3.7 summarize data on scale-free networks, which we discussed in this chapter. In fact, these are data on networks (or subgraphs) reported as having scale-free (power-law) degree distributions. Notice that the observed values of γ do not exceed a boundary due to the finite sizes of networks. We show (very approximately) this boundary by the grey region in Fig. 3.32.

From Fig. 3.32 one can also appreciate the wide range of sizes of real scale-free nets, which is from 10^3 to 10^9 vertices. Table 3.7 is more informative: in addition, data on clustering (the characteristic of the closest environment of a vertex) and shortest-path lengths (the characteristic of the global structure of a net) are represented. (See also Figs 3.33 and 3.34 where the clustering coefficient and the average shortest-path length with respect to the corresponding values for classical random graphs are shown versus the sizes of networks.) However, notice that the right-hand side of the table contain many blank spaces. Several natural reasons exist for such 'blankness': (1) It is slightly more difficult to obtain a clustering coefficient or a shortest-path length than a degree distribution. (2) The clustering coefficient, in principle, is defined only for undirected networks. (3) A number of lines in Table 3.7 contain data for subgraphs, whose clustering and shortest-path lengths are far less interesting than for entire networks.

Table 3.7 (overleaf) *Sizes (the total numbers of vertices and edges) and values of the γ exponent for the networks or subgraphs reported as having power-law (in-, out-)degree distributions; the average clustering coefficients; and the average shortest-path length in these scale-free networks. Also, the clustering coefficients and average shortest-length paths are given with respect to the corresponding values of classical random graphs with the same numbers of vertices and edges. For each network (or class of networks), data are presented in more or less chronological order, so that the recent exciting progress is visible. Errors are not shown (see the caption of Fig. 3.32). They essentially depend on the size of a network and on the value of γ. We recommend that our readers look over this section before using these values.*

[1]The average undirected-shortest-path length (and the corresponding ratio with respect to that of the classical random graph) is represented in parentheses.

[2]The data for the network of operating AS were obtained for one day in December 1999. The clustering coefficient and the average shortest-path length are given for the 'average' net in 1999 ($N = 5287$ and the total number of edges is $L = 10\,100$).

[3]Stanford Public Information Retrieval System.

[4]In fact, the data were collected from a small set of vertices of the web of human sexual contacts. These vertices almost surely have no connections between them.

[5] The clustering coefficient is represented for the part of the network of metabolic reactions of *E. coli* (282 substrates with average number of connections 7.35), which was obtained by ignoring the directness of edges (Wagner and Fell 2001).

[6]The value of the clustering coefficient is given for the largest ('giant') connected component (466 vertices with mean degree 2.3) of the protein physical interaction network consisting of 985 vertices with mean degree 1.83 (Wagner 2001a).

[7]For the Word Web, two values of the γ exponent are given: below and above the crossover point of the degree distribution (see Fig. 3.24). The second value was obtained with much less precision than the first. The values of the clustering coefficient of the Word Web noticeably depend on the construction procedure. In the table, they are given for two known constructions.

[8]Food webs are very small, so reliable conclusions are hardly possible. In subsequent studies (Dunne, Williams, and Martinez 2002a) the degree distributions of such food webs were interpreted as exponential-like.

[9]The out-degree distribution of the telephone call graph cannot be fitted by a power-law dependence.

[10]Instead of the number of edges, for this subgraph, we show one-half of the product of the number of vertices and the average degree. The clustering and the mean shortest-path length are given for a larger e-mail net ($N = 59\,969$, $L = 0.863 \times 10^5$) which contains the net of student accounts as a subnet.

	Network or subgraph	Number of vertices	Number of edges	γ
1	Complete map of the nd.edu domain of the Web	325 729	1 469 680	$\gamma_i = 2.1$ $\gamma_o = 2.45$
2	Pages of the WWW scanned by Altavista [1] in October 1999	2.711×10^8	2.130×10^9	$\gamma_i = 2.1$ $\gamma_o = 2.7$
3	'————' (another fitting of the same data)			$\gamma_i = 2.10$ $\gamma_o = 2.82$
4	Map of sites of a domain of the WWW, spring 1997	2.60×10^5	—	$\gamma_i = 1.94$
5	Undirected map of sites in a domain of the WWW	153 127	2.70×10^6	—
6	A set of public company home pages	4923	1.335×10^7	$\gamma_i = 2.05$
7	A set of US newspaper home pages	—	—	$\gamma_i = 2.05$
8	A set of university home pages	—	—	$\gamma_i = 2.63$
9	A set of computer scientist home pages	56 880	—	$\gamma_i = 2.66$
10	Interdomain level of the Internet, December 1998	4389	8256	2.2
11	Interdomain level of the Internet, December 1999 [2]	6374	13 641	2.2
12	Router level of the Internet in 1995	3888	5012	2.5
13	Router level of the Internet in 2000	$\sim$ 150 000	$\sim$ 200 000	$\sim$ 2.3
14	Citations in the ISI database 1981–June 1997	783 339	6 716 198	$\gamma_i = 3.0$
15	'————' (another fitting of the same data)			$\gamma_i = 2.9$
16	'————' (another estimate from the same data)			$\gamma_i = 2.5$
17	Citations in *Phys. Rev. D* **11–50** (1975–1994)	24 296	351 872	$\gamma_i = 3.0$
18	'————' (another fitting of the same data)			$\gamma_i = 2.6$
19	'————' (another estimate from the same data)			$\gamma_i = 2.3$
20	'————' (another estimate from the same data)			$\gamma_i = 1.9$
21	Collaboration network of movie actors	212 250	61 085 555	2.3
22	'————' (another fitting of the same data)			3.1
23	Collaboration network of MEDLINE	1 388 989	1.028×10^7	2.5
24	Coauthorships in the SPIRES [3] e-archive	56 627	4.898×10^6	1.2
25	Collaboration net collected from math. journals	70 975	0.132×10^6	2.1
26	Collaboration net collected from neurosci. journals	209 293	1.214×10^6	2.4
27	Web of human sexual contacts [4]	2810	—	3.4
28	Networks of metabolic reactions [5]	$\sim$ 200–800	$\sim$ 600–3000	$\gamma_i = 2.2$ $\gamma_o = 2.2$
29	Net of protein–protein interactions (yeast proteome) [6]	1870	2240	$\sim$ 2.5
30	Word Web [7]	470 000	17 000 000	1.5 / 2.7
31	Food web of Silwood park [8]	154	366	$\sim$ 1
32	Java Developement Framework (largest component)	1376	2174	2.5
	'————' (second largest component)	1364	1930	2.65
33	Computer game graph	1989	4.78×10^3	2.85
34	Large digital electronic circuits	2×10^4	4×10^4	3.0
35	Telephone call graph [9]	47×10^6	8×10^7	$\gamma_i = 2.1$
36	E-mail net (student accounts in Kiel University) [10]	5165	6.57×10^4	1.3, γ_i=1.5
37	Energy landscape network for a 14-atom cluster	4196	87 219	2.78

	C	C/C_r	$\bar{\ell}$	$\bar{\ell}/\bar{\ell}_r$	References
1	—	—	11.2	—	Albert *et al.* 1999
2	—	—	16 (6.8)	(1.0)	Kumar *et al.* 2000a, Broder *et al.* 2000
3					Newman *et al.* 2001
4	—	—	—	—	Adamic *et al.* 2000
5	0.108	0.47×10^3	3.1	0.93	Adamic 1999
6	—	—	—	—	Pennock *et al.* 2002
7	—	—	—	—	Pennock *et al.* 2002
8	—	—	—	—	Pennock *et al.* 2002
9	—	—	—	—	Pennock *et al.* 2002
10	—	—	4	0.6	Faloutsos *et al.* 1999
11	0.24	3.3×10^2	3.7	0.58	Pastor-Satorras *et al.* 2001
12	—	—	12.1	1.39	Faloutsos *et al.* 1999
13	—	—	10	0.8	Govindan *et al.* 2000
14	—	—	—	—	Redner 1998
15					Tsallis *et al.* 2000
16					Krapivsky *et al.* 2000
17					Redner 1998
18					Tsallis *et al.* 2000
19					Krapivsky *et al.* 2000
20					Vázquez 2001b
21	—	—	4.54	1.25	Barabási *et al.* 1999
22					Albert *et al.* 2000a
23	0.066	6×10^3	4.6	0.9	Newman 2001e
24	0.726	0.24×10^3	4.0	1.88	Newman 2001e
25	0.59	1.1×10^4	9.5	1.16	Barabási *et al.* 2002
26	0.76	1.4×10^4	6	1.2	Barabási *et al.* 2002
27	—	—	—	—	Liljeros *et al.* 2001
28	0.32	12	3.2	0.95	Jeong *et al.* 2000, Wagner *et al.* 2001
29	0.022	4.4	6.8	0.8	Jeong *et al.* 2001, Wagner 2001a
30	0.69/0.44	$4.4\times10^3/2.8\times10^3$	2.65	0.87	Ferrer i Cancho *et al.* 2001a
31	0.15	5	3.4	1.05	Montoya *et al.* 2001, Solé *et al.* 2002
32	0.06	25	6.39	1.02	Valverde *et al.* 2002
	0.08	40	6.91	1.01	Valverde *et al.* 2002
33	0.08	35	6.2	1.28	Valverde *et al.* 2002
34	3×10^{-2}	1.5×10^2	~ 6	~ 1	Ferrer i Cancho *et al.* 2001b
35	—	—	—	—	Aiello *et al.* 2000
36	0.156	3.25×10^3	4.95	0.48	Ebel *et al.* 2002
37	0.073	7.4	2.32	1.04	Doye 2002

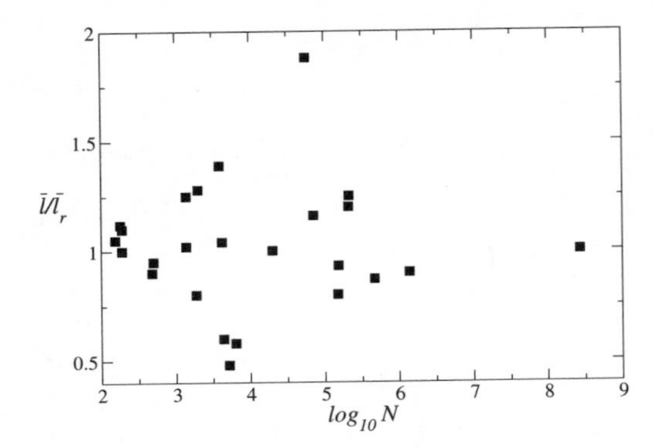

FIG. 3.34. Log–linear plot of the relative average shortest-path lengths of scale-free networks with respect to the average shortest-path lengths of the corresponding classical random graphs versus the sizes of the networks. The points are plotted using data from Table 3.7. In addition, the data on several non-scale-free networks are included. Notice that all the points are inside the band $0.5 < \overline{\ell}/\overline{\ell}_r < 2$.

In addition, in Table 3.7 we compare the clustering and shortest-path lengths of networks with the values for the corresponding classical random graphs. One can see that the values of clustering coefficients in the networks in the table ($0.03 < C < 0.76$) are much greater than those of the corresponding classical random graphs. Contrastingly, all the shortest-path lengths collected in the table turn out to be close to those for classical random graphs.

The relative difference is really small, and, roughly speaking, the average shortest-path lengths of real scale-free nets are the same as in classical random graphs. So, *the 'compactness' of a network does not depend crucially on a fat tail in the distribution of connections.*

We should say that classical random graphs are not the best choice for comparison. Rather, they demonstrate the contrast between networks with fat-tailed distributions of connections and those with rapidly decreasing distributions. So, large ratios of the clustering coefficients of real nets and those of classical random graphs in Table 3.7 do not necessarily mean a strong clustering of real networks.

If we want to be more realistic, it is better, for comparison, to use graphs with the same degree distributions as the real ones. In Section 1.3 (see Fig. 1.8) we discussed a simple construction procedure from graph theory which provides networks with a given degree distribution but, unlike real networks, without correlations. Necessary analytical results for these uncorrelated equilibrium graphs are already available. The clustering coefficient and the average shortest-path length of such a graph are given by eqns (6.1) and (6.10) respectively from Section 6.1. These formulae contain only the total number of vertices in a net and the first and the second moments of its degree distribution. One can find the second moment from empirical degree distributions of real graphs and substitute into these formulae. In many cases, the resulting values are much closer to the empirical clustering coefficients and average shortest-path lengths than those of classical random graphs (Newman 2002a, Ebel, Mielsch, and Bornholdt 2002). However, as one might expect, these values do not coincide, and the difference can be large, especially for the clustering.[32]

More recent data on scale-free networks are appearing rapidly, and so, soon, our list will turn out to be very incomplete. On the other hand, one or two of the smallest nets will, perhaps, be excluded.

[32]One can easily find the second moment of an empirical degree distribution. However, empirical researchers rarely compute this important value, and so a systematic comparison with equilibrium uncorrelated graphs with the same degree distributions is still absent. What can we get from such a comparison? Let us consider the clustering (as we saw, the average shortest-path lengths in various types of graphs are relatively close in any case). Two situations are possible. (1) A real network and its equilibrium uncorrelated analogue with the same degree distribution have relatively close clustering. Then, the clustering coefficient of the real net is close to its, roughly speaking, minimal possible value for random networks of a given size with the same degree distribution. It is the clustering produced simply by random connections, as in Fig. 1.8 from Section 1.3. In fact, this clustering is a size effect: the value of the clustering coefficient approaches zero as the size of a network tends to infinity. (2) The clustering coefficients of these nets differ strongly: the clustering of the real net is much higher than the minimum possible value. In such an event, it is certainly worthwhile to search for some special cause of such strong clustering (see Sections 4.7 and 5.13).

4

EQUILIBRIUM NETWORKS

The subject of this book is the statistical mechanics of networks. Following traditional statistical mechanics, we will try to consider equilibrium and non-equilibrium systems separately, as far as possible. This separation can be managed fairly well as long as degree distributions and related simple properties are studied.

4.1 Statistical ensembles of random networks

First we shall recall some basic notions from Chapter 1 at a further level. Recall that, in fact, we are studying not networks themselves but statistical ensembles of random networks.

A graph g is a set of $N(g)$ vertices connected by $L(g)$ edges. It is reasonable to consider labelled graphs; that means that each vertex of a graph has its own label i. Such a graph is represented by the $N \times N$ adjacency matrix $\hat{a}$, whose elements a_{ij} are numbers of edges connecting vertices i and j. For simplicity, in this chapter we consider undirected graphs. This means that the matrix is symmetric: $a_{ij} = a_{ji}$. If closed one-edge loops (we call them *tadpoles*, see Fig. 4.1(a) and (b)) are absent then $a_{ii} = 0$. If multiple connections between pairs of vertices (for them we use the standard term *melons* from field theory, see Fig. 4.1(c)) are absent, then a_{ij} is either 0 or 1 for $i \neq j$.[1]

The degree of a vertex is $k_i = \sum_j a_{ij}$, hence the total degree of the network is equal to the sum of all elements of the adjacency matrix: $\sum_i k_i = \sum_{ij} a_{ij} = 2L = K$.

[1]Networks with tadpoles and melons are no less important and interesting as networks without such 'degeneracy'. In principle, even in the Web, tadpoles and melons are possible: you can make a reference to your own page, you can make many references to the same site (for example, linking the list of your publications with online journals and electronic archives), opposing hyperlinks can connect a pair of Web pages.

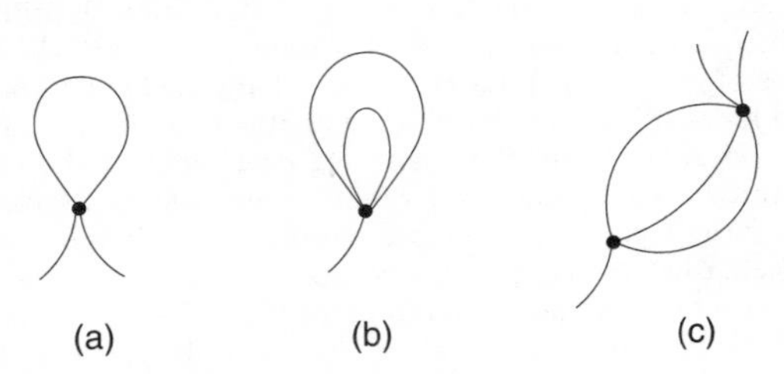

FIG. 4.1. 'Tadpoles' (a) and (b) and a 'melon' (c).

A statistical ensemble of graphs is defined by the following:

(1) a set G of graphs, and
(2) a rule that associates some *statistical weight* $\mathcal{P}(g) > 0$ with each graph g from this set: $g \in G$.

One may say that G is a set of adjacency matrices. As for the rule, there are various possibilities (see below):

(a) We can directly ascribe a specific statistical weight to each graph $g \in G$.
(b) We can define some process on the set G (some rules of evolution) which results in statistical weights for all these graphs.

The average of any quantity $S(g)$, which depends on the properties of a graph, is given by

$$\langle S \rangle = \frac{1}{Z} \sum_{g \in G} S(g)\mathcal{P}(g)\,, \tag{4.1}$$

where the sum is over all graphs from the set G and Z is the *partition function* of the ensemble, $Z \equiv \sum_{g\in G} \mathcal{P}(g)$.

For example, what does a degree distribution mean? By definition, $P(k) = \langle N(k) \rangle / N$, where $\langle N(k) \rangle$ is an average number of vertices with degrees k in the statistical ensemble that we consider. From this point on, in the present chapter we consider only a simple situation: all members of ensembles have the same number of vertices, N.

We see that the quantity of special interest is the number of vertices with degrees k in a graph g of this ensemble, $N(k, g)$. Suppose that we know the degree sequence for all vertices of the graph, $\{k_i(g)\}$ (we specially indicate that these degrees are for vertices of just this graph). To compute $N(k, g)$ we must scan the sequence and find how many times this degree k occurs. This scanning can be done using the Kronecker symbol $\delta[k_i(g) - k] \equiv \delta_{k_i(g),k}$.[2] Then we see that $N(k, g) = \sum_{i=1}^{N} \delta[k_i(g) - k]$. Therefore,

$$P(k) \equiv \frac{\langle N(k) \rangle}{N} = \frac{1}{N} \left\langle \sum_{i=1}^{N} \delta[k_i - k] \right\rangle = \langle \delta[k_1 - k] \rangle\,, \tag{4.2}$$

We have taken into account that all the averages $\langle \delta[k_i - k] \rangle$ are equal. Thus we see that the degree distribution is the average of the Kronecker symbol. Similarly, one can express more complex quantities (for example, many-degree correlators) in terms of the averages of combinations of the Kronecker symbols. After this is done, one needs the full set of statistical weights for the ensemble to find the average with these weights. If the weights are given (case (a) from above), we

[2] In this book we use this useful notation for the Kronecker symbol, $\delta[k - k'] \equiv \delta_{kk'}$ each time the arguments have small indices. In order not to confuse the Kronecker symbol with the δ-function, we always use parentheses for the latter, $\delta(x)$.

can use them immediately, otherwise we must first obtain them. How can this be done?

In case (b) above, we have a dynamical process on a set of graphs (rules of the evolution of networks for a given ensemble). In the situations discussed in this chapter, this process leads to a limiting equilibrium state.[3] The statistical weights of equilibrium statistical ensembles are limiting results of this process.

In principle, these weights can be obtained from detailed balance conditions, that is the absence of 'currents' in the equilibrium state. The detailed balance conditions contain the (given) transition probabilities for pairs of graphs from the set. Knowing the statistical weights, we can (in principle) find the averages of any quantities on the ensemble, and so obtain a complete physical description of the ensemble. Unfortunately, for networks this program can be strictly fulfilled only in very simple cases.

However, often it is not necessary to know statistical weights to find some simple quantities. Often, it is enough to use a less strict approach: one can write master or rate equations (or balance equations in an equilibrium situation) directly for a quantity of interest. For example, this quantity may be a degree distribution or the average number of vertices of degree k. The main known results were obtained in just this simpler way, which we use in our book.

Before we proceed with our networks, let us briefly recall basic ensembles from statistical mechanics.

(1) *Microcanonical ensemble*: energy and number of particles are fixed.
(2) *Canonical ensemble*: temperature and number of particles are fixed.
(3) *Grand canonical ensemble*: temperature and chemical potential are fixed.

Statistical weights for each of these equilibrium ensembles are defined by the following condition: the entropy is maximal under the constraint that the two corresponding quantities are fixed. In other words, the system is maximally random under this restriction.

In the *thermodynamic limit*, that is for infinite systems, these ensembles are equivalent. In this limit, the relative fluctuations of the number of particles in a grand canonical ensemble become invisible.

The difficulty in network theory is that it is hard to define the Hamiltonian in a unique way and introduce standard thermodynamic variables.[4] Moreover, any theory from first principles is impossible: networks are very diverse. This is why we can speak only about analogies of these standard ensembles for networks, using simple demonstrative examples.

4.2 Classical random graphs

The first introduced random graph was the Erdős–Rényi network (Erdős and Rényi 1959, 1960). There are two main constructions of such graphs with a fixed number of vertices, N:

[3]Note that this equilibrium state may be absent.

[4]See, however, the paper of Berg and Lässig (2002).

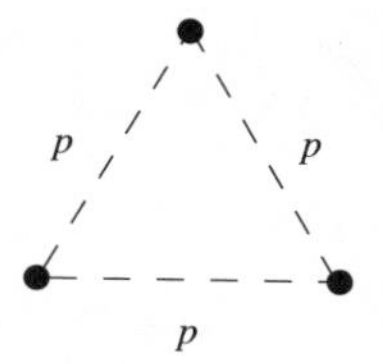

FIG. 4.2. A simple classical random graph constructed by procedure (i). $N = 3$.

(1) Each two vertices of the network are connected by an edge with probability p. Naturally, this edge is absent with probability $1 - p$.
(2) A given number L of edges connects randomly chosen pairs of vertices. One can realize this construction procedure by adding new edges one by one and repeatedly connecting randomly chosen pairs of vertices. In graph theory, this is called a *random graph process*.

Construction (2) provides us with one graph but, repeated many times, it generates a set of random graphs, each one with equal weight.

Note that construction (2) produces graphs with tadpoles and melons if one does not forbid this by special restriction. It is possible to prove that in the large graph limit the averages of physical quantities on both these graphs, with and without the restriction, are the same. Therefore, at this point, we are not concerned with the possible degeneracy of edges. Construction (1) does not have this problem: tadpoles and melons are absent (see Fig. 4.2).

These two constructions define two equivalent statistical ensembles of graphs.

The set of graphs in construction (1) is all $2^{N(N-1)/2}$ graphs with any number of edges smaller than or equal to $N(N-1)/2$ (recall that the number of vertices, N, is fixed). The statistical weight of a graph g with $L(g)$ edges taken from this set is $P_{GC}(g) = p^{L(g)}(1-p)^{N-L(g)}$. $L(g)$ varies from graph to graph. So, one can naturally call this ensemble *grand canonical* (compare with standard statistical definition from above).

The set of graphs in construction (2) consists of all possible graphs with a given number L of edges and N vertices. Statistical weights of graphs from this set are equal. One can call this ensemble *microcanonical* (see the discussion below for random networks with arbitrary degree distributions).

We will not discuss the well-studied classical random graphs in detail, and so we obtain a degree distribution for them in a very intuitive way. It is easier to use construction (1).

Each vertex in the graph with N vertices is in the same situation. It can have any number of edges attached, from zero (a 'bare' vertex) to $N-1$. If the vertex is of degree k, that is it has k edges attached, these k edges can occupy $N - 1$ possible positions. Then standard combinatorics readily lead to the following degree distribution of the classical random graph:

$$P(k) = \binom{N-1}{k} p^k (1-p)^{N-1-k} , \tag{4.3}$$

that is the binomial distribution, so that the average degree is $\overline{k} = p(N-1)$ and the network contains, on average, $pN(N-1)/2$ edges. For large N and fixed $\overline{k}$, it takes the Poisson form

$$P(k) = e^{-\overline{k}}\,\overline{k}^{\,k} / \, k! \, . \tag{4.4}$$

Notice that this limit ($N \to \infty$, $pN \to$ const) is the most interesting. For example, we shall see that the percolation threshold in these graphs is at $\overline{k} = 1$.

One can calculate more complex many-degree distributions and check that these graphs are uncorrelated. Intuitively, this is evident from the construction at least for infinitely large graphs.

4.3 How to build an equilibrium net

The construction of Erdős and Rényi is restricted: it produces only Poisson degree distributions. How can equilibrium uncorrelated random graphs with more complex degree distributions be constructed? We have demonstrated in Section 1.3 that, in principle, they can be created using a formal 'geometrical' procedure ('configuration construction') from graph theory. But do equilibrium graphs with arbitrary degree distributions exist in reality? Is it possible to create such a graph using some natural dynamical procedure?

The point of our interest is the fat tails of degree distributions. We saw in Chapter 2 that in networks growing under the mechanism of preferential linking, such distributions are widespread. So, *can equilibrium networks have fat-tailed degree distributions?*

Microcanonical ensemble

First let us look at the construction of mathematicians (see Section 1.3) from the point of view of statistical mechanics (Dorogovtsev, Mendes, and Samukhin 2002b).[5] What they made were maximally random graphs under the restriction that their degree distribution is equal to a given distribution $P(k)$. Now we formulate this construction more rigorously.

One can call this a *microcanonical ensemble of random networks*:[6]

(*set*) Let $N(k)$ be a sequence of non-negative integers such that $\sum_k N(k) = N$. G_M is the set of all possible graphs of size N with the following property: in each of these graphs the number of vertices of degree k is $N(k)$.

(*rule*) We ascribe equal statistical weight to each graph $g \in G_M$.

Actually this is a static construction (one can also say a geometrical construction or the 'configuration model'). Statistical weights are directly ascribed to graphs.

Unlike a standard situation in statistical mechanics (see above) this microcanonical ensemble is determined not by fixing the values of only two quantities

[5]In this chapter, for simplicity, we fix the number N of vertices in graphs.

[6]Since direct relation to a standard situation is impossible, the names of the ensembles for networks are rather conventional.

but by fixing the sequence $\{N(k)\}$, that is by the degree distribution $P(k)$. Automatically this also fixes the average degree $\overline{k} = \sum_k kP(k)$ and so the total number of edges $L = \overline{k}N/2$.

In the infinite graph limit correlations between different vertices of these networks evidently disappear: the networks are uncorrelated.

In principle, one may not exclude tadpoles and melons from this construction. Let us estimate roughly the numbers of tadpoles and melons in such a case if a degree distribution is a rapidly decreasing function. Our estimate will show what this word 'rapidly' means.

First consider tadpoles. If a vertex is of degree k and the total number of edges in the graph is $L = K/2$, where K is the total degree, the probability that the vertex has at least one tadpole is of the order of $(k/K)k$. Then the total number of tadpoles in a network is estimated to be of the order of $N\sum_k P(k)k^2/K = \langle k^2\rangle/\overline{k}$. Here we accounted for the fact that the occurrence of two or more tadpoles at a vertex is a much less probable event than one tadpole. If the second moment of the degree distribution is finite, we see that the total number of tadpoles is finite even in the infinite network.

A similar estimate can be obtained for the total number of double connections. The probability that two vertices of degrees k and k' are connected is of the order of $k(k'/K) = N^{-1}kk'/\overline{k}$. The probability that they are connected twice is of the order of $(N^{-1}kk'/\overline{k})^2$. Then the total number of doubly connected pairs of vertices is of the order of $N^2\sum_{k,k'} P(k)P(k')(N^{-1}kk'/\overline{k})^2 \sim (\langle k^2\rangle/\overline{k})^2$. This number is also asymptotically independent of N if the second moment of the degree distribution is finite. More than doubly connected pairs of vertices disappear as $N \to \infty$.

Thus, in 'safe' situations, the numbers of tadpoles and melons in these graphs are still finite when the network is infinitely large. Intuitively, it is clear, then, that the averages of physical quantities are the same for this construction with tadpoles and melons as without them. The former graphs are usually called *degenerate* or *non-Mayer*, the latter are called *non-degenerate* or *Mayer*. This equivalence can be proved for self-averaging physical quantities, for example for a degree distribution.

On the other hand, if the second moment of the distribution diverges, as happens, for example, when a single vertex attracts a finite part of connections ('condensation' of edges), the above arguments are not valid, and nothing definitely can be said.[7]

Very similar arguments show that the probability that a pair of nearest-neighbour vertices in these graphs is connected to the same third vertex is small, and thus small loops are rare. So, the clustering is low (see Section 1.4), and the local structure of the networks is tree-like (see the discussion in Section 6.1).

[7]Actually, even when the second moment diverges, the equivalence may remain valid if the divergence is not too strong and the relative number of tadpoles and melons decreases with growing N.

Canonical ensemble

For a physicist, the above construction seems rather formal. Now let us build equilibrium graphs of this type by using a dynamic procedure. In other words, let us describe a process that leads, after the relaxation, to an equilibrium ensemble of uncorrelated random networks equivalent to the above in the thermodynamic limit.

We now need to find a way to transform one graph of a given set (N is fixed) to another. This can be done naturally by rewiring or deleting/adding edges or by moving them from one position to another. A simple, although not unique, possibility is the following. First we choose an edge, for example, at random, and then find for it a new position by implementing the idea of preferential linking. This means that the probability that this edge becomes attached to some vertex is proportional to a function of its degree, $f(k)$. In particular, one may use the following construction (Dorogovtsev, Mendes, and Samukhin 2002b, see Fig. 4.3).[8]

Canonical ensemble of random networks:

(*set*) The set G_M contains all graphs with a given number L of edges.

(*rule*) At each step of the evolution, one of the ends of a randomly chosen edge is rewired to a preferentially chosen vertex of degree k. The rate of this process is $f(k)$.[9] The limiting state of this process is the equilibrium canonical ensemble.

The ensemble is described by a given sequence of rates $\{f(k)\}$ and L (or, equivalently, $\overline{k} = 2L/N$).

It is clear that we are exploiting the idea of preferential linking. Moreover, here we have used it twice. As for choosing a vertex to attach to, this is preferential attachment by definition. But first we have to select an edge to rewire. We choose it at random, so that, at first sight, there is no preference. However, we must also choose, at random, one of its end vertices from which to rewire the edge. There is $2L = N\overline{k}$ ends of edges, so that the probability that some vertex j will be chosen is equal to $k_j/(2L)$, where k_j is the degree of the vertex. Thus, this choice is again preferential, and the preference is proportional to the degrees of vertices (*proportional preference*).

In the thermodynamic limit, this construction evidently generates uncorrelated random networks with degree distributions dependent on particular $\{f(k)\}$ and $\overline{k}$. How can this be related to the microcanonical ensemble of random networks? To answer this question we must obtain the degree distribution of the canonical ensemble.

[8] A similar construction was proposed by Burda, Correia, and Krzywicki (2001).

[9] Equivalently, one can move a randomly chosen edge to a new position between a pair of vertices of degrees k and k' at the rate $f(k)f(k')$. The result (in the thermodynamic limit) is the same.

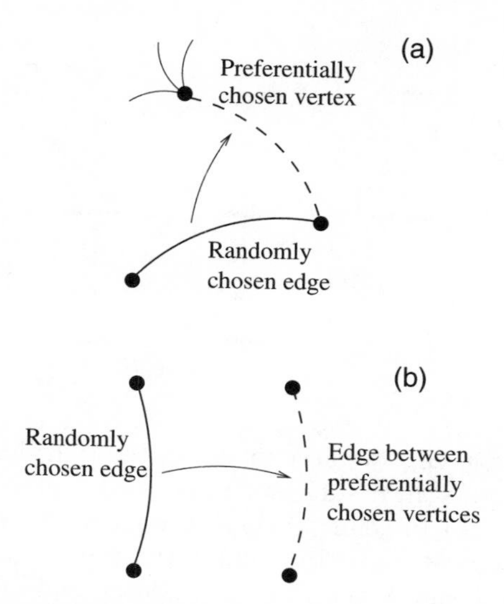

FIG. 4.3. Construction procedures for the canonical ensemble of random networks. (a) A randomly chosen end of a randomly chosen edge is rewired to a preferentially chosen vertex. (b) A randomly chosen edge is removed to a new position between a pair of preferentially chosen vertices. These procedures are equivalent in the limit of a large net.

Here we especially did not forbid tadpoles and melons. We have mentioned that in 'safe' situations (the absence of a condensate and the finite second moment of the degree distribution), large equilibrium networks with and without tadpoles and melons are actually equivalent.

To find the degree distribution we use not a strict but a simple convenient approach. We write a rate equation for the average number of vertices of degree k at time t, $\langle N(k,t)\rangle = NP(k,t)$. $\sum_k \langle N(k,t)\rangle = N$. In other words, we express $\langle N(k,t+1)\rangle$ in terms of these averages in the previous instant. In fact, we describe the change of the state of the system, which is caused by rewiring one of the ends of a randomly chosen edge. This equation for the *large* network is of the form

$$\langle N(k,t+1)\rangle = \langle N(k,t)\rangle - \frac{f(k)}{N\langle f(k)\rangle}\langle N(k,t)\rangle + \frac{f(k-1)}{N\langle f(k)\rangle}\langle N(k-1,t)\rangle$$
$$- \frac{k}{N\overline{k}}\langle N(k,t)\rangle + \frac{k+1}{N\overline{k}}\langle N(k+1,t)\rangle \,. \tag{4.5}$$

Here $\langle f(k)\rangle = \sum_k P(k) f(k)$.

This is the first occurrence of such an equation in our book. Therefore, let us focus upon it. In Fig. 4.4 we show all the processes which are accounted for in the right-hand side of eqn (4.5).

(1st) The first term on the right-hand side of the equation is clear.

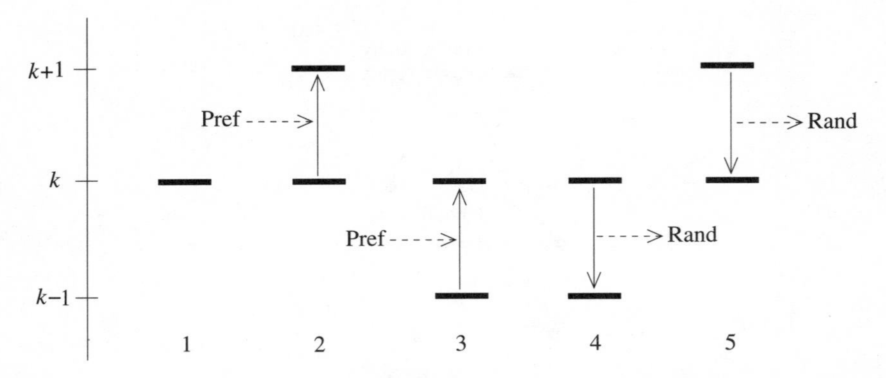

FIG. 4.4. The processes that we account for in the right-hand side of eqn (4.5) (see text). 1 represents the first term, 2 the second, and so on. The solid arrows show how the number of connections of a vertex changes. The dashed lines display the rewirings of edges. In 2 and 3, this is rewiring to a preferentially chosen vertex. In 4 and 5, the randomly chosen end of a randomly chosen edge is removed from a vertex.

(2nd) When the edge end becomes (preferentially with the probability $f(k)/N\langle f(k)\rangle$) attached to a vertex of degree k, the number of vertices of degree k decreases by one. This loss is described by the second term on the right-hand side.

(3rd) The third (gain) term is due to vertices of degree $k-1$ that get an edge (with the probability $f(k-1)/N\langle f(k)\rangle$) and arrive at a state with k edges.

(4th) The fourth term is due to the loss of one of the edges by vertices of degree k. As we saw above, this happens with the probability $k/2L$ and reduces $\langle N(k,t)\rangle$.

(5th) Finally, the last term is due to the loss by vertices of degree $k+1$ of one of their edges (the probability is $(k+1)/(2L)$). This increases $\langle N(k,t)\rangle$.

This equation is very simple. Actually, what we are playing with is the redistribution of a given number of balls (edge ends) among a fixed number of boxes (vertices). At each step, a ball is taken from a box chosen with the preference function proportional to k and thrown into a box chosen with the other preference function $f(k)$.[10] We emphasize that this is equivalent only to networks with tadpoles and melons. Equilibrium networks without tadpoles and melons can be reduced to this 'one-particle' problem only in 'safe' situations (see above).

Now we can readily write the equation for the degree distribution:

[10]This resembles the backgammon model ('balls-in-boxes') (see Bialas, Burda, and Johnston 1997, 1999, Bialas, Bogacz, Burda, and Johnston 2000, and Burda, Correia, and Krzywicki 2001). In this model L balls are distributed among N boxes. Each of the boxes is considered to have the same (functionally) statistical weight dependent only on the number of balls in the box. Therefore, our construction presents a physical realization of the backgammon model: the construction allows us to find the above statistical weights of the boxes.

$$N\frac{\partial P(k,t)}{\partial t} = \frac{1}{\langle f(k)\rangle}[-f(k)P(k,t) + f(k-1)P(k-1,t)]$$

$$+\frac{1}{\overline{k}}[-kP(k,t) + (k+1)P(k+1,t)]\,. \tag{4.6}$$

This is a very simple linear homogeneous equation. Its stationary solution $P(k) \equiv P(k, t \to \infty)$ satisfies

$$\frac{f(k)}{\langle f(k)\rangle}P(k) - \frac{k+1}{\overline{k}}P(k+1) = \text{const} = 0\,. \tag{4.7}$$

We set const $= 0$ since $kP(k) \to 0$ as $k \to \infty$, otherwise the degree distribution is not normalizable. Then we have

$$f(k) = \frac{(k+1)P(k+1)}{P(k)}\frac{\langle f(k)\rangle}{\overline{k}}\,. \tag{4.8}$$

This relation (1) allows us to find rates $\{f(k)\}$ from a given degree distribution $P(k)$ but (2) also makes it possible to obtain $P(k)$ from $f(k)$. Let us discuss these two possibilities focusing on the interesting classes of distributions.

(1) Let $P(k)$ be given.

As we are interested only in equilibrium properties, the rates $f(k)$ are defined up to an arbitrary constant factor, which is important only for the approach to equilibrium. We can always choose this constant such that $\langle f(k)\rangle/\overline{k} = 1$. Then, from a given $P(k)$ one can find the preference function $f(k)$:

$$f(k) = \frac{(k+1)P(k+1)}{P(k)}\,. \tag{4.9}$$

This relation establishes a one-to-one correspondence between microcanonical and canonical ensembles of random networks, or, which is the same, between two construction procedures. In fact we have constructed a canonical ensemble of uncorrelated equilibrium networks, which is equivalent to a microcanonical one.

Here we do not discuss in detail for which classes of a degree distribution this is possible. For example, for an exponential degree distribution, where $P(k+1)/P(k) = \text{const}$, it should be $f(k) = k+1$ up to an arbitrary constant factor.

(2) Now let the preference function $f(k)$ be given.

When does it yield fat-tailed degree distributions? Is this possible at all? Indeed, we saw that linear preferential linking in growing networks produces fat-tailed (often, even scale-free) structures in a wide range of parameters, which may be called self-organized criticality. And for the equilibrium networks? We recall the question with which we started this section: when does this equilibrium network (loosely speaking, the canonical ensemble of random networks) have a fat-tailed degree distribution?

From eqn (4.8) one can see that for this it is necessary (but not sufficient!) that $f(k)\overline{k}/\langle f(k)\rangle = k + o(k)$, as $k \to \infty$ ($o(k) \ll k$ by definition). Let us choose an arbitrary constant multiple of $f(k)$ such that $f(k) = k + o(k)$, as $k \to \infty$. In the present section we assume that the condensation of edges is absent. Then we see from eqn (4.8) that *a fat-tailed degree distribution in an equilibrium network can be realized only at a single (critical) value of the average degree* $\overline{k}_c$, where $\overline{k}_c$ is taken from the condition

$$\overline{k} = \langle f(k)\rangle \,. \tag{4.10}$$

Here the averaging is taken with the distribution obtained from eqn (4.8), so that this condition is actually self-consistent. In the next section, we shall show that $\overline{k}_c$ is the condensation point.

Below this critical point, the 'critical' fat-tailed distribution is cut off by the factor x_s^k, where $x_s \equiv \overline{k}/\langle f(k)\rangle < 1$. Notice the difference from the growing networks, where a linear preference actually ensures a fat-tailed degree distribution.

The specific form of the (fat-tailed) degree distribution at the critical point depends on how $f(k)$ approaches its asymptote $f(k) \cong k$ as $k \to \infty$. For example, what will $f(k)$ be when the degree distribution is scale-free: $P(k) \sim k^{-\gamma}$?

Let us substitute $P(k) \sim k^{-\gamma}$ into eqn (4.9). At large k we obtain

$$f(k) \cong k(1 + k^{-1})\frac{k^{-\gamma}(1-\gamma k^{-1})}{k^{-\gamma}} \cong k + 1 - \gamma + \mathcal{O}(k^{-1})\,. \tag{4.11}$$

This preference produces scale-free networks when the average degree approaches its critical value. For $\overline{k} < \overline{k}_c$, we have $P(k) = [x_s(\overline{k})]^k k^{-\gamma}$.

A more general construction. Note that we can use a more general construction procedure than in Fig. 4.3(a). At each time step, let an edge end be rewired from a vertex that is chosen with a preference function $h(k)$ to a vertex chosen with a preference function $f(k)$. For simplicity, we assume $h(k = 0) = 0$. Indeed, without this assumption one has to take into account that it is impossible to remove connections from bare vertices. Then instead of eqn (4.8) we obtain

$$f(k) = \frac{h(k+1)P(k+1)}{P(k)}\frac{\langle f(k)\rangle}{\langle h(k)\rangle}\,. \tag{4.12}$$

Grand canonical ensemble

We will consider the situation above the critical point, $\overline{k} > \overline{k}_c$, in two subsequent sections. Here we will briefly discuss a way to construct the grand canonical ensemble of equilibrium networks. Our canonical ensemble was constructed by preferential rewiring. This procedure conserves the total number of edges, as it is necessary for the canonical ensemble. For large networks, this is the same as (1) deleting an arbitrary chosen edge and (2) creating simultaneously a new edge between two preferentially chosen vertices (each of them is chosen with probability proportional to the function $f(k)$ of its degree). One can imagine this as the edge jumps from one position to another (see Fig. 4.3(b)). One edge jumps per time step.

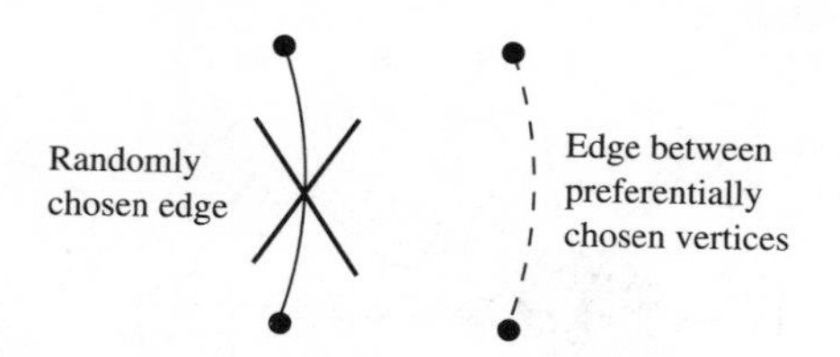

FIG. 4.5. The construction procedure for the grand canonical ensemble of random networks. Two processes proceed in parallel: (1) the deletion of randomly chosen edges and (2) the emergence of new edges between preferentially chosen vertices.

In the grand canonical ensemble, the total number of edges is not conserved. To construct the grand canonical ensemble, we change one element in the previous picture: let randomly chosen edges be deleted at one rate and new edges connect pairs of preferentially chosen vertices at another rate, so that their total number is not conserved (see Fig. 4.5). The construction looks as follows:

Grand canonical ensemble of random networks:

(*set*) The set G_{GC} contains all graphs with any number of edges.[11]

(*rule*) Randomly chosen edges are deleted at a rate λN (λ is the inverse lifetime of an edge; λ is fixed as $N \to \infty$, that is in the thermodynamic limit). Edges between vertices i and j emerge at a rate $f(k_i)f(k_j)$.

The grand canonical ensemble is determined by the set $(\lambda, \{f(k)\})$.

Correspondence between ensembles

In the thermodynamic limit, all three ensembles are equivalent if their parameters are properly adjusted.[12] Equation (4.8) fixes the correspondence between microcanonical and canonical ensembles. Let us establish a one-to-one correspondence between canonical and grand canonical ensembles assuming that rates $f(k)$ are equal in both these cases. When $N \to \infty$ in the canonical ensemble, the total number of edges which 'evaporate' per unit time must be equal to the total number of edges which are 'absorbed' in this time: $\lambda NL = \lambda N(\overline{k}N)/2 = \langle f(k)\rangle^2 N(N-1)/2$. Then we have

$$\lambda = \frac{\langle f(k)\rangle^2}{\overline{k}} \,. \tag{4.13}$$

[11]This is if we allow tadpoles and melons. If they are forbidden, the set G_{GC} consists of all Mayer graphs with any number of edges smaller than or equal to $N(N-1)/2$ (recall that the number of vertices N is fixed).

[12]A rigorous proof of this claim is non-trivial. In particular, one must use the ergodic hypothesis and prove the vanishing smallness of the fluctuations of the total number of edges in the thermodynamic limit of the grand canonical ensemble. Meanwhile, just this smallness means that in these networks, basic physical quantities are self-averaging.

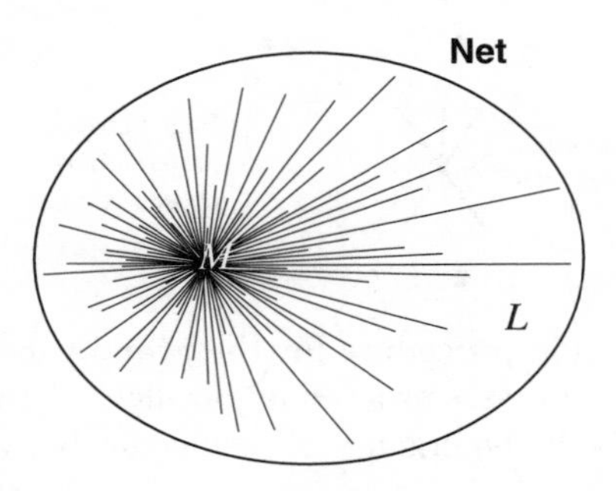

FIG. 4.6. When the average number of connections of a vertex in the equilibrium network exceeds a critical value $\overline{k}_c$, a finite fraction of all edges is condensed on a single vertex.

This relates λ for the grand canonical ensemble to the average degree $\overline{k}$ for the canonical ensemble and fixes the one-to-one correspondence between these ensembles of random networks.

4.4 Econophysics: condensation of wealth

We explained that below the critical point ($\overline{k} < \overline{k}_c$), if it exists, our ensembles of uncorrelated random networks without tadpoles and melons (Mayer graphs) are equivalent to ensembles of graphs with tadpoles and melons (non-Mayer graphs) and the problem is actually reduced to that of distributing balls among boxes. This equivalence is violated (1) in finite networks and (2) above $\overline{k}_c$, where, as we will soon see, a finite fraction of edges turns out to be attached to a few vertices.

The situation for Mayer graphs, as we will see in the next section, is unclear. Here we discuss a much simpler case of networks with tadpoles and melons, where we can forget about edges themselves, and consider only their ends. This is really a simple problem but also a convenient case to discuss some other applications of these ideas.

We return to our dynamical construction procedure, namely to the canonical ensemble of random networks. One can see that eqn (4.7) has no thermodynamically stable normalizable solution $P(k)$ if $\overline{k} > \langle f(k) \rangle$; that is, for $\overline{k} > \overline{k}_c$. The introduction of the condensation allows us to get rid of this problem.

Let us assume that one or a few vertices attract a finite fraction of edges (see Fig. 4.6). More rigorously, this means that the fraction of such vertices approaches zero as $N \to \infty$. Assume that M of the total number L edges is condensed in such a way. The stationary situation is realized when

$$\frac{M}{N\overline{k}} = \frac{f(M)}{N\langle f(k) \rangle} \, . \tag{4.14}$$

The meaning of this balance condition is obvious: the number of edges that are rewired from the centre(s) of condensation is equal to the number of edges that become condensed. Here $\overline{k} = 2L/N$ and $\langle f(k)\rangle$ are the averages for the entire network. Then $\langle f(k)\rangle/\overline{k} = f(M)/M$.

One can also introduce averages $\overline{k}\,' = (2L - M)/N$ and $\langle f(k)\rangle' = [N\langle f(k)\rangle - f(M)]/N$ for the non-condensed fraction. One can see that

$$\frac{M + N\overline{k}\,'}{M} = 1 + \frac{N\overline{k}\,'}{M} = \frac{f(M) + N\langle f(k)\rangle'}{f(M)} = 1 + \frac{N\langle f(k)\rangle'}{f(M)}\,, \tag{4.15}$$

so that $\langle f(k)\rangle'/\overline{k}\,' = \langle f(k)\rangle/\overline{k} = f(M)/M$.

The formal form of the balance equation for the degree distribution $P'(k)$ of 'normal' vertices is the same as previously, that is eqn (4.7). Let the preference function be $f(k) = k + o(k)$ as $k \to \infty$. Then $\langle f(k)\rangle/\overline{k} = f(M)/M \to 1$, and we see that the degree distribution of 'normal' vertices in the condensed phase is the same as at the critical point, $P_c(k)$. This means that (1) it is fat-tailed and (2) the total degree of 'normal' vertices is the same as at the critical point, $N\overline{k}_c$. Hence

$$M = N(\overline{k} - \overline{k}_c)\,. \tag{4.16}$$

The same results were obtained in the framework of the backgammon model (Burda, Johnston, Jurkiewicz, Kaminski, Nowak, Papp, and Zahed 2001). How many centres of condensation there are in a network, depends on some subtleties. For example, finite-size effects are essential in this situation (Bialas, Burda, and Johnston 1997). A specific form of $f(k)$ for large k may also be important.

Where else, apart from networks, can this be used?

The straightforward application is in *econophysics*, namely wealth distribution. In fact what we are discussing is wealth distribution in a stable ('equilibrium') society (the total large number of agents N is fixed). There is no mortality, and the total wealth L is conserved. The wealth (degree) is permanently redistributed among agents (vertices). At each moment some amount of money is spent. In the situation that we consider this amount is proportional to the wealth one has, although this is only a possibility. This money comes to somebody else with a probability proportional to some preference function of their wealth. The situation is simple, and we will not worry about the exclusion of tadpoles (a transfer from one pocket to another) and melons (a multiple transfer of money from one agent to another—Onassis likes your money).

What happens as the society grows richer? Let this growth be so ('infinitely') slow that the situation is constant equilibrium and the above consideration is valid.

From the previous section we see that if $\overline{k} = 2L/N$ is small (the 'poor' equilibrium society), the wealth distribution is of exponential type. If the type of preference that we implement allows wealth condensation, then starting from some critical average wealth $\overline{k}_c$ a lucky, infinitely small part of the population captures a finite fraction of the total wealth (the 'rich' equilibrium society). The wealth distribution of the rest is the same as at the condensation point, that is fat-tailed.

Moreover, the growth of the total wealth above the critical value does not increase the wealth of the population except for several lucky stars. The entire extra wealth $L - N\overline{k}_c$ appears in the pockets of a very few people. The others have their total supremum $N\overline{k}_c$ even if the average wealth approaches infinity. That's life...

A 'shark' in equilibrium

In the previous section we did not account for many important factors such as mortality, inheritance, etc. Such schematic pictures cannot account for everything. Nevertheless, one factor we have missed may be really crucial. We assumed that our society is absolutely homogeneous: the preference function $f(k)$ is the same for everybody, so that their chance to accumulate money is determined only by their present wealth. It does not depend on their relations, on whether they are clever, or silly, strong, or weak, etc. Of course, the society is inhomogeneous, and we may account for this by introducing different preferences (different preference functions) for different 'vertices' (individuals).

To avoid going into detail, it is better to discuss a 'natural' linear type of preference, $Bk+A$. One can suggest that variation of the constant A from person to person does not change the situation crucially. In Section 5.10 we will discuss the effects of such inhomogeneity in growing networks. These effects are not striking. The effect of the inhomogeneity of the factor B evidently must be much stronger.[13]

The behaviour of the system (network, society, etc.) depends on the particular form of the distribution of B. For brevity, here we consider a trivial particular case, which demonstrates the effect.

Let there be only one strong person ('shark'), and all the others have the same simple preference $k+1$. As was shown above (see Section 4.3), in a homogeneous case, this preference produces the exponential 'wealth' distribution:

$$P(k+1)/P(k) = \overline{k}/(\overline{k}+1)\,, \tag{4.17}$$

and any condensation is absent. Let the preference of the strong guy be $g(k+1)$, where the constant g (the 'fitness' or the 'strength' of the strong guy) is greater than 1.

We introduce a personal degree distribution $P''(k)$[14] for this guy and a degree distribution $P'(k)$ for all others. Then the detailed balance conditions are

$$\frac{(k+1)P'(k)}{(N-1)(\overline{k}'+1)+g(\overline{k}''+1)} = \frac{(k+1)P'(k+1)}{N\overline{k}}\,, \tag{4.18}$$

[13] In growing networks, this inhomogeneity was originally introduced by Bianconi and Barabási (2000a).

[14] Recall at this point that we do not have a single random network and a single society but a statistical ensemble of random networks or a statistical ensemble of societies. Consequently, for the 'shark' we must introduce his personal distribution and his average wealth.

$$g\frac{(k+1)P''(k)}{(N-1)(\overline{k}'+1)+g(\overline{k}''+1)} = \frac{(k+1)P''(k+1)}{N\overline{k}},$$

where $\overline{k}''$ is the average wealth of the strong guy and $\overline{k}'$ is the average wealth of the others; $(N-1)\overline{k}' + \overline{k}'' = N\overline{k}$.

First suppose that the condensation is absent. Then both the left denominators approach $N(\overline{k}+1)$ as $N \to \infty$. The first equation is the same as in the absence of the strong guy, and so the degree distribution is the same as in the homogeneous situation. The equation for the 'shark' turns out to be

$$P''(k+1)/P''(k) = g\overline{k}/(\overline{k}+1)\,. \tag{4.19}$$

One can see that if $g < (\overline{k}+1)/\overline{k}$, that is $\overline{k} < \overline{k}_c = 1/(g-1)$, there are no problems, and $P(k) \propto [\overline{k}/(\overline{k}+1)]^k$. Otherwise the wealth distribution $P''(k)$ of the 'shark' diverges, and the solution is absent. As one might expect, this indicates the condensation on this vertex.

Let us assume that this is just the case. Suppose that M dollars (balls, edges, etc.) are condensed in the pockets of the strong guy. The condensation means that M/N approaches a finite limit as $N \to \infty$; $(N-1)\overline{k}' + M = N\overline{k}$. Analogously to eqn (4.14), the number M is stationary when

$$g\frac{M+1}{(N-1)(\overline{k}'+1)+g(M+1)} = \frac{M}{N\overline{k}}\,. \tag{4.20}$$

Taking into account that $M \gg 1$ gives

$$g = 1 + \frac{1}{\overline{k}} + (g-1)\frac{M}{N\overline{k}}\,, \tag{4.21}$$

and so above the condensation point, $\overline{k} > \overline{k}_c = 1/(g-1)$,

$$M = N\overline{k} - \frac{N}{g-1}\,. \tag{4.22}$$

But with this condensate, the first line of eqn (4.18) gives for the wealth distribution of 'normal poor' people (the degree distribution of 'normal' vertices)

$$\frac{P'(k+1)}{P'(k)} = \frac{1}{g}\,. \tag{4.23}$$

This provides the wealth distribution of poor people, $P'(k) \propto g^{-k} = e^{-k\ln g}$, which just coincides with the wealth distribution at the condensation point. The total 'wealth' of poor people above the condensation point is the same all the time, $M - N\overline{k} = N/(g-1)$.

To characterize the evolution of the exponential wealth distribution of poor people we introduce a function $r(\overline{k}, g)$ such that $P'(k) \propto r^k(\overline{k}, g)$. The behaviour of r versus $\overline{k}$ is shown in Fig. 4.7.

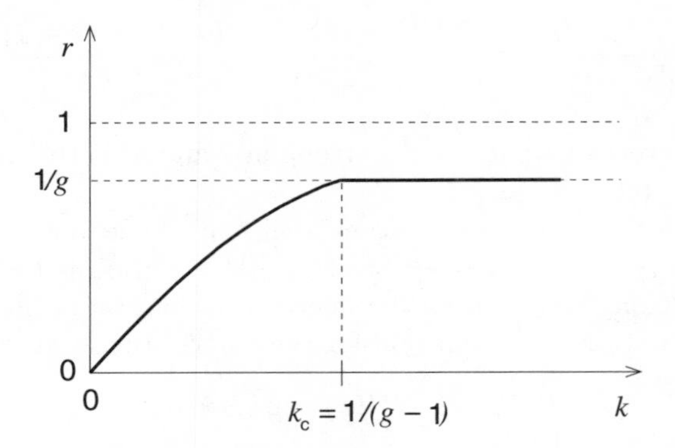

FIG. 4.7. The dependence r ($P(k) \propto r^k$) versus the average wealth in the 'equilibrium society' with a single 'strong' person of a strength $g > 1$, if the wealth distribution is exponential in a homogeneous situation. The parameter $r < 1$ characterizes the wealth distribution $P(k)$. The smaller r is, the faster the exponential decay. The wealth condensation occurs at the average wealth $\overline{k}_c = 1/(g-1)$. For $\overline{k} > \overline{k}_c$, the wealth distribution is the same as at the condensation point.

On the other hand, we can look at the situation when the total wealth is fixed. Then, if our strong guy is not strong enough, $g < 1 + 1/\overline{k}$, his global effect is zero. When g exceeds the critical value, the strong guy is like Bill Gates. It is a simple exercise to consider the effect of several 'sharks' with various strengths. One easily finds that in such a case, only the strongest person can produce the global effect and condense our money.

We see that, unlike the condensation in a homogeneous situation, the condensation in an equilibrium inhomogeneous system (a society or a network) is not accompanied by a fat-tailed distribution.

One can study a more complex situation. Let a network have a condensation point even without a strong guy, in a homogeneous situation. Hence, its degree distribution is fat-tailed at this point (for Mayer and non-Mayer graphs)[15] and above it (at least for non-Mayer graphs). One can easily check that in such a case, the introduction of a single 'strong' vertex diminishes the value of the critical average degree. Moreover, the degree distribution necessarily gets an exponential size-independent cut-off.

We did not account for mortality. At first sight, mortality inevitably removes the condensation since someday the strong guy will die. However, this issue is not so obvious: the answer depends on the time, which is necessary to approach equilibrium. In addition, mortality is always accompanied by inheritance.

[15] We emphasize that the value of $\overline{k}_c$ is the same for Mayer and non-Mayer networks irrespective of whether the net is homogeneous or not.

4.5 Condensation of edges in equilibrium networks

In the previous section we have actually described the structure of the condensed phase in networks with tadpoles and melons. The general situation for these networks (homogeneous and inhomogeneous) is clear. However, the problem of the condensed phase for Mayer graphs is more complex and is still unsolved. In this event, the approach of the thermodynamic limit is non-trivial. Therefore, here we can present only very intuitive arguments for networks without tadpoles and melons and briefly discuss natural possibilities.

First we will look at the condensation in the inhomogeneous networks. We explained above that at the condensation point, the degree distribution of both the Mayer and non-Mayer inhomogeneous graphs has a size-independent exponential cut-off. It would be very strange if the degree distribution of a network had the cut-off at the condensation point and had no cut-off in the condensed phase. In this respect, the situations for Mayer and non-Mayer inhomogeneous graphs more or less coincide.

Now consider condensation in homogeneous networks without tadpoles and melons; that is, when the degree distribution at the condensation point was fat-tailed. One can check that the assumption that there is only one centre of condensation in the network leads to a contradiction: the thermodynamically stable state is absent. The same follows if we suppose that there are several such centres. The thermodynamically stable condensed state is possible if the total number of the centres of condensation n is not finite but tends to infinity in an infinitely large network, although the ratio $n/N \to 0$. The number of tadpoles and melons grows slower than N, and so, again, this case seems to be similar to that for non-Mayer graphs.[16]

At this point we show our cards and finish the section with a résumé: the comparison of equilibrium and non-equilibrium networks, which are constructed by preferential linking.

(1) *Non-equilibrium networks, namely networks growing under the mechanism of preferential linking*:
If the preference is linear, at least at large degrees, the degree distribution is fat-tailed (and, often, even scale-free) over a wide range of parameters without any condensate. In fact, this is self-organized criticality. Condensation in homogeneous growing networks[17] is absent. Inhomogeneities of vertices do not remove these fat tails but may result in condensation.

(2) *Equilibrium networks with preferential linking*:
If the preference is (asymptotically) linear, a degree distribution can be fat-tailed only in the presence of condensate and at the condensation point. The condensation may take place in a homogeneous network, above some critical value of mean degree. The inhomogeneity of a network may diminish this critical value and simultaneously remove fat-tails.

[16]... but the problem is open.

[17]Here we mean a linear preference.

The detailed discussion of non-equilibrium networks is the subject of the next chapter, so the reader still has to take (1) on faith.

4.6 Correlations in equilibrium networks

The constructions that we considered above naturally generate uncorrelated networks at least in the thermodynamic limit. In other words, correlations between their vertices were absent. However, such correlations are important in real networks. We will see that the correlations are inevitable in growing networks (see Section 5.12) but in equilibrium nets they may also be important. Let us discuss how correlations can be characterized and generated.

The difficulty is that there may be several types of correlations in networks.[18] For example, clustering (that is, the presence of loops of length 3 in a network) means specific correlations (see Sections 1.4 and 4.7). The presence of more long loops also means correlations.[19] Finally, degrees of vertices may be correlated. Here we will discuss the last type of correlations.

How can we describe correlations between degrees of vertices?

In Section 4.1 we explained that a degree distribution is, in fact, the average of the Kronecker symbol over the ensemble of networks, $P(k) = \langle N(k) \rangle / N = \langle \delta[k_1 - k] \rangle = \langle \sum_{i=1}^{N} \delta[k_i - k] \rangle / N$, where 1 is an arbitrary vertex and $N(k)$ is the total number of vertices of degree k in a network from the ensemble. A direct way to describe correlation is to introduce a more complex distribution by averaging a combination of Kronecker symbols.

Various such correlators may be considered, but usually the following distribution is used. Correlations are characterized by the distribution of degrees of the nearest-neighbour vertices:

$$P(k, k') = \frac{1}{N^2} \left\langle \sum_{i,j=1}^{N} \delta[k_i - k] a_{ij} \delta[k_j - k'] \right\rangle . \tag{4.24}$$

Here, the factor a_{ij} (elements of the adjacency matrix) ensures that only nearest neighbours are accounted for in the sum. One can give a natural physical interpretation of $P(k, k')$: it is the probability that a randomly chosen edge connects vertices of degrees k and k'. So, $P(k, k')$ plays the role of a 'degree–degree distribution for edges'. If correlations in a network are absent, $P(k, k')$ factors and can be expressed in terms of the degree distribution:

$$P(k, k') = \frac{kP(k)k'P(k')}{[\sum_k kP(k)]^2} . \tag{4.25}$$

Indeed, in any network, correlated and uncorrelated, the probability that a randomly chosen end of a randomly chosen edge is of degree k is proportional to

[18] Note that these types are not independent.

[19] This is why correlations are actually present in *finite* nets from the above constructions.

$kP(k)$. We have already used this in Section 4.3. The reason is simple: a vertex of degree k keeps $k/(2L)$ edges. If the network is uncorrelated, and so the end vertices of an edge are independent, $P(k,k')$ must be proportional to $kP(k)k'P(k')$. The denominator $[\sum_k kP(k)]^2$ is introduced for normalization.

Let us construct an equilibrium random network with such pair degree–degree correlations of the nearest-neighbour vertices. Here we propose a construction procedure for a graph with a given (arbitrary) degree–degree distribution for nearest neighbours, $P(k,k')$, and a fixed number of vertices, N.

The distribution $P(k,k')$ is all we need for the complete description of the network, since from the degree–degree distribution one can easily get the ordinary degree distribution.[20] Indeed, applying the above arguments readily yields

$$\sum_k P(k,k') = \frac{kP(k)}{\sum_k kP(k)} . \tag{4.26}$$

This relation gives the function $P(k)$ up to a constant factor, which, in turn, can be obtained from the normalization condition $\sum_k P(k) = 1$. Thus we get

$$\overline{k} = \left[\sum_{k,k'} P(k,k')/k\right]^{-1} , \tag{4.27}$$

and so the total number of edges is known, $L = \overline{k}N/2$.

Then, we have N and L fixed and can formulate the construction procedure for such correlated graphs in the spirit of the construction from graph theory for uncorrelated graphs with a given degree distribution (see Fig. 1.8 from Section 1.3).

Let, for simplicity, the (fixed) number N of vertices tend to infinity.

(1) Using the given $P(k,k')$ find the total number of edges, L.
(2) Create the 'infinite' number L of pairs of integers (k,k') distributed as $P(k,k')$. These are L edges with ends of degrees k and k'. Therefore we have edges labelled by degrees of their ends but have no vertices yet.
(3) Select, at random, groups of ends of degree k, each one consisting of k ends. Tie them in bunches, each of k tails. These bunches are vertices of degree k.

And that is all: the correlated network is ready. One can check that this procedure is possible in the thermodynamic limit, and that it generates tree-like graphs.

Speaking more loosely, this corresponds to *the microcanonical ensemble of random graphs with a given correlation*:

[20] It is worthwhile to emphasize this point. In these networks $P(k,k')$ is a basic object. It completely determines their structure, and an ordinary degree distribution is only a consequence of secondary importance.

(*set*) G_M is the set of all possible graphs with N vertices, in which the number $N(k, k')$ of edges with ends of degrees k and k' is taken from a given degree–degree distribution.

(*rule*) The statistical weights of these graphs are equal.

This is a geometric (configuration) construction (compare with Section 4.3) but as in Section 4.3 we can also construct such graphs dynamically. For example, this can be done by removing a randomly selected edge to a new position chosen preferentially, that is to a pair of vertices chosen with a preference $f(k, k')$.

Here we introduce a new category, *pair preference*, and a pair preference function. This is a more general and, in some sense, a more natural thing than a one-vertex preference, where the probability to become attached, $f(k)$, depends only on the degree of a vertex. Indeed, an edge has not one but two ends. In general, they must become attached not to one but to two vertices, and so it is a combination of two degrees that is a natural form of preference.[21]

In this dynamical construction (in fact, this is a canonical ensemble of correlated random networks), the given preference $f(k, k')$ and the total number of edges, L, determine $P(k, k')$.

4.7 Small-world networks

Another type of correlation in networks is clustering (see Section 1.4). Watts and Strogatz (1998) proposed a very simple construction procedure that generates equilibrium networks with correlations of this type.

These graphs are obtained from regular lattices by rewiring bonds or by making shortcuts between randomly chosen vertices.[22] The resulting random structures, which have been named *small-world networks*,[23] are, in fact, superpositions of lattices and classical random graphs (see Fig. 4.8). One can say that in these networks, the crossover from lattices to random graphs is realized. So, they combine the high clustering of regular lattices and the smallness (small-world effect) usual for 'normal' random networks.

The fact that such a crossover exists in a wide range of parameters (even at a rather small concentration of shortcuts) is not so trivial. The popular term 'small-world network' for very specific networks is a reflection of the surprise over the appearance of the small-world effect in slightly spoiled lattices, which are still locally perfect.

[21] Preferential attachment was introduced by Barabási and Albert (1999) by using a rather specific network, a citation graph: the Barabási–Albert model. In such graphs, new vertices become attached to old ones, and actually only one of the ends of each edge becomes attached preferentially. In this situation the usual form of preference, $f(k)$, is the only possibility.

[22] Similar constructions with disordered lattices are also possible.

[23] Do not confuse them with 'small worlds' that are, by definition, networks with the small-world effect.

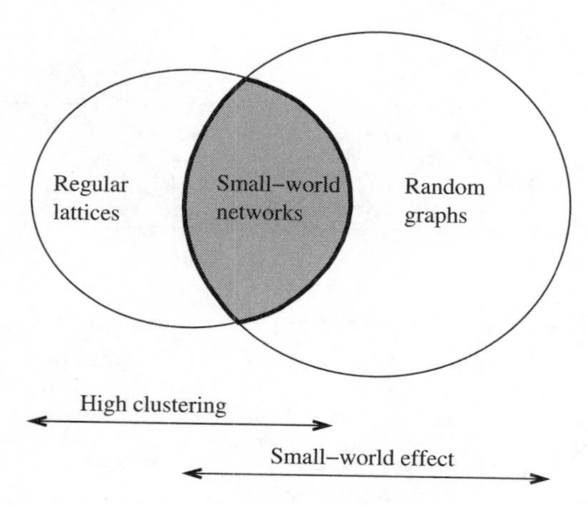

FIG. 4.8. Small-world networks are the superposition of regular lattices and random graphs (see discussion below). Usually, the latter are classical random graphs but, in principle, any random nets with the small-world effect can be used in these constructions. The small-world networks inherit their high clustering from the regular lattices. The small-world networks inherit the small-world effect from the random graphs.

4.7.1 *The Watts–Strogatz model and its variations*

For half a century, physicists have studied the effect of a local disorder on lattices: all the time regular structures were spoiled or deformed locally, that is over one or ten or a hundred lattice cells. Watts and Strogatz asked: 'what will happen if we spoil a lattice globally?' Let even this 'global' damage be, at first sight, small. What will be the effect?

The original network of Watts and Strogatz is constructed in the following way (see Fig. 4.9(a)). Initially, a regular one-dimensional lattice with periodic boundary conditions is present. Each of $\mathcal{L}$ vertices has $z \geq 4$ nearest neighbours ($z = 2$ was not appropriate for Watts and Strogatz since, in this case, the clustering coefficient of the original regular lattice is zero). Then one takes all the edges of the lattice in turn and with probability p rewires them to randomly chosen vertices. In such a way, a number of long-range connections appear. Obviously, when p is small, the situation has to be close to the original regular lattice. For a large enough p, the network is similar to the classical random graph. Note that the periodic boundary conditions are not essential.

Watts and Strogatz studied the crossover between these two limits. Their main interest was in the average shortest path, $\overline{\ell}$, and the clustering coefficient (recall that each edge has unit length). The simple but exciting result was the following. Even for the small probability of rewiring, when the local properties of the network are still nearly the same as for the original regular lattice and the clustering coefficient does not differ essentially from its initial value, the average

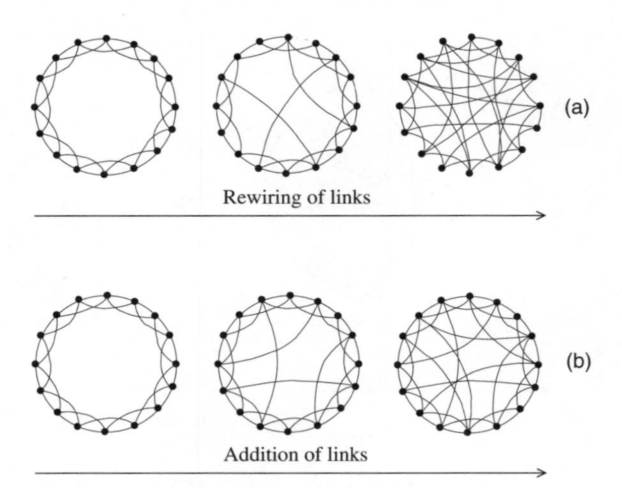

FIG. 4.9. Small-world networks. (a) The original Watts–Strogatz model (Watts and Strogatz 1998) with the rewiring of connections. (b) The network with the addition of shortcuts to a regular lattice.

shortest-path length is already of the order of that for classical random graphs (see Fig. 4.10).

We have mentioned that this result seems quite natural: the average shortest-path length is very sensitive to the shortcuts. One can see that it is enough to make a few random rewirings to decrease $\overline{\ell}$ by several times. On the other hand, several rewired edges cannot crucially change the local properties of the entire network. This means that the global properties of the network already change strongly at $pz\mathcal{L} \sim 1$, when there is one shortcut in the network, that is at $p \sim 1/(\mathcal{L}z)$, when the local characteristics are still close to the regular lattice. One of these local characteristics is the clustering coefficient C,[24] so just the behaviours of $\overline{\ell}$ and C were compared.

Instead of rewiring edges, one can add shortcuts between randomly chosen vertices of a regular lattice (see Fig. 4.9(b)). The main features of the network do not change. For example, one can check that in both these cases, the degree distribution is of a Poisson type: it decreases rapidly at high degrees. In both the cases, the dependence of the average shortest-path length on the extent of the damage displays a scaling behaviour in the crossover region.

The nature of this scaling of the crossover is simple. One can start with a regular lattice of arbitrary dimension d where the number of vertices $N = \mathcal{L}^d$.[25] $\mathcal{L}$ is the linear size of the lattice. Let z be the average number of the

[24] Another local characteristic is, of course, degree. However, in the original procedure of Watts and Strogatz, the average degree of the network does not change when links become rewired. So, it cannot be used for comparison.

[25] Note that many regular lattices (for example, a square lattice, $z = 4$, or a simple chain, $z = 2$) have zero clustering coefficient. If one is interested in the evolution of clustering, it is

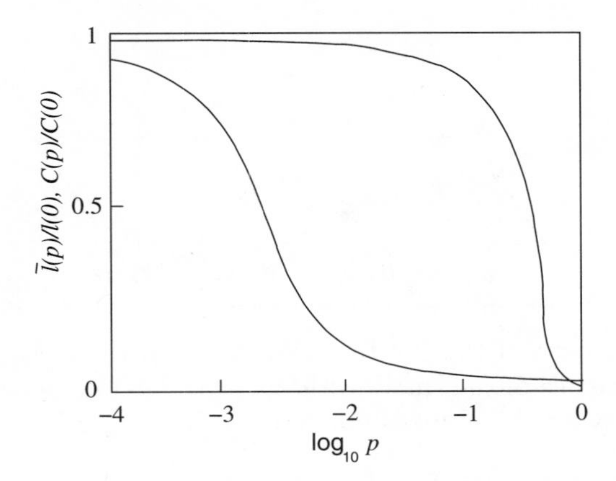

FIG. 4.10. Average shortest-path length $\bar{\ell}$ and clustering coefficient C of the Watts–Strogatz model versus the fraction of the rewired connections p (after Watts and Strogatz 1998). Both these quantities are normalized to their values for the original regular lattice. The network contains 1000 vertices, and the average degree is 10. Notice that the clustering coefficient is practically constant in the range where the average shortest-path length diminishes sharply.

nearest neighbours of a vertex in this lattice. In this case, the number of edges in the regular lattice is $z\mathcal{L}^d/2$. To keep the correspondence to the Watts–Strogatz model, let us define p in such a way that for $p = 1$, there is the same number $z\mathcal{L}^d/2$ of random shortcuts as the total number of edges in the regular lattice.

Let us look at what length scales we have in the network:

(1) Of course, we have the regular lattice spacing, that is the unit length.

(2) One can introduce the characteristic length $\xi = \mathcal{L}/N_s^{1/d} = (pz)^{-1/d}$, where N_s is the total number of shortcuts (see, e.g., Ozana 2001). Then $N_s/\mathcal{L}^d = \xi^{-d}$ plays the role of the average 'density of the ends of shortcuts', and ξ is the average distance between the closest ends of shortcuts (see Fig. 4.11). This characteristic length also has another meaning. Suppose we measure the average shortest-path length $\bar{\ell}(l)$ between pairs of vertices with a separation by a Euclidean distance l on the regular lattice. If the Euclidean separation of the vertices is smaller than the Euclidean distances between the vertices and the nearest shortcuts, then the situation is the same as for a regular lattice, and $\bar{\ell}(l \ll \xi) \cong l$ (see Figs 4.11 and 4.12). Otherwise the average length of the shortest path connecting these two vertices coincides with the average shortest-path length of the network: $\bar{\ell}(l \gg \xi) \cong \bar{\ell}$. That is, starting from this scale, the network 'forgets' about the regular lattice.

(3) We have the average shortest-path length $\bar{\ell}$.

better to use, for example, a honeycomb lattice etc. However, the distribution of the shortest-path lengths and other 'global' properties of the network do not depend on the structure of the specific 'mother' lattice.

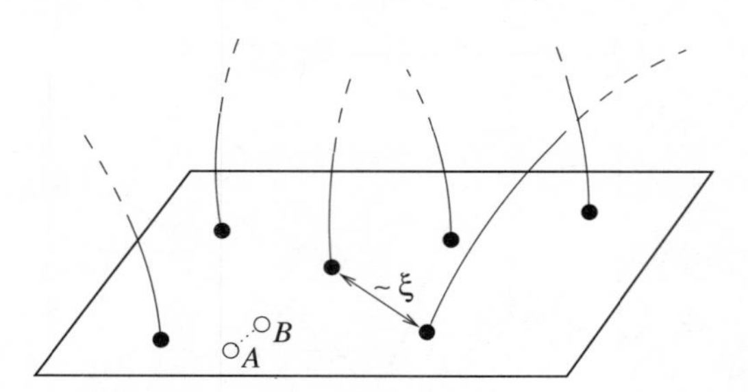

FIG. 4.11. The length of the shortest path between the close points A and B on a mother lattice of the small-world network ($AB \ll \xi$) coincides with the Euclidean distance AB. This is not the case if $AB \gtrsim \xi$.

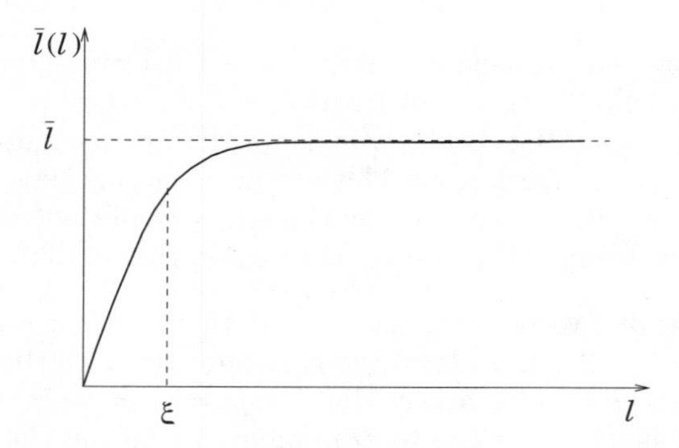

FIG. 4.12. Typical average length of the shortest path between two vertices versus their Euclidean separation l. ξ is the average distance between the closest ends of shortcuts.

(4) Finally, we have the linear size of the lattice, $\mathcal{L}$.

In the regular lattice ($N_s = 0$) the average shortest-path length is of the order of the linear size of the lattice, $\overline{\ell} \sim \mathcal{L}/z$. In the region interesting for us, that is when the number N_s of shortcuts is small enough, $\overline{\ell}$ does not depend on a lattice spacing. So, we can forget about the characteristic length (1).

Both the dimensionless ratios $\overline{\ell}/\mathcal{L}$ and $\overline{\ell}/\xi = (\overline{\ell}/\mathcal{L})N_s^{-1/d}$ can only be functions of dimensionless number N_s. Indeed, we have no other dimensionless quantities in the problem. Then we can write $\overline{\ell}/\mathcal{L} = f(2N_s)$, and so

$$\overline{\ell}(p, \mathcal{L}) = \mathcal{L} f(pz\mathcal{L}^d)\,. \tag{4.28}$$

Equally, using $\overline{\ell}/\xi = \overline{\ell}(pz)^{1/d} = N_s^{-1/d} f(2N_s) = g(\mathcal{L}/\xi) = g(\mathcal{L}(pz)^{1/d})$, we can see that

$$\overline{\ell}(p, \mathcal{L}) = (pz)^{-1/d} g(\mathcal{L}(pz)^{1/d})\,. \tag{4.29}$$

Here $f(x)$ and $g(x) = xf(x^d)$ are scaling functions.

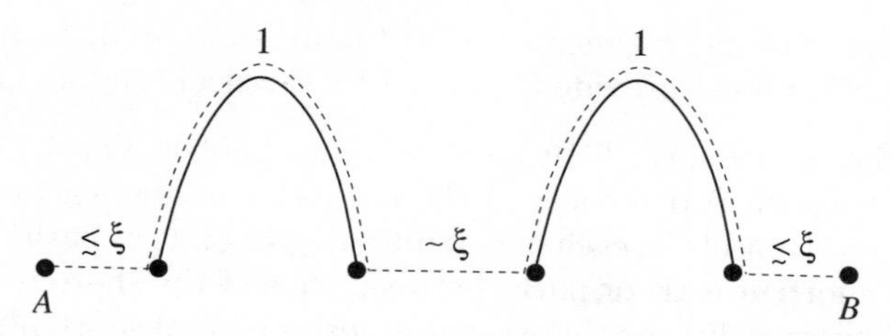

FIG. 4.13. Typical structure of the shortest path between a pair of vertices if their Euclidean separation exceeds the average distance between ends of shortcuts, ξ.

Equations (4.28) and (4.29) follow one from the other and present various forms of the same scaling: the average shortest-path length of the small-world networks depends on the combination of their linear sizes $\mathcal{L}$ and the probability p of the rewiring or the addition of shortcuts.

The usual scaling regime, where these relations are valid, is $\mathcal{L} \to \infty$, $p \to 0$, while the number of shortcuts $N_s = pz\mathcal{L}^d/2$ is finite. The scaling simplifies the situation. Indeed, to describe completely the $\overline{\ell}(p, \mathcal{L})$ dependence, we must know only the scaling function $f(x)$ (or $g(x)$) of a single variable. Many efforts have been made to find this scaling function, which describes the crossover. The exact form is yet unknown. However, we can easily obtain its limiting behaviours:

(1) For a regular 'mother' lattice, we have $\overline{\ell}(p = 0) \sim \mathcal{L}/z$, and so $f(0) = \text{const}$ (the value of this constant depends on the subtleties of the definition of $f(x)$).

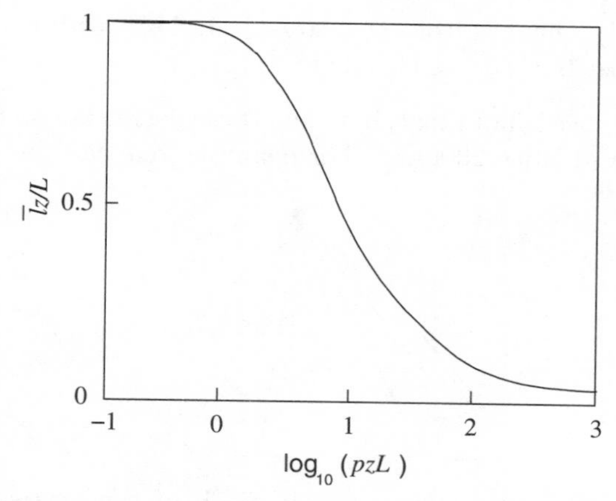

FIG. 4.14. Scaling of the average shortest-path length of small-world networks (after Newman and Watts 1999b). The combination $\overline{\ell}z/\mathcal{L}$ versus pzL for the network which is constructed by the addition of random shortcuts to a one-dimensional lattice. The mother lattice has a size L and coordination number z.

(2) In the other limiting regime, $x \to \infty$ (many shortcuts), at first sight the scaling function must provide $\overline{\ell}$ typical of a classical random graph.

However, the situation is a little more complex. Look at Fig. 4.13. The typical shortest path between two vertices in the network contains parts of two kinds: (1) the shortcuts themselves, each one of unit length; and (2) paths from vertices to the nearest shortcut ends or paths between ends of the shortcuts, each of the order of ξ (the average distance between the closest ends of shortcuts). Therefore, to obtain the estimate for $\overline{\ell}$ in this region, we must multiply the typical average shortest-path length of the classical random graph of size $\sim N_s$ by ξ: $\overline{\ell} \sim \xi \ln N_s$. Here we show only the important size-dependent factor $\ln N_s$. Then we have $g(x \gg 1) \sim \ln x$, and so $f(x) \sim \ln x / x^{1/d}$. The resulting scaling function $f(x)$ is shown in Fig. 4.14.

4.7.2 *The smallest-world network*

One can spoil a regular lattice globally in another way. Let us select vertices of the lattice at random and connect each of them to an additional vertex, that is a hub (see Fig. 4.15). This is more or less the same as connecting all the selected vertices together. In this network, the crossover from the regular lattice to a random network with a hub is realized.

The effect is, in principle, the same, and so all the scaling relations from above are valid for this situation. The only difference is in the form of the scaling function $f(x)$:

- We saw that in the small-world networks of Watts and Strogatz $\overline{\ell}$ changes from $\sim \mathcal{L}$ to $\sim \xi \ln N_s$, which gives $f(x \ll 1) = \text{const}$ and $f(x \gg 1) \sim \ln x / x^{1/d}$.
- Now $\overline{\ell}$, obviously, changes from $\sim \mathcal{L}$ to $\sim \xi \cdot 1$. This readily yields again $f(x \ll 1) = \text{const}$ but $f(x \gg 1) \sim 1/x^{1/d}$.

Thus the difference is not great, but then these networks can be solved exactly (Dorogovtsev and Mendes 2000a).[26] For example, one can find exactly the dis-

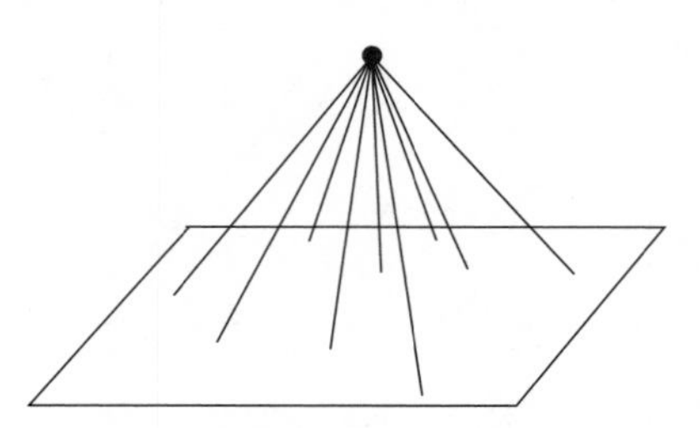

FIG. 4.15. The 'smallest-world' network. Each of the vertices of the regular lattice has some probability of connecting to a single hub. This hub controls all the long-range connections in the network. One may say that shortcuts are 'condensed' on the hub.

[26]In fact, this is a problem *à la* mean-field theory.

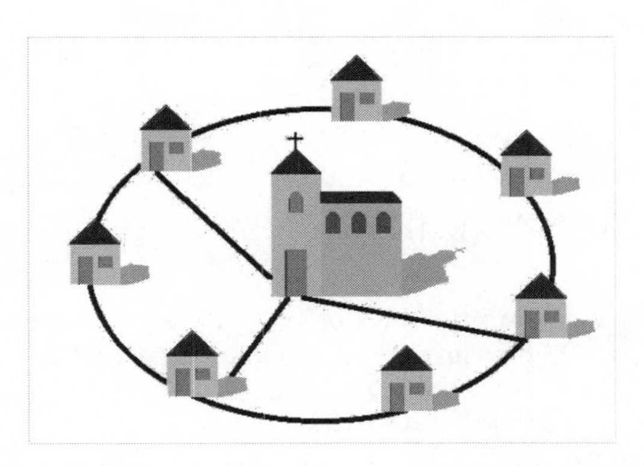

FIG. 4.16. The real 'smallest-world' network. Asocial inhabitants live in this village: usually, they keep in contact only with their neighbours but some of them also attend the church.

tribution of the shortest-path length, which rapidly decreases with the growing shortest-path length, as in the Watts–Strogatz model (Barrat and Weigt 2000) and many other characteristics.

We should add that in reality, networks with such a structure of connections are widespread. Figure 4.16 demonstrates only one of many possible examples.

5

NON-EQUILIBRIUM NETWORKS

Here we consider the basic principles of the structural organization of growing networks at a slightly higher level than in Chapters 1 and 2.

5.1 Growing exponential networks

In Section 2.1, we explained that if new edges in a growing network become attached to vertices at random, without any preference, the degree distribution is exponential. Let us show how this result may be derived. For the demonstration we choose an undirected citation graph to compare the results with those from the next section and not to repeat Section 2.1 word for word.

At each time step, we add one vertex and attach it to a randomly selected old vertex. We start the growth from two doubly connected vertices, $s = 1$ and $s = 2$, at time $t = 2$ (see Fig. 5.1). The variable $s = 1, 2, 3, \ldots$ marks vertices and is actually the birth time of a vertex. At time t, there are t vertices and t edges in the graph, and so the average degree is constant all the time, $\overline{k} = 2$. The double connection at the initial instant has been introduced to make formulae slightly more elegant and is, in fact, not essential.

Now we recall that when we speak about a random network, we actually mean a statistical ensemble of random nets, which contains numerous members (realizations). Then, if we are interested in the state of a vertex s, we must consider the following probability which relates to the whole ensemble: $p(k, s, t)$; that is, the probability that a vertex s has degree k at time t.

The master equation, which describes the evolution of this probability, is very simple:

$$p(k, s, t+1) = \frac{1}{t} p(k-1, s, t) + \left(1 - \frac{1}{t}\right) p(k, s, t) \,. \tag{5.1}$$

We have met similar equations in the previous chapter, so the structure of the equation is clear. The probability that the vertex gets a new connection is $1/t$.

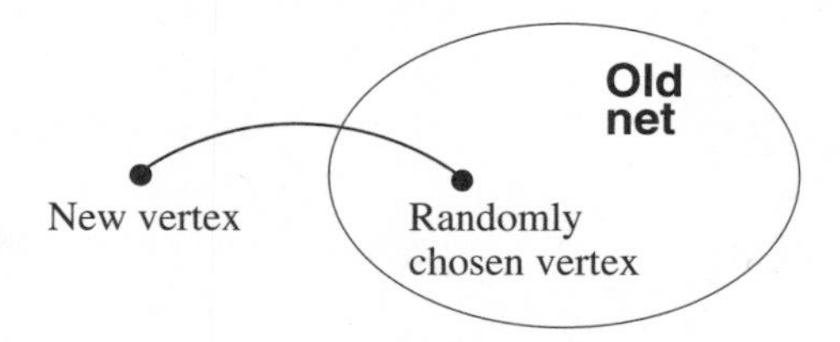

FIG. 5.1. Scheme of the growth of exponential citation network. At each time step, a new vertex become attached to a randomly chosen vertex.

The probability that the vertex remains in a former state is $1-(1/t)$. There exist only two possibilities to get a vertex of degree k:

(1) The first possibility is that in the previous instant it was of degree $k-1$ and attached a new edge. This yields the first term on the right-hand side.

(2) The second possibility is that in the previous instant the vertex was of degree k and did not attach anything. From this we obtain the second term on the right-hand side.

The initial condition is also obvious: there are two vertices, each one of degree 2, so $p(k,\, s=1,2,\, t=2)=\delta_{k,2}$. The matter which makes the equation non-trivial is the boundary condition $p(k,\, s=t,\, t>2)=\delta_{k,1}$.

The total degree distribution of the entire network follows from the above probability for individual vertices:

$$P(k,t)=\frac{1}{t}\sum_{s=1}^{t}p(k,s,t)\,. \tag{5.2}$$

Using this definition and applying $\sum_{s=1}^{t}$ to both sides of eqn (5.1), we immediately obtain the following master equation for the total degree distribution:

$$(t+1)P(k,t+1)-tP(k,t)=P(k-1,t)-P(k,t)+\delta_{k,1}\,. \tag{5.3}$$

Obviously, the initial condition is $P(k,t=2)=\delta_{k,2}$.

The reader who is willing to get this equation in an even simpler way may recall how we used rate equations for the average number $\langle N(k,t)\rangle$ of vertices of degree k at time t in Section 4.3. For the network under consideration, such an equation can be written immediately:

$$\langle N(k,t+1)\rangle=\langle N(k,t)\rangle+\frac{1}{t}\langle N(k-1,t)\rangle-\frac{1}{t}\langle N(k,t)\rangle+\delta_{k,1}\,. \tag{5.4}$$

(1st) The first term on the right-hand side of the equation is obvious.

(2nd) When the edge end becomes attached to a vertex of degree $k-1$, the number of vertices of degree k increases by one. This gain is described by the second term on the right-hand side.

(3rd) When the edge becomes attached to a vertex of degree k, the number of vertices of this degree decreases by one.

(4th) The new term! The network grows, and at each time step one vertex with one connection is added. This gives the Kronecker symbol in the equation.

Recalling that $P(k,t)=\langle N(k,t)\rangle/t$ (the number of vertices is equal to time) gives eqn (5.3). The continuum (for t) limit of this equation is

$$\frac{\partial P(k,t)}{\partial t}+P(k,t)=\frac{\partial[tP(k,t)]}{\partial t}=P(k-1,t)-P(k,t)+\delta_{k,1}\,. \tag{5.5}$$

Let us compare the structure of this equation for the growing network with that of the corresponding master equation (4.6) for equilibrium networks. The principal differences between these linear equations are the following:

(1) Equation (4.6) is homogeneous (its right-hand side contains only terms linear in $P(k,t)$), and eqn (5.5) is inhomogeneous (the additional Kronecker symbol is present).
(2) Equation (4.6) contains the term $\partial[tP(k,t)]/\partial t$ on the left-hand side, and eqn (5.5) contains only $\partial P(k,t)/\partial t$.

The first difference is great. The origin of the inhomogeneity of eqn (5.5) is the fact that the network is a non-equilibrium one, and there is a 'current', that is the permanent injection of edges.

The second difference is not so striking. The reason for $\partial[tP(k,t)]/\partial t$ in eqn (5.5) is the growth of the network. In principle, a non-equilibrium state may be realized in a network with a constant number N of vertices and edges (see Section 5.18). In such a situation, as in eqn (4.6), we have $N\partial P(k,t)/\partial t$.

The equation for the stationary degree distribution $P(k) \equiv P(k, t \to \infty)$ is

$$2P(k) - P(k-1) = \delta_{k,1} \tag{5.6}$$

without any additional conditions. We can check this: multiply both the sides of the equation by k and sum over k. This gives the proper result: $\overline{k} = 2$. The solution of the equation is an exponential degree distribution:

$$P(k) = 2^{-k} \tag{5.7}$$

(check by substitution). We have obtained a similar result in a slightly simpler manner in Section 2.1. Note that the normalization is proper.

The average degree of vertex s at time t is

$$\overline{k}(s,t) = \sum_{k=1}^{\infty} kp(k,s,t)\,. \tag{5.8}$$

Applying $\sum_{k=1}^{\infty} k$ to both sides of eqn (5.1), we get the equation for this quantity,

$$\overline{k}(s,t+1) = \overline{k}(s,t) + \frac{1}{t}\,, \tag{5.9}$$

with the boundary condition $\overline{k}(s=t, t>2) = 1$. Of course, we could guess this without any derivation. The resulting average degree of individual vertices is

$$\overline{k}(s, t>2) = 1 + \sum_{j=1}^{t-s} \frac{1}{s+j} \cong 1 - \ln(s/t)\,, \tag{5.10}$$

where the asymptotic form is valid for large s and t. We see that the average degree of individual vertices of this network *weakly* diverges for the oldest vertices. Hence, the oldest vertex is the 'richest' (of course, only in the statistical sense, that is with high probability).

We started from eqn (5.1) for the degree distributions of individual vertices, $p(k, s, t)$. It is one more simple exercise to obtain the solution of this equation for large s and t and fixed s/t:

$$p(k, s, t) = \frac{s}{t} \frac{1}{(k + 1)!} \ln^{k+1} \left(\frac{t}{s} \right) , \tag{5.11}$$

so that this distribution decreases very rapidly with growing degree.

Some readers may say that it is too hard for them to solve linear discrete equations, especially when they are more complex than eqn (5.6). But there is a way to make life easier. Recall Chapter 2. We may pass to the continuum limit for degree. In this limit, master equations become very simple. At first sight, this must work for large degrees, but mathematicians know that such limiting is an extremely dangerous operation. Sometimes it works, sometimes not, and while using the continuum approximation, you have to check your work all the time. However, for simple growing networks this approximation usually yields exact results for most useful quantities or produces unimportant deviations.

For example, let us pass to the continuum degree limit in the master equation (5.6) for $P(k)$. We get

$$\frac{dP(k)}{dk} = -P(k) , \tag{5.12}$$

and so $P(k) \propto e^{-k}$. Compare this result with the exact degree distribution $P(k) = 2^{-k}$. The difference is of secondary importance.

To finish, we note that this network displays the small-world effect, as do most of the networks in this book.

5.2 The Barabási–Albert model

Unfortunately, the exponential degree distribution is not much better than a Poisson one. Both distributions are rapidly decreasing functions, and we desire fat-tailed ones. The next step was made by Barabási and Albert (1999) who combined the growth and preferential linking. Their model is a citation graph where new vertices become attached to old ones with a probability proportional to their degrees (see Fig. 5.2 and Section 2.2). So, this is the natural proportional preference. Let us also take the next step and discuss the Barabási–Albert model.

We change only one detail in the graph from the previous section: now the attachment of new vertices is with proportional preference. The probability that a vertex of degree k gets a new connection is $k/(2t)$ (recall that $2t = t\overline{k}$ is the total degree of the network at time t, $\overline{k} = 2 = \text{const}$). The probability that it remains in its state is $1 - k/(2t)$. Then the master equation for the degree distributions of the individual vertices is

$$p(k, s, t + 1) = \frac{k - 1}{2t} p(k - 1, s, t) + \left(1 - \frac{k}{2t} \right) p(k, s, t) \tag{5.13}$$

with the same initial condition, $p(k, s = 1, 2, t) = \delta_{k,2}$, and boundary one, $p(k, s = t, t > 2) = \delta_{k,1}$, as in the previous section. In a similar way, as above,

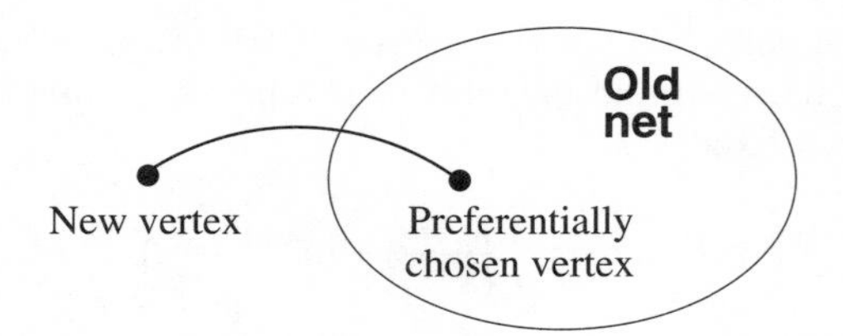

FIG. 5.2. The scheme of the growth in the Barabási–Albert model. The case in which each new vertex has a single connection is shown. At each time step, a new vertex become attached to a preferentially chosen vertex.

we obtain from eqn (5.13) the equation which governs the evolution of the average degrees of vertices and has a clear physical meaning:

$$\overline{k}(s,t+1) = \overline{k}(s,t) + \frac{\overline{k}(s,t)}{2t} \tag{5.14}$$

with the same boundary condition $\overline{k}(s=t, t>2) = 1$. The second term on the right-hand side describes the attachment with proportional preference (compare with eqn (5.9)). In Section 2.2 we have seen that equations of this type produce a power-law singularity of $\overline{k}(s,t) \sim (s/t)^{-\beta}$ in the region of the oldest vertices. Here we introduce a new exponent β which describes this singularity. We see from eqn (5.14) that for the Barabási–Albert model, $\beta = 1/2$.

We have explained in Section 2.2 that it is the region of old vertices that determines the tail of the degree distribution in growing networks. It is no great deal to find an asymptotic solution of the master equation (5.13) in the region $1 \ll s \ll t$. Fixing the ratio $k/\overline{k}(s,t) = k\sqrt{s/t}$, that is a natural scale in the problem, yields

$$p(k,s,t) = \sqrt{\frac{s}{t}} \exp\left(-k\sqrt{\frac{s}{t}}\right) . \tag{5.15}$$

Then the degree distributions of individual vertices depend on k, s, and t in a scaling fashion. Soon we will see that this scaling behaviour is a rather general feature of scale-free networks.

As in the previous section, to derive the master equation for the total degree distribution one may use eqn (5.13) or, if one wants, the rate equation for the average number $\langle N(k,t)\rangle = tP(k,t)$ of vertices of degree k at time t:

$$\langle N(k,t+1)\rangle = \langle N(k,t)\rangle + \frac{k-1}{t\overline{k}}\langle N(k-1,t)\rangle - \frac{k}{t\overline{k}}\langle N(k,t)\rangle + \delta_{k,1} \tag{5.16}$$

(compare with eqn (5.4) from the previous section). In the same manner, we obtain in the continuum time limit

$$\frac{\partial[tP(k,t)]}{\partial t} = \frac{1}{2}[(k-1)P(k-1,t) - kP(k,t)] + \delta_{k,1} \tag{5.17}$$

(compare with eqn (5.5)), whence the stationary equation is of the form

$$P(k) + \frac{1}{2}[kP(k) - (k-1)P(k-1)] = \delta_{k,1}\,. \tag{5.18}$$

One can immediately write the solution of this equation, that is the degree distribution of the Barabási–Albert model:

$$P(k) = \frac{4}{k(k+1)(k+2)} \tag{5.19}$$

(check by substitution). At large degrees, this gives, as in Section 2.2, $P(k) \propto k^{-\gamma}$: the fat tail of the distribution with $\gamma = 3$. The continuum degree limit of the stationary master equation (5.17) is

$$\frac{1}{2}\frac{d[kP(k)]}{dk} = -P(k)\,. \tag{5.20}$$

Its solution is $P(k) \propto k^{-3}$. So, the continuum approximation gives the exact exponent of the degree distribution, and this coincidence is quite standard for growing networks.

The meaning of the continuum approximation

Let us pass to the continuum degree limit in eqn (5.13) for individual vertices. We obtain the equation

$$t\frac{\partial p(k,s,t)}{\partial t} = -\frac{1}{2}\frac{\partial[kp(k,s,t)]}{\partial k} \tag{5.21}$$

with the boundary condition $p(k, s = t, t) = \delta(k-1)$. The solution of this equation is $kp(k,s,t) = \delta(\ln k - \frac{1}{2}\ln t + C(s)) = k\delta(k - C(s)\sqrt{t})$. Thus we get

$$p(k,s,t) = \delta(k - \overline{k}(s,t)) \tag{5.22}$$

where $\overline{k}(s,t) = (s/t)^{-1/2}$ is, naturally, the average degree of the vertex s at time t. Note that we take $C(s) = 1/\sqrt{s}$ to satisfy the boundary condition. Relation (5.22) is a general property of the continuum approximation for such networks (check this for the exponential network).

Thus, *the continuum approximation, in fact, means the δ-function ansatz for the degree distributions of individual vertices.* At first sight, it seems awkward since the real distributions $p(k,s,t)$ are very far from δ-functions, but in growing networks this, usually, works.[1]

Suppose that for some strange reason we are afraid of master equations. Then, as in Chapter 2 and in the original work of Barabási and Albert, we can dispense with them.

[1]The reason for high quality of this ansatz is, in principle, clear. The result (the slowly decreasing dependence of the degree distribution) is determined by a slow variation of $\overline{k}(s)$ and not by a specific form of the rapidly decreasing $p(k)$.

For the Barabási–Albert model, the δ-ansatz (5.22) immediately leads to the equation for the average degree of vertices:

$$\frac{\partial \overline{k}(s,t)}{\partial t} = \frac{\overline{k}(s,t)}{\int_0^t du\, \overline{k}(u,t)} \,. \tag{5.23}$$

The initial condition is $\overline{k}(0,0) = 0$, and the boundary one $\overline{k}(t,t) = 1$: each new vertex has one edge attached. One can check that this is consistent. Indeed, applying $\int_0^t ds$ to eqn (5.23) we obtain

$$\frac{\partial}{\partial t} \int_0^t ds\, \overline{k}(s,t) = \int_0^t ds\, \frac{\partial}{\partial t} \overline{k}(s,t) + \overline{k}(t,t) = 1 + 1 \,, \tag{5.24}$$

from which the proper relation follows, $\int_0^t ds\, \overline{k}(s,t) = 2t$; that is, the total degree in this case equals double the number of edges. Therefore, eqn (5.23) takes the form

$$\frac{\partial \overline{k}(s,t)}{\partial t} = \frac{1}{2} \frac{\overline{k}(s,t)}{t} \tag{5.25}$$

(compare with the exact eqn (5.14)). Its general solution is $\overline{k}(s,t) = C(s) t^{1/2}$, where $C(s)$ is an arbitrary function of s. Taking into account the boundary condition, $\overline{k}(t,t) = 1$, one has

$$\overline{k}(s,t) = \left(\frac{s}{t}\right)^{-1/2} \,. \tag{5.26}$$

If one introduces the scaling exponent β: $\overline{k}(s) \propto s^{-\beta}$, then in the Barabási–Albert model, $\beta = 1/2$.

Using the δ-ansatz produces the following general relation for the degree distribution in the continuum approach:

$$P(k,t) = \frac{1}{t} \int_0^t ds\, \delta(k - \overline{k}(s,t)) = -\frac{1}{t} \left(\frac{\partial \overline{k}(s,t)}{\partial s}\right)^{-1} [s = s(k,t)] \,. \tag{5.27}$$

In the right side of this formula we must substitute $s(k,t)$, which is the solution of the equation $k = \overline{k}(s,t)$. For the Barabási–Albert model, $s(k,t) = t/k^2$, and so $P(k) = -t^{-1} \partial s(k,t)/\partial k = -t^{-1} \partial(t/k^2)/\partial k = 2/k^3$. Notice that the factor 2 differs from that of the exact degree distribution (5.19). In return, the 'normalization' $\int_1^\infty dk\, (2/k^3) = 1$ is valid. This procedure may easily be realized for any network growing under the mechanism of preferential linking, and there are a few cases when it fails.

5.3 Linear preference

The proportional preference which is used in the Barabási–Albert model is only a particular kind of preference that produces fat-tailed and scale-free degree

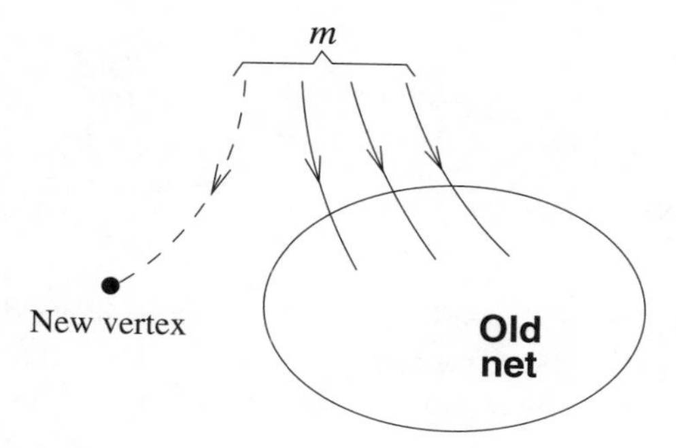

FIG. 5.3. Scheme of the growth of the directed network. At each time step a new vertex and m directed edges are added. The positions of their source ends are not essential. The target ends of these edges become attached to vertices of the network according to the rule of preferential linking. A new vertex can also attach an edge.

distributions. The resulting value 3 of the γ exponent is only a very particular value. In Section 2.2 we have introduced a more general *linear preference* and demonstrated its effect at a basic level. The linear preferential linking was also discussed in Chapter 4 when we tried to obtain fat-tailed degree distributions in an equilibrium network.

Let us consider a *directed* growing network (see Fig. 5.3). Let only incoming connections of vertices and in-degree distributions concern us. In this case, we consider only the attachment of target ends of edges, and can forget about source ends.[2] For brevity, we use the notations $q \equiv k_i$ for in-degree and γ instead of γ_i for in-degree distribution exponent.

The network grows in the following way:

(1) At each time step, a new vertex is added to the network.
(2) Simultaneously, m new directed edges emerge in the network. We do not worry about their source ends. They go out of non-specified vertices or even from outside of the network.
(3) The target ends of the new edges become attached to vertices chosen with *linear preference.* This means that the probability that a new edge becomes attached to some vertex of in-degree q is proportional to $q + A$, where A is a positive constant.

The parameter A plays the role of *additional attractiveness* of vertices.[3] It is convenient to introduce a new notation: $A = am$.

[2]Of course, all that we discuss in this section is equally relevant to out-degrees if we consider only the attachment of source ends of edges.

[3]Notice that, for this directed network, the positive addition A in the rule of preference is necessary. Without it, vertices without incoming connections would never attach new edges. To avoid this difficulty, one can also introduce an additional channel of random linking (see Section 5.7).

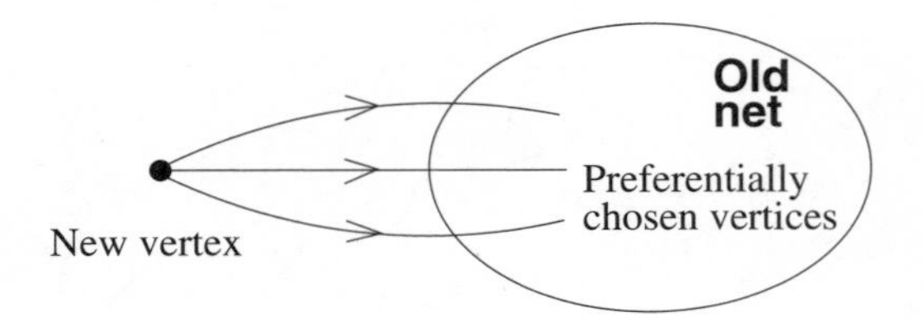

FIG. 5.4. If at each time step all the source ends of new edges in Fig. 5.3 are attached to a new vertex, the network is a citation graph. In this case, the numbers of incoming connections of each vertex are equal.

How does this network relate to the Barabási–Albert model? The answer does not depend on the position of the source ends of new edges. Thus, at each time step, we can attach all of them to a new vertex (see Fig. 5.4), so that old vertices obtain only incoming links. Now we have a citation graph (as in the Barabási–Albert model) where each vertex has exactly m outgoing connections. Then the degree of an arbitrary vertex is $k = q+m$, where q is its in-degree. If we set $A = m$, we get the same rule of preferential attachment as in the Barabási–Albert model: the probability of attaching a new edge will be proportional to $q + A = q + m = k$, that is to the degree of a vertex. Thus, when $A = m$ or, equivalently, $a = 1$ we get the Barabási–Albert model.

Let us discuss the general case. This network can be analysed exactly (Dorogovtsev, Mendes, and Samukhin 2000a), However, the continuum approach will be sufficient for us. The arguments similar to those of Sections 2.2 and 5.2 immediately lead to the following equation for the mean in-degree of individual vertices:

$$\frac{\partial \overline{q}(s,t)}{\partial t} = m \frac{\overline{q}(s,t) + A}{\int_0^t du \, [\overline{q}(u,t) + A]} \tag{5.28}$$

with the initial condition $\overline{q}(0,0) = 0$ and the boundary one $\overline{q}(t,t) = 0$ (new vertices have no incoming edges). The denominator on the right-hand side is obviously $\int_0^t du\,[\overline{q}(u,t) + A] = mt + At = m(1 + a)t$, so the solution of the equation is

$$\overline{q}(s,t) + A = A \left(\frac{s}{t}\right)^{-\beta} , \tag{5.29}$$

where $\beta = 1/(1+a)$. We see that indeed, when $a = 1$, we arrive at the result for the Barabási–Albert model. Using this power-law $\overline{q}(s,t)$ and taking into account the relation (5.27) allows us to obtain the tail of the degree distribution; $s(q,t) \sim tq^{-1/\beta}$, and so $P(q) = -t^{-1}\partial s(q,t)/\partial q \sim -t^{-1}\partial[tq^{-1/\beta}]/\partial q \propto q^{-(1+1/\beta)} = q^{-\gamma}$. Thus we have $\gamma = 2 + a = 2 + A/m$ but, moreover, we obtain a relation for scaling exponents β and γ:

$$\beta(\gamma - 1) = 1\,. \tag{5.30}$$

Even at first sight this relation seems general, and soon we will see that this is indeed the case.

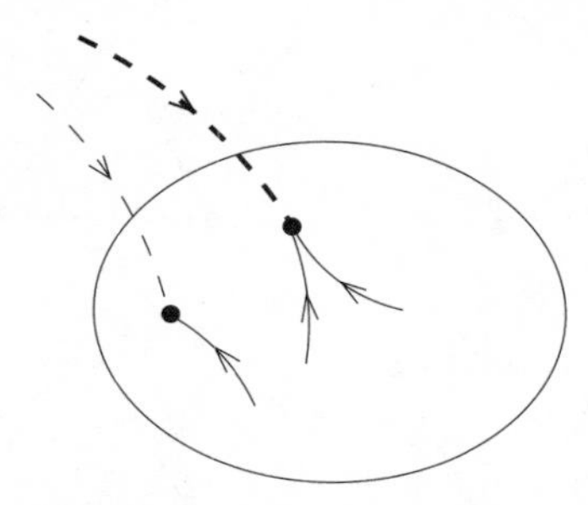

FIG. 5.5. If the attaching of the target ends of new edges to the target ends of existing edges is random, than the right vertex (two incoming edges) has a two times higher probability of receiving a new connection than the left vertex (one incoming edge).

We can change the parameter A in the preference rule over a wide range, from 0 to ∞, and all the time the network is scale-free. This is quite typical of the self-organized criticality. γ increases from 2 to ∞, and β decreases from 1 to 0 as A grows from 0 to ∞. If $A \to 0$, all connections will be to the oldest vertex. If A approaches ∞, preference is absent, and we must obtain an exponential growing network. Exponent γ runs over the range $(2, \infty)$, which is allowed for networks with a constant average degree (see Section 1.3).

Looking ahead, we note that the linear preference is the only type of preference that produces the fat tails of degree distributions.

5.4 How the preferential linking emerges

In Sections 5.2 and 5.3, the rule of preferential attachment was introduced directly. We postulated the form of the preference function $f(k)$ (the probability that a new edge becomes attached to a vertex with k connections). The function $f(k)$ determines the structure of the network, but where does this form come from? Of course, preferential attachment seems to be natural—one can recall our verbal arguments from Section 2 ('popularity is attractive')—but the problem is not as simple as it seems. Why are just the proportional and linear preferences realized in numerous networks? What is the nature of these kinds of preference? Evidently, a general answer is hardly possible. So, here we present simple arguments and a demonstrative example.

First of all, the most natural type of preference is the absence of any preference, that is random attachment of edges. Suppose that in a directed network, the target ends of new edges become attached to the target ends of randomly chosen edges (see Fig. 5.5). On the one hand, this is attachment (*of ends to ends*) without any preference. However, on the other hand, one sees that in this process, the target ends become attached *to vertices* with probability proportional to in-degrees of vertices. This is proportional preference.

In Section 4.3 we have seen that, in undirected networks, attachment to a randomly chosen end of a randomly chosen edge actually corresponds to attach-

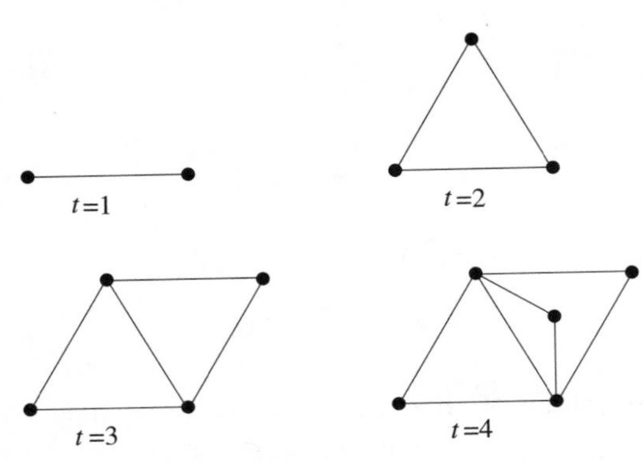

FIG. 5.6. The network grows by attaching each new vertex to both the ends of a randomly chosen edge (Dorogovtsev, Mendes, and Samukhin 2001c). At each time step, a single vertex is added.

ment to vertices with probability proportional to their degrees. Using this idea, one can construct a growing network with the degree distribution similar to that in the Barabási–Albert model but with strong clustering (Dorogovtsev, Mendes, and Samukhin 2001c). The undirected network is defined as follows (see Fig. 5.6):

(1) Initially ($t = 1$), two vertices connected by an edge are present.
(2) At each time step, a new vertex is added.
(3) It is connected to both ends of a randomly chosen edge by two edges.

The degree distribution and some other characteristics of this network were obtained explicitly for any size of network. The stationary degree distribution practically coincides with that of the Barabási–Albert model, eqn (5.19). Of course, the γ exponent is 3. At this point, for us, the time-dependent (or one may say, size-dependent) degree distribution $P(k,t)$ is interesting. We omit details of the tedious calculations of $P(k,t)$ (see Dorogovtsev, Mendes, and Samukhin 2001c). Actually, this is a direct analysis of the discrete master equation for the time-dependent degree distribution. Only the following result of the calculations is important for us: at long times, the time-dependent degree distribution takes the scaling form

$$P(k,t) = P(k)\, F(k/\sqrt{t})\,, \tag{5.31}$$

where $P(k)$ is the stationary degree distribution, and the scaling function $F(k/\sqrt{t})$ is as shown in Fig. 5.7. In Section 5.2 we have presented a scaling form for the degree distributions of individual vertices (see eqn (5.15)).

Relation (5.31) gives another insight into the scaling properties of the growing network.[4] Figure 5.7 displays the following two features of the time-dependent degree distribution. First, the distribution has a cut-off at $k_{cut} \sim \sqrt{t}$. This coincides with the estimate from Section 1.3, $k_{cut} \sim t^{1/(\gamma-1)}$ ($\gamma = 3$). Secondly, there

[4]Note that in this network mortality is absent. For scaling relations, this is essential.

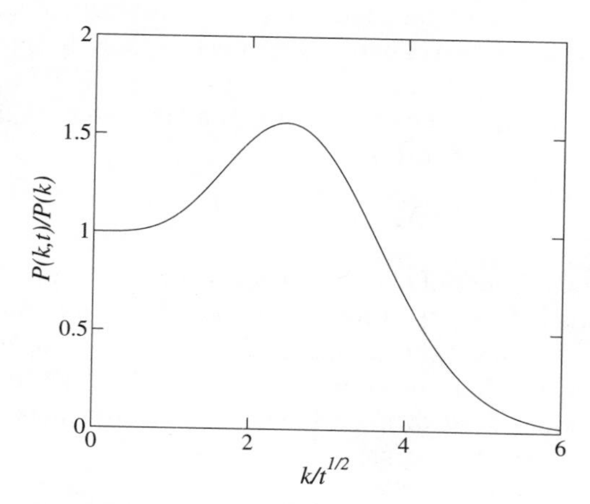

FIG. 5.7. Scaling form of the deviation of the degree distribution of the network from Fig. 5.6 versus $k/\sqrt{t}$. t is the size of the network. The form of the hump depends on the initial configuration of the network.

is a hump in the time-dependent degree distribution near the cut-off. This hump does not disappear as the network grows but moves to large degrees, decreasing in height. Furthermore, one can check that the hump is the trace of initial conditions: if the growth starts not from a single edge but from a more complex configuration, then the form of the hump will be different.[5]

Below we shall discuss the scaling properties of growing networks in a more general context. Here we note that, although the degree distribution of the network practically coincides with the Barabási–Albert model, there is an essential difference. The growth generates numerous triple loops in the network (see Fig. 5.6), so that the structure is very far from tree-like. One may even say that the network consists of triple loops. From Section 1.4 we know that this means high clustering (a high value of the clustering coefficient). Contrastingly, the clustering of the Barabási–Albert model is low (see Section 5.13).

Cliques

The network that we discuss here is very convenient for analytical study. In Section 3.2.1 we mentioned cliques in the WWW whose indexing is important for the optimization of a search. The cliques in a directed network, by definition, are bipartite subgraphs that have all possible edges going from hubs to authorities. The statistics of cliques in our growing network are very simple (see Appendix C). It is possible to make this network directed: let each new vertex have two outgoing edges. Then each pair of nearest-neighbour vertices plays the role of authorities

[5]One can easily eliminate the hump. For this, for example, we may introduce additional rewiring of edges in the old part of the network (see below). After a while, the rewiring wipes out any memory of the initial conditions, and the hump disappears. Ageing and mortality of vertices and edges also suppress the effect of initial conditions.

of a bipartite subgraph based on them. Hence the number of authorities in each clique in the net is exactly two. Then, the problem is only to find the distribution of cliques over the numbers of hubs.

Simple calculations in Appendix C show that the average number of cliques with h authorities in the network is

$$N_{\text{cliq}}(h,t) = 2^{-h}t\,. \tag{5.32}$$

One can see that the number of cliques is high. It grows proportionally to the size of the network. This is in sharp contrast to a negligibly small number of cliques in classical random graphs.[6] The exponential number-of-hubs distribution (5.32) correlates with empirical data obtained from the WWW (see Section 3.2.1). The high number of cliques in growing, strongly correlated networks makes these objects important in real networks.

5.5 Scaling

In previous sections we found that a number of quantities of particular growing scale-free networks may be written in a scaling form, and the scaling exponents involved may be connected by a simple relation. Can these forms and relations be applied to other growing scale-free networks?

To answer this question we use general considerations. In this section, it is not essential whether we consider degree, in-degree, or out-degree. Hence we use one general notation, k. When one speaks about scaling properties, a continuum treatment is sufficient, so that we can use the following expressions: *(1)*

$$P(k,t) \cong \frac{1}{t}\int_{t_0}^{t} ds\, p(k,s,t) \tag{5.33}$$

(hereafter, we mark all points we use to derive the scaling relations)[7] and *(2)*

$$\overline{k}(s,t) = \int_0^\infty dk\, k p(k,s,t)\,. \tag{5.34}$$

In addition, we will need the normalization condition for $p(k,s,t)$: *(3)*

$$\int_0^\infty dk\, p(k,s,t) = 1\,. \tag{5.35}$$

Next we suppose that the stationary degree distribution $P(k)$ exists *(4)*. Usually this holds for linearly growing networks, that is for networks where the

[6] Here we do not prove that the number of cliques in classical random graphs is small. The reader whose curiosity is piqued should refer to the paper of Kumar, Raghavan, Rajagopalan, Sivakumar, Tomkins, and Upfal (2000b). One can make an even more general premise: the number of cliques is small in equilibrium uncorrelated graphs.

[7] In relation (5.33) we suppose that, initially, the network consists of t_0 vertices. This assumption is actually not essential.

total number of edges is proportional to the total number of vertices, and so the mean degree is constant.

Then from eqn (5.33), we immediately see that $p(k,s,t)$ has to be of the form $p(k,s,t) = \rho(k,s/t)$.

Substituting this into the normalization condition (5.35) yields $\int_0^\infty dk\, \rho(k,x) = 1$, and so it should be $\rho(k,x) = g(x)f(kg(x))$, where $g(x)$ and $f(x)$ are arbitrary functions.[8]

Now we assume that the stationary distribution $P(k)$ and the average degree $\overline{k}(s,t)$ exhibit scaling behaviour, that is $P(k) \propto k^{-\gamma}$ for large k *(5)*, as they should do for scale-free networks, and *(6)* the average degree of a vertex is a power-law function of its birth time, $\overline{k}(s,t) \propto s^{-\beta}$, for $1 \ll s \ll t$.

Then, from eqn (5.34), one can see that $\int_0^\infty dk\, k\rho(k,x) \propto x^{-\beta}$. Substituting $\rho(k,x) = g(x)f(kg(x))$ into this relation, one obtains $g(x) \propto x^\beta$. Of course, without loss of generality, one may set $g(x) = x^\beta$, and so we obtain the following scaling form of the degree distribution of individual vertices:

$$p(k,s,t) = (s/t)^\beta f(k(s/t)^\beta)\,. \tag{5.36}$$

Finally, from the power-law form of $P(k)$, that is $\int_0^\infty dx\, \rho(k,x) \propto k^{-\gamma}$, and using eqn (5.36), we obtain $\gamma = 1 + 1/\beta$. Thus we see that relation (5.30) between the exponents is *universal* for such scale-free growing networks.[9] Notice that, in fact, we used *(7)* the rapid decrease of $\rho(k,x)$ at large x. We have observed this rapid diminution while discussing the Barabási–Albert model (see eqn (5.15) from Section 5.2).

We are presenting details of this simple derivation to show that scaling relations are based only on *(1)*–*(7)*. Notice that *(1)* also implies that mortality is absent in the network.

Exponent β also has another meaning. Let us treat the degree distribution in terms of *Zipf's law*, which is a standard form of presenting empirical data. One can arrange the vertices of a network in decreasing order of their degrees, so that the vertex of the highest degree is the first (the rank $r = 1$), the second-highest-degree vertex has the rank $r = 2$, and so on. If Zipf's law is valid, the degree of a vertex is a power of its rank, $k \propto r^{-\nu}$, where ν is the corresponding exponent.

On the other hand, the rank of a vertex is easily expressed in terms of the cumulative distribution: $r(k) \propto \int_k^\infty dk\, P(k) \equiv P_{cum}(k)$. For a power-law degree distribution $P(k) \propto k^{-\gamma}$, we see that $r(k) \propto k^{1-\gamma}$. Then $k \propto k^{-1/(\gamma-1)} = k^{-\beta}$, and thus our β exponent coincides with ν from Zipf's law.

[8] This general relation is valid even for exponential networks.

[9] Owing to this relation, sometimes, a single glance at the master equation or the equation for the average degree is quite enough to describe the degree distribution of a growing network. The structure of these equations reveals exponent β, from which the γ exponent of the degree distribution follows.

Proceeding in a similar way as above, one can obtain the following general form of the total degree distribution $P(k,t)$ for growing scale-free networks in the scaling regime:

$$P(k,t) = k^{-\gamma}F(kt^{-\beta}) = k^{-\gamma}F(kt^{-1/(\gamma-1)})\,. \tag{5.37}$$

Here $F(x)$ is a scaling function. We have obtained this form for the exactly solvable network with $\gamma = 3$ from Section 5.4 (see eqn (5.31) and Fig. 5.7).

5.6 Generic scale of 'scale-free' networks

This is not the first time that we have met the scale $k \sim k_0 t^{1/(\gamma-1)}$ in the degree distribution of scale-free networks.[10] Recall our first estimate (1.8) of the cut-off k_{cut} of the degree distribution from Section 1.3. Also recall the hump from Fig. 5.7. Both the cut-off and hump are present just at this point, $k_{cut} \sim k_0 t^{1/(\gamma-1)}$.

This size-dependent k_{cut} is a generic scale of all 'scale-free' networks. The cut-off as well as the hump sets strong restrictions for observations of power-law distributions since there are very few really large networks in nature.

The measurement of degree distributions is always hindered by strong fluctuations at large degrees. The reason for such fluctuations is the poor statistics in this region. One can easily estimate the characteristic value of degree, k_f, above which the fluctuations are strong. If $P(k) \sim k^{-\gamma}$, $tk_f^{-\gamma} \sim 1$. Therefore, $k_f \sim t^{1/\gamma}$. One may improve the situation by using the cumulative distributions, $P_{cum}(k) \equiv \int_k^\infty dkP(k)$, instead of $P(k)$ (see Appendix B). Also, in simulations, one may make many runs to improve the statistics. Nevertheless, one cannot exceed the cut-off, k_{cut}, that we are discussing. This cut-off is the real barrier for observation of the power-law dependence.

No scale-free networks with large values of γ were observed. $\gamma = 4, 5$, or 6 is absent. The reason for this is clear. Indeed, the power-law dependence of the degree distribution can be observed only if it exists for at least two or three orders of degree.[11] For this, the networks must be large: their size should be, at least, $t > 10^{2.5(\gamma-1)}$. Then, if γ is large, one practically has no chance of finding the scale-free behaviour.

Recall Fig. 3.32 from Section 3.11, where in the log–linear scale, we presented the values of the γ exponents of all the networks reported as having power-law degree distributions versus their sizes (recall also Table 3.7). One can see that almost all the plotted points are inside the region restricted by the lines $\gamma = 2$, $\log_{10} t \sim 2.5(\gamma - 1)$, and by the logarithm of the size of the largest scale-free network, that is the WWW, $\log_{10} t \sim 9$.

[10] Notice that the scaling form (5.37) allows a constant factor k_0, which has a simple meaning. This degree is a low boundary of the power-law region of the degree distribution. Sometimes $k_0 \sim 1$, sometimes it is essentially higher.

[11] The number of these orders shows the extent of the impudence of researchers. Several years ago, usually it was about three; two years ago, it was down to two; one year ago, it was about one or less. What will this number be in a year's time?

5.7 More realistic models

The models of growing networks from previous sections demonstrate how scale-free distributions arise. Here we discuss more realistic models of scale-free networks. Of course, even these 'realistic' models only just illustrate reality. At best, they can be used for rough estimations.

Linking in the body of a net

In principle, regarding the operation of the preferential mechanism, there is no difference between connecting a pair of old vertices and connecting a new vertex and an old one. Even our first popular discussion of the 'attractiveness of popularity' (Section 2.2) was already illustrated by the growing network where any vertices, new or old, may be connected. The network from Section 5.3 is also of this type.

However, very often, the growth of the network combines the attachment of new vertices to old ones and the linking of old vertices. Suppose that in such a network, the mechanism of preferential linking operates. Then, how do these two channels of growth interfere?

Let us consider an illustrative example. The undirected network grows as follows (see Fig. 5.8):

(1) At each time step, a new vertex is added.
(2) It becomes preferentially attached to m old vertices. The probability of the attachment to a vertex is proportional to its degree.
(3) Simultaneously, cm new edges emerge between preferentially chosen old vertices. Here, c is a non-negative constant. The probability that such an edge becomes attached to a pair of vertices of degrees k_1 and k_2 is proportional to the product $k_1 k_2$.

One can see that when $c = 0$, this is the Barabási–Albert model. We immediately write the equation for the evolution of the average degree of vertices in this network:

$$\frac{\partial \overline{k}(s,t)}{\partial t} = m(1+2c)\frac{\overline{k}(s,t)}{2m(1+c)t} \,. \tag{5.38}$$

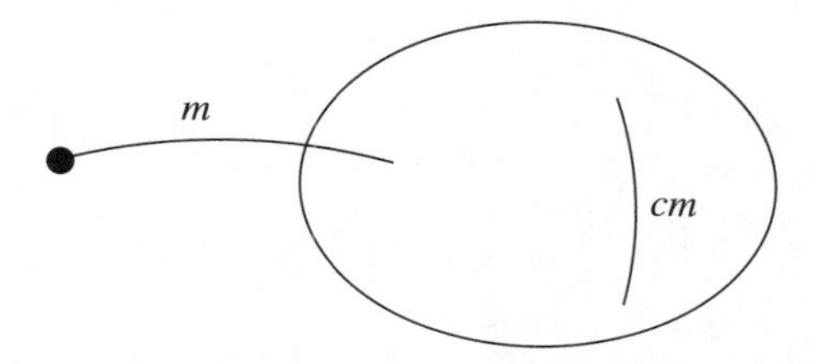

FIG. 5.8. Scheme of the growth of a network with the creation of connections between already existing vertices. At each time step, the preferential attaching of m new vertices is combined with the emergence of cm new edges between preferentially chosen old vertices.

The boundary condition is $\overline{k}(t,t) = m$. The structure of the equation is evident. As we are discussing distributions of one-vertex quantities, our problem is equivalent to the attachment of independent ends of edges. At each time step, we must attach $m + 2mc$ ends of edges. The total degree of a network is equal to $2(m + mc)t$. Each end becomes attached to a vertex of degree k with probability $k/[2(m + mc)t]$. This yields eqn (5.38), from which we readily obtain $\beta = (1 + 2c)/[2(1 + c)]$. Using relation (5.30) gives

$$\gamma = 2 + \frac{1}{1 + 2c} \,. \tag{5.39}$$

Thus, linking in the body of the network diminishes exponent γ. γ decreases from 3, ($c = 0$) to 2 with growing c. Recall that 2 is the lower boundary for γ, which is allowed for networks with finite average degree.

Above, we made new connections in the network's body. Now let us discuss what happens when we destroy them. In the above rules, we change only one line. Now instead of (3), we have:

(3′) Simultaneously (at each time step) $c'm$ randomly chosen edges are deleted.

We have already used the fact that a random selection of an edge means a (proportional) preferential choice of two vertices (for example, see Sections 4.3 and 5.4). This means that at each time step we detach $2c'm$ ends from vertices chosen with probability $k/[2(m{-}mc')t]$. The only change in eqn (5.38) is $c \to -c'$. From this we see immediately that, in this case, $\gamma = 2 + 1/(1 - 2c')$, and as the rate of the deletion increases, γ grows.

Rewiring

Rewirings in the old part of a network (Albert and Barabási 2000a) produce a very similar effect. Suppose that a network grows as shown in Fig. 5.9. Again we change the last line in the above rule:

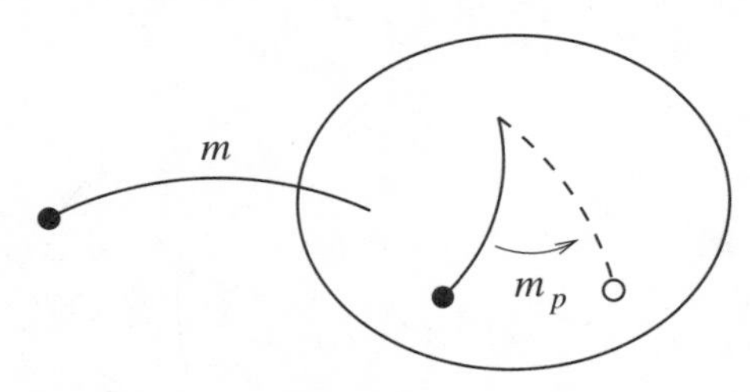

FIG. 5.9. Scheme of the growth of a network with the rewiring of connections in the old part of the net. At each time step, (1) a new vertex is added, (2) it is attached to m preferentially chosen old vertices, (3) m_p vertices are chosen at random, and one of the edges from each of these vertices is rewired to a preferentially chosen vertex.

(3″) At each time step, we select at random m_p vertices and rewire one of the edges from each of them to preferentially chosen vertices. The preference is proportional.

If m_p is essentially smaller than m, we may neglect the emergence of bare vertices. Then the equation for the average degrees is very simple:

$$\frac{\partial \overline{k}(s,t)}{\partial t} = (m+m_p)\frac{\overline{k}(s,t)}{2mt} - m_p\frac{1}{t} \tag{5.40}$$

with the boundary condition $\overline{k}(t,t) = m$. The second term in the equation is due to the rewirings from m_p randomly chosen vertices; $2mt$ is the total degree of the network at time t. One can see that $\beta = (m+m_p)/(2m)$, and so

$$\gamma = 2 + \frac{m-m_p}{m+m_p}\,. \tag{5.41}$$

Exponent γ decreases with growing rate of rewirings (compare with eqn (5.39)) but we repeat that m_p must be small enough.

Preference and indifference

In real growing networks, as a rule, many channels of linking are present. Even for very rough estimates we have to account for several basic contributions. Let us consider the minimal model of the growth of directed networks, which already allows us to obtain first estimates. As usual we use the following notation for in-degree, $q \equiv k_i$. We do not care where the source ends of edges are.

The rules of the growth (see Fig. 5.10; compare with the scheme of the growth of the WWW, Fig. 3.4 from Section 3.2.1) are as follows:

(1) At each time step, a new vertex is added.

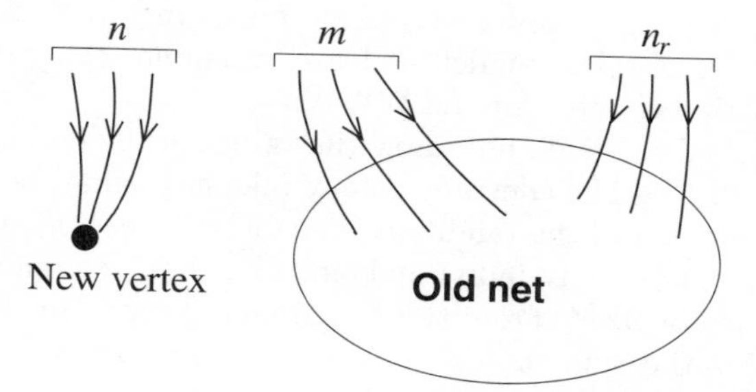

FIG. 5.10. Scheme of the growth of the network with a mixture of preferential and random linking. At each time step a new vertex with n incoming connections is added. Simultaneously the target ends of m new edges become attached to preferentially chosen vertices, and, in addition, the target ends of n_r new edges become attached to randomly chosen vertices. We are not concerned with the location of the source ends of edges.

(2) It has n incoming connections.
(3) Simultaneously, m extra edges are distributed with preference. This means that a target end of each of them becomes attached to a vertex chosen with linear preference: the probability of choosing a vertex of in-degree q is proportional to $q + A$.
(4) In addition, at each time step, the target ends of n_r edges become attached to randomly chosen vertices.

We shall see that the reasonable values of the constant A, that is *additional attractiveness*, are $A > -(n+n_r)$. The growth combines preferential linking with random attachment. What is the result of such a combination?

The equation for the average in-degrees of vertices has the form

$$\frac{\partial \overline{q}(s,t)}{\partial t} = n_r \frac{1}{t} + m \frac{\overline{q}(s,t) + A}{\int_0^t du[\overline{q}(u,t) + A]} \tag{5.42}$$

with the initial condition $\overline{q}(0,0) = 0$ and the boundary one $\overline{q}(t,t) = n$. The first term on the right-hand side is due to the 'indifferent' attachment, the second term accounts for linking with linear preference. The latter is the same as for the network from Section 5.3 (see eqn (5.28)). In this case, the total in-degree of the network is equal to $(n_r + m + n)t$, and so $\beta = m/(m + n_r + n + A)$. We see that $0 < \beta < 1$, and

$$\gamma_i = 2 + \frac{n_r + n + A}{m} \,. \tag{5.43}$$

This relation shows that all the three factors, the indifferent attachment (n_r), the presence of incoming connections of new vertices (n), and the initial attractiveness A, play the same role. Each of them increases the exponent, so that it varies in the range between 2 and ∞.

5.8 Estimations for the WWW

We have already explained how new pages appear in the WWW (see Fig. 3.4 from Section 3.2.1). The last model, at least, resembles this process. So, let us try to obtain rough estimates for the WWW.

The difficulty is that we do not know the values of the quantities on the left-hand side of eqn (5.43). The constant A may take *any* values between $-(n_r+n)$ and infinity, the number of the randomly distributed edges, n_r, in principle, may not be small (there are many individuals making their references practically at random), and n is not fixed. From the experimental data (see Section 3.2.1) we know more or less the sum $m + n + n_r \sim 10 \gg 1$ (between 7 and 10, more precisely), and that is all.

The only thing we can do is to fix the scales of the quantities. The natural characteristic values for $n_r + n + A$ in eqn (5.43) are (a) 0, (b) 1, (c) $m \gg 1$, and (d) infinity. In the first case, all new edges are attached to the oldest vertex since only this one is attractive for linking, and so $\gamma_i \to 2$. In the last case, there is no preferential linking, and the network is not scale-free, $\gamma_i \to \infty$. Let us consider the truly important cases (b) and (c).

In-degree distribution

Let us assume that the process for the appearance of each document in the Web is as simple as the procedure for the creation of your personal home page described in Section 3.2.1. If only one reference to the new document ($n = 1$) appears, and if one forgets about the terms n_r and A in eqn (5.43), then, for the γ_i exponent of the in-degree distribution, we immediately obtain the estimate $\gamma_i - 2 \sim 1/m \sim 10^{-1}$. This estimate (Bornholdt and Ebel 2001, Dorogovtsev, Mendes, and Samukhin 2000b) indeed coincides with the empirical value $\gamma_i - 2 = 0.1$ (see Section 3.2.1). Therefore, the estimation looks good. Nevertheless, we should repeat that this estimate follows only from the fixation of the scales of the involved quantities, and many real processes are not accounted for in it.

Out-degree distribution

Above we discussed the distribution of incoming links. Equation (5.43) may also be applied for the distribution of links which go out from Web documents, since the model of the previous section can easily be reformulated for outgoing edges of vertices. In this case all the quantities in eqn (5.43) take other values which are again unknown. However, we can estimate them. As we explained in Section 3.2.1, there are usually several citations (n) in each new WWW document. In addition, one may think that the number of links distributed without any preference, n_r, is not small now. Indeed, even beginners proceed by linking their pages. Hence, $n+n_r \sim m$ (we have no other available scale), and so $\gamma_o - 2 \sim m/m \sim 1$. We can compare this estimate with the empirical value, $\gamma_o - 2 = 0.7$ (see Section 3.2.1).

Thus, we see that the estimates are reasonable, and that it is hardly possible to improve them.

5.9 Non-linear preference

Now we return to our favourite question: *when do distributions of connections have fat tails?* From Section 5.3, we learned that linear preferential linking in growing networks provides a fat-tailed (and even scale-free) degree distribution. In respect of behaviour at large degrees, evidently, only the asymptotic part of the preference function is important. So, asymptotically linear preference $f(k) \cong k+A$ as $k \to \infty$ is quite sufficient for fat-tailed distributions in growing networks.

What happens when the preference is non-linear at large degrees? Let us discuss the particular case: power-law preference $f(k) = k^y$, where y is some non-negative constant. Here we only present the main results; the derivation can be found in Appendix D.

For brevity, let the network be an undirected citation graph, and let each new vertex have a single connection to a preferentially selected vertex. Then the case with $y = 1$ is exactly the Barabási–Albert model, $P(k) \propto k^{-3}$. $y = 0$ corresponds to indifferent linking, that is an exponential network (see Section 5.1).

At first, we consider the region $y < 1$ and go down a little from $y = 1$. In this event, highly connected old vertices become less attractive for attachment as compared with the linear or proportional preference. The singularity of $\overline{k}(s)$

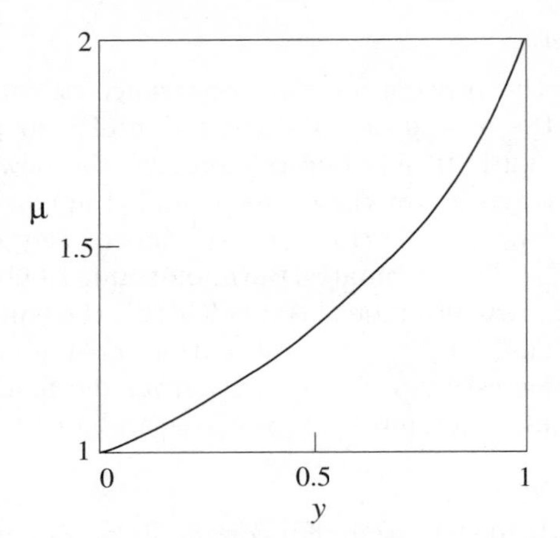

FIG. 5.11. Dependence of the coefficient μ in the exponential of the degree distribution (5.44) on exponent y. The preference is the power law $\propto k^y$ (after Krapivsky, Redner, and Leyvraz 2000, Krapivsky and Redner 2001).

at the oldest vertex, $s = 0$, in this case is weaker than a power-law one. One may expect a direct crossover from the k^{-3} dependence to the exponential one as y decreases from $y = 1$ to 0. This is actually the case but the crossover is not quite trivial. The resulting degree distribution is of the form[12]

$$P(k) \propto k^{-y} \exp\left(-\frac{\mu}{1-y} k^{1-y}\right), \tag{5.44}$$

where μ depends on y, see Fig. 5.11. The reader can check that this expression reduces to k^{-3} when $y \to 1$ and it yields, of course, pure exponent $\propto e^{-k}$ for $y = 0$.

Contrastingly, in the complementary case $y > 1$, the attractiveness of the old highly connected vertices increases. Furthermore, if $y > 2$, there is a finite probability that the oldest vertex attracts *all* connections, and so each other vertex has a single connection. Let, indeed, the oldest 'guy' have all t connections at time t with probability $P_{old}(t)$. Then, readily, $P_{old}(t+1) = P_{old}(t)\{t^y/[t^y + t \cdot 1^y]\}$, and so for the infinite network we obtain

$$P_{old}(t \to \infty) = \prod_{t=1}^{\infty} [1 + t^{1-y}]^{-1} . \tag{5.45}$$

This product is non-zero for $y > 2$.

The effect for $1 < y \leq 2$ is slightly less exciting but very similar: The total number of vertices of degrees higher than 1 is negligible (see Appendix D).[13]

[12]For the exact form and a more rigorous derivation see Krapivsky, Redner, and Leyvraz (2000) and Krapivsky and Redner (2001).

[13]In fact, this is condensation but of a quite different nature than that of the condensation in equilibrium networks (compare with Sections 4.4 and 4.5).

So, *only linear preference produces scale-free growing networks.* On the other hand, it is more than sufficient to provide a wide spectrum of fat-tailed degree distributions.

5.10 Types of preference providing scale-free networks

Indeed, only linear preferential linking produces scale-free and fat-tailed distributions, but there are numerous versions of linear preference. Generally speaking, the linear preference means that the probability that a new edge, at time t, becomes attached to a vertex s of degree $k(s,t)$ is proportional to a linear combination

$$f(k,s,t) = G(s,t)k + A(s,t)\,. \tag{5.46}$$

We call $A(s,t)$ the additional attractiveness. $G(s,t)$ is, often, called 'fitness'. One can choose these two coefficients in the linear dependence on degree in a thousand various forms and obtain a thousand various networks.

For example, in Section 5.7, we discussed the combination of preferential and indifferent (random) linking. Note that this is a particular case of the above formula. It corresponds to $G =$ const and the fluctuating additional attractiveness $A = A(t)$ that at random takes two values: a finite constant (preferential linking) and the infinite value (indifferent linking).

There is another possibility: one can make older vertices less attractive than youngsters by choosing appropriate forms of the vertex–age-dependent fitness $G = G(t-s)$ and additional attractiveness $A = A(t-s)$. This ageing crucially changes the structure of the network (see Fig. 5.12). Meanwhile, such ageing is widespread in networks. Recall nets of scientific citations: we rarely cite old papers.

And this is not all. We can force new edges to play the role of fans with different passions for the popularity of their idols, vertices, by taking $G = G(t)$ and $A = A(t)$. One may consider any other possibility, and very often obtain fat tails and scale-free networks.

We saw in Chapter 4 that the inhomogeneity of networks can produce a number of exciting effects. Here, inhomogeneity means different properties of distinct vertices; that is, G and A in eqn (5.46) depend on the label of a vertex, s. In Section 5.3 we showed that if G and A are independent of s, the oldest vertices have a great chance to become the 'richest' ones. Variations of $A(s)$ and, especially, $G(s)$ violate this rule. A new vertex with, for example, high G has a good chance to become the most connected vertex in the network. In particular, this explains the appearance of new, incredibly popular sites in the WWW.

We have learned from Chapter 4 that variations of G (multiplicative factor), in principle, may produce stronger effects than variations of the additional attractiveness A. We saw that they may even yield the condensation of edges on the 'strongest' vertices in equilibrium networks (see Sections 4.4 and 4.5).

We defer our discussion of the condensation in growing networks until the next section. Here we dwell on less exciting effects of the inhomogeneity (Bianconi and Barabási 2001a), where degree distributions are not changed so crucially. A

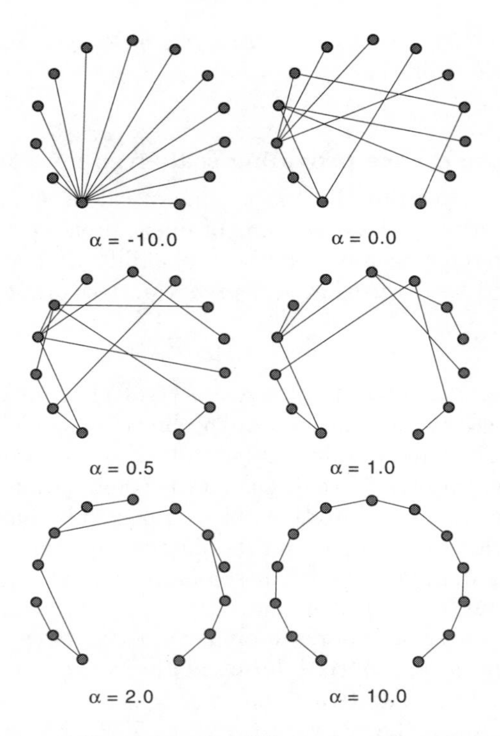

FIG. 5.12. Structure of the network with ageing of vertices (Dorogovtsev and Mendes 2000b). The linking is preferential, but, in addition, the probability that a new vertex becomes attached to an old one is proportional to $\tau^{-\alpha}$, where τ is the age of the old vertex. The variations of the structure of the network with increasing exponent α are shown. If α is large, the network takes a linear form.

more detailed discussion of these 'weak effects' of inhomogeneity can be found in Appendix E.

The results for the growing network with inhomogeneous 'fitness' are easily understandable. If a growing scale-free network is homogeneous, we know how to get its degree distribution. In the continuum approach (see Section 5.2) we have $s(k,t) \sim tk^{-1/\beta} = tk^{1-\gamma}$, and so $P(k) \sim -t^{-1}\partial s(k,t)/\partial k \sim -t^{-1}\partial[tk^{1-\gamma}]/\partial k \sim \gamma k^{-\gamma}$. If this network is 'spoiled' by fluctuations of 'fitness' G, which are described by a distribution $P(G)$, the evolution of degree of each distinct vertex is determined by its own G, and so one may introduce for each vertex its own exponent $\gamma(G)$. The averaging of the above expression over the distribution $P(G)$ produces the degree distribution of the inhomogeneous net that looks as follows (Bianconi and Barabási 2001a):

$$P(k) \propto \int dG P(G)\gamma(G)k^{-\gamma(G)} . \tag{5.47}$$

The right-hand side of this expression is a superposition of weighted power-law dependences. Exponents of these power laws, $\gamma(G)$, can be found from a self-consistency condition. We explain this procedure in Appendix E. For some distributions $P(G)$, it works, and then nothing serious happens to the degree distribution: it is close to a power law.

For example, when one spoils the Barabási–Albert model in such a way, and G is homogeneously distributed between 0 and 1, the result is

$$P(k) \propto \frac{k^{-2.255\ldots}}{\ln k} \,. \tag{5.48}$$

However, for many distributions $P(G)$, the solution $\gamma(G)$ is absent, and the above considerations fail. We discuss what happens in such a case in the next section.

5.11 Condensation of edges in inhomogeneous nets

We will follow our favourite principle: do not prove but demonstrate. So, for illustration, we use the simplest distribution $P(G)$ that produces the effect. However, these results are valid in much more general situations. As in Section 4.4, only one 'guy' (vertex, 'shark') in our growing network will be 'stronger' than others. Its fitness ('strength') $g > 1$ is greater than the unit fitness of the other vertices.[14] What will be the effect of this 'strong guy'?

Let us consider the illustrative example, namely a directed growing network with the growth rules that provide elegant results:

(1) At each time step, a new vertex is added. It has n incoming connections.
(2) Simultaneously, m new edges emerges. Their target ends become preferentially attached to vertices.
(3) The preference function of each vertex $s \neq w$ is proportional to its in-degree $q(s)$. The preference function of the vertex $s = w$ is equal to its in-degree $q(w)$ multiplied by a factor $g > 1$.

In the continuum approach, we have two equations for the average in-degrees: for the strong guy and for the others,

$$\begin{aligned}
\frac{\partial \overline{q}(w,t)}{\partial t} &= m\frac{g\,\overline{q}(w,t)}{(g-1)\overline{q}(w,t) + \int_0^t ds\,\overline{q}(s,t)}\,,\\
\frac{\partial \overline{q}(s,t)}{\partial t} &= m\frac{\overline{q}(s,t)}{(g-1)\overline{q}(w,t) + \int_0^t ds\,\overline{q}(s,t)}\,,
\end{aligned} \tag{5.49}$$

where the boundary condition is $\overline{q}(t,t) = n$. One can see that at long times, the total in-degree of the network is $\int_0^t ds\,\overline{q}(s,t) \cong (m+n)t$.

At long times, two distinct situations are possible.

[14] One can check that the variation of additional attractiveness of a single vertex would produce unnoticeable effects.

(1) Suppose that the in-degree of the strongest vertex grows slower than t, and so the denominators in the equation approach $(m+n)t$. This means that the strong guy is unimportant, and the second equation coincides with that for the corresponding homogeneous case. Thus we obtain the exponents $\beta = m/(m+n) \equiv \beta_0$ and $\gamma = 2 + n/m \equiv \gamma_0 = 1 + 1/\beta_0$, where $0 < \beta_0 < 1$, $2 < \gamma_0 < \infty$. Here, we introduce the exponents γ_0 and β_0 of the network in which all vertices have equal 'strength' (fitness), $g = 1$.

This situation corresponds to that of the previous section with the only difference that now we have only one *rara avis*, and its effect on exponents is negligible.

But when does this happen? When does the degree of the strong vertex grow more slowly than t? To find this, let us consider the first line of eqn (5.49) in this case:

$$\frac{\partial \overline{q}(w,t)}{\partial t} = \frac{gm}{m+n} \frac{g\overline{q}(w,t)}{t} \, . \tag{5.50}$$

The solution of this equation at long times is $\overline{q}(w,t) = \text{const}\, t^{gm/(m+n)}$. Therefore, we see that, indeed, the slow growth can be realized, but only when g is smaller than some critical point:

$$g < g_c \equiv 1 + \frac{n}{m} = \gamma_0 - 1 \, . \tag{5.51}$$

(2) Now, let the strength of the strong guy be greater than the critical value, $g > g_c$. How does the strong vertex's degree grow at long times? Can it grow more slowly than $\propto t$? Certainly it cannot, since this contradicts eqn (5.50). Can it grow more rapidly than $\propto t$? No, this is impossible in principle, since the maximum number of incoming connections which a vertex can get is mt. Then we have the only possibility: at long times the in-degree of the strong vertex grows proportionally to the total number of vertices with a factor smaller than m, that is $q(s,t) = d\,t$, $d < m$.

This means that, for $g > g_c$, a finite fraction of all preferentially distributed edges is captured by the strong vertex (Bianconi and Barabási (2001b) called this the Bose–Einstein condensation). Therefore, g_c is the condensation point. Then we see that a single strong vertex may produce a macroscopic effect. In this case, the denominators in eqn (5.49) are $[(g-1)d + m + n]t$ and it takes the form[15]

$$\begin{aligned} \frac{\partial \overline{q}(w,t)}{\partial t} &= \frac{gm}{(g-1)d + m + n} \frac{\overline{q}(w,t)}{t} \, , \\ \frac{\partial \overline{q}(s,t)}{\partial t} &= \frac{m}{(g-1)d + m + n} \frac{\overline{q}(s,t)}{t} \, . \end{aligned} \tag{5.52}$$

[15]Notice that, since the in-degree of the 'strong' vertex is now macroscopic, the bar over $q(w,t)$ is superfluous.

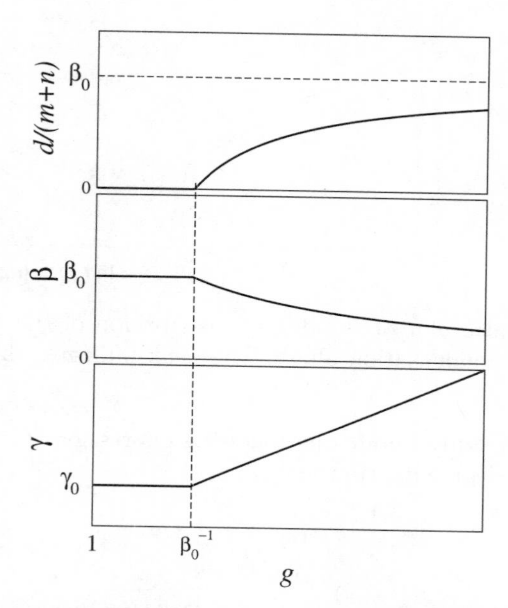

FIG. 5.13. Condensation of edges in growing networks. The fraction of all edges, $d/(m+n)$, captured by a single 'strong' vertex and scaling exponents β and γ versus relative fitness ('strength') of the strong vertex (Dorogovtsev and Mendes 2001b). The network is large. The condensation point is $g_c = 1/\beta_0 = \gamma_0 - 1$, where β_0 and γ_0 are scaling exponents for the network without a strong vertex. $d/(m+n) \to \beta_0$ and $\beta \to 0$ as $g \to \infty$.

The first line is for the strong vertex and the second one is for the others. Now we recall that $q(w,t)$ is proportional to t. Then, from the first of eqns (5.52), we obtain the condition

$$\frac{gm}{(g-1)d + m + n} = 1\,, \tag{5.53}$$

and we see that, for $g > g_c$, the following fraction of all edges in the network is captured by the strongest vertex:

$$\frac{d}{m+n} = \frac{d}{m}\frac{m}{m+n} = \frac{g-g_c}{g-1}\frac{1}{g_c} \tag{5.54}$$

(see Fig. 5.13). Note that the resulting value of d at large times is independent of the initial conditions.

What is the in-degree distribution of the other vertices? Notice that the factor on the right-hand side of the second equation of eqns (5.52) is smaller than that in the first equation ($g > 1$). Using the condition (5.53) and the second equation

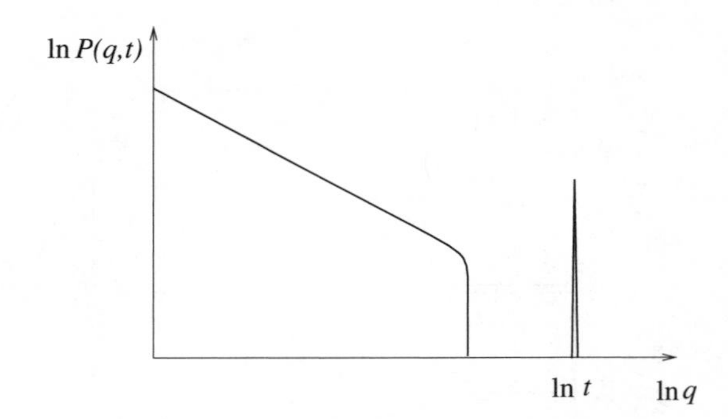

FIG. 5.14. Schematic plot of the degree distribution of the growing scale-free network in the condensation phase. The peak indicates the condensation of edges.

of eqns (5.52) we readily obtain the following expressions for the exponents of this part of the in-degree distribution:

$$\beta = \frac{1}{g} < \beta_0 \,, \quad \gamma = 1 + g > \gamma_0 \,. \tag{5.55}$$

Figure 5.14 schematically shows the complete (in-)degree distribution of the network in the condensed phase. The fraction of all edges captured by the strongest vertex and the β and γ exponents versus g are shown in Fig. 5.13. Note that γ increases as the strength ('fitness') of the strong vertex grows.

Therefore, the situation is quite different from the condensation on a 'strong' vertex in equilibrium networks with linear preference[16] (see Sections 4.4 and 4.5). Let us compare:

- *Equilibrium networks.*
  The condensation is possible even in homogeneous equilibrium networks. If the condensation is due to the inhomogeneity of a net, the degree distribution at the condensation point is not fat-tailed. It has a network-size-independent cut-off. In the condensed phase, the degree distribution of 'normal' vertices coincides with that at the condensation point.
- *Growing networks.*
  Without inhomogeneity, the condensation is impossible. At the condensation point, the degree distribution is fat-tailed. In the condensed phase, the degree distribution of 'normal' vertices is also fat-tailed, although it does not coincide with that at the point of condensation.

Here we allow ourselves to make more general statements than those following from our illustrative examples, but these claims are true.[17]

[16] Recall that only the linear preference can provide us with fat tails of degree distributions.

[17] We also allow ourselves to forget about the problem of tadpoles and melons from Chapter 4. For the condensation in growing networks, this problem also exists.

Econophysics

As for equilibrium networks (see Section 4.4), we can interpret the condensation phenomenon in growing nets in terms of wealth distribution, this time in growing societies. In our example above, incoming connections play the role of money, in-degree is wealth, 'normal' vertices are us, and the 'strong' vertex is Bill Gates. The network is a happy growing society with Bill Gates. The reader can express all the results of the present section in these terms.

Exponent γ of the wealth distribution characterizes the fairness of the society. A society with higher γ is more 'fair', since it has relatively fewer riches. In this sense, the exponential distribution, that is $\gamma = \infty$, is the apotheosis of 'fairness'. On the other hand, a power-law wealth distribution with $\gamma \leq 2$ provides a few riches with a finite fraction of the total wealth, which is, of course, unfair (we do not discuss this case here).

The condensation of a finite fraction of wealth by the single 'shark' we speak about, is certainly unfair. Notice, however, the effect of this condensation on the distribution of money in the pockets of other, not so lucky people. Look at Fig. 5.13: in the phase with 'condensate', the 'fairness' γ grows with increasing strength g of the strongest guy. If the world is captured by a Bill Gates or a czar, the distribution of wealth of the others becomes more fair!

One should note that, in our scenario, the 'strong guy' does not take money away from other people but only *intercepts* them. From eqn (5.51) we see that the closer exponent γ_0 is to 2, the smaller 'strength' g of the strong guy is necessary to exceed the threshold. In other words, it is easier to lay your hands on a less fair society.

Anomalous relaxation

In this section we spoke about one strong guy, one strong vertex. Now suppose that in our growing network, there are several such 'supermen' of distinct strength. In this situation, if there is condensation, then at long times (large sizes) only the 'strongest' vertex attaches a finite fraction of edges.

The words 'long times' for condensation phenomena in growing networks may mean *very long* times indeed. We discussed above only the final initial-condition-independent state, which is realized in the limit of large networks. This state is approached very slowly (Dorogovtsev and Mendes 2001b). Let there be a single 'strong guy' with $g > g_c$. Using eqn (5.52), one can analyse the approach of $q(w,t)$ of the in-degree of this 'strong' vertex to the asymptotic value dt for the infinite network ($t \to \infty$). The result is non-trivial: in the entire condensate phase, relaxation to the final state is of a power-law kind:

$$\frac{q(w,t) - d\,t}{t} \propto t^{-(g-g_c)/g} \,. \tag{5.56}$$

So, the fraction of all edges captured by the strong vertex relaxes to the final value by a power law. Its exponent $(g-g_c)/g$ approaches zero at the condensation point $g = g_c$. We stress that *this slow relaxation is realized for any $g > g_c$; that is, in* *the entire condensed phase. At the threshold, the relaxation proceeds slower than any power law.* This is in sharp contrast to standard critical phenomena. Recall that standard relaxation is exponential both above and below a phase transition, and is a power law only at the critical point. The origin of this difference is the growth of the network.[18]

5.12 Correlations in growing networks

In Chapters 1 and 4 we explained that the basic construction of equilibrium networks is a graph in which any correlations are absent. Contrastingly, *there are no non-equilibrium networks without correlations.* In the present section and in the next one, we explain how correlations arise in growing networks.

Degree–degree correlations

The important type of correlations in a network is correlations between degrees of its nearest-neighbour vertices. Recall that these correlations are described by the probability $P(k,k')$ that a randomly chosen edge has end vertices of degrees k and k' (see Section 4.6). In fact, this is the joint degree–degree distribution of the nearest neighbours. We explained that if such degree–degree correlations are absent, then $P(k,k')$ factorizes: $P(k,k') \propto kP(k)k'P(k')$.[19] In particular, if the network is scale-free with degree distribution $P(k) \propto k^{-\gamma}$, the absence of such correlations means that $P(k,k') \propto k^{1-\gamma}k'^{\,1-\gamma}$.

The last simple relation is not valid for growing networks (Krapivsky and Redner 2001). Indeed, in growing networks (scale-free and non-scale-free), a natural asymmetry between 'old' and 'young' vertices is always present. 'Parents' and 'children' are in quite different situations. Distributions of their connections and the dynamics of these distributions are very different. So, vertices in such networks are statistically non-equivalent, the structure of their connections depends on their degrees, and this indicates the presence of correlations in such networks.

The above arguments sound too abstract, so we will illustrate the effect using a convenient example. We consider a scale-free citation graph. Let us begin with the answer. The asymptotic form ($k \gg k' \gg 1$) of the distribution of degrees of nearest neighbours is

$$P(k,k') \sim k^{-(\gamma-1)}k'^{\,-2}\,. \tag{5.57}$$

We see that a factorization is absent.[20]

[18]Look at the list of differences between master equations for growing and equilibrium networks (see Section 5.1). Master equations for the degree distribution of growing networks contain the logarithmic derivative over t, $t\partial/\partial t$, the corresponding equations for equilibrium nets contain $\partial/\partial t$. This difference leads to the anomalously slow relaxation in the growing networks.

[19]Notice that $P(k,k')$ does not factorize into the product $P(k)P(k')$!

[20]The only (occasional) exception here is $\gamma = 3$. In particular, this is the case of the Barabási–Albert model, where the pair degree–degree correlations are anomalously week.

Let us discuss where these correlations come from. At first, we fix 'small' k'. In this event, the first power factor of $P(k,k')$ is the degree distribution of nearest neighbours of a 'young' vertex (indeed, it is these vertices that have 'small' degrees). In particular, it is the probability that a new vertex becomes attached to a vertex of degree k. Now we use the fact that the scale-free networks are the result of the mechanism of *linear* preferential linking. Then this probability is proportional to $(k+\text{const})P(k)$, which is just what we need, namely $\sim k^{-(\gamma-1)}$, at large degrees.

Now we fix large k in $P(k,k')$. In this case, the k'-dependent factor coincides with the degree distribution of the nearest neighbours of the oldest and one of the most connected vertices. In citation graphs, this distribution can be easily estimated. In such graphs, only new vertices become attached to old ones. Hence, to find the probability that the oldest vertex is connected to a vertex s it is sufficient to consider only the moment of its birth: $t = s$. In addition, we account for the fact that the linear preferential linking mechanism operates and the average degree of a vertex grows with time as t^{β}, $\beta = 1/(\gamma-1)$. Therefore, the probability that a vertex s is connected to the oldest vertex is estimated as $(s^{\beta}+\text{const})/s \sim s^{\beta-1}$. In the continuous approach (see Section 5.2), the degree of a vertex s is $k' \sim (s/t)^{-\beta}$. Then, the probability that the oldest vertex is connected to a vertex of degree k' is $\sim {k'}^{-(\beta-1)/\beta} = {k'}^{-2+\gamma}$. From this, we readily see that the degree distribution of nearest neighbours of the oldest vertex is $\sim {k'}^{-2+\gamma}{k'}^{-\gamma} = {k'}^{-2}$. Thus, the second factor in eqn (5.57) is also clear.

The form of the joint degree–degree distribution depends strongly on details of the linking procedure. For instance, the emergence of new connections or rewirings in the old part of a growing network crucially changes $P(k,k')$. On the other hand, our illustrative example with linear preferential linking is only an example. Even without any preferential linking, growing networks display strong correlations.

In Section 3.2.3, we mentioned that reliable measurements of the joint degree–degree distribution of neighbouring vertices are difficult. The reason for this is poor statistics. Indeed, if a network has L edges, the empirical data consist of only L pairs of degrees. To find the degree–degree distribution with high precision, one should have a sufficient number of edges with the same pairs of degrees. If $L \ll N^2$, this is practically impossible.

There is a way to get round this difficulty and observe correlations (Pastor-Satorras, Vázquez, and Vespignani 2001, see Section 3.2.3). One can measure not the complete distribution $P(k,k')$ but the average degree of the nearest neighbours of a vertex of degree k, that is $\overline{k}_{nn}(k)$. At least this characteristic indicates the presence of such correlations.

For the scale-free citation graph, $\overline{k}_{nn}(k)$ can be easily estimated. If $\gamma < 3$, from eqn (5.57), we obtain the dependence $\overline{k}_{nn}(k) \propto k^{-(3-\gamma)}$ for small enough degrees k, while for larger k, the dependence is a slow, logarithm-like function. Contrastingly, if γ of a citation graph exceeds 3, $\overline{k}_{nn}(k)$ is an increasing function of k. In particular, this is the case for the exponential citation graph, which grows under the mechanism of random linking.

In- and out-degree correlations

In directed growing networks, the correlation may even be between in- and out-degrees of the same vertex (Krapivsky, Rodgers, and Redner 2001). The origin of these correlations is quite obvious. If the target and source ends of each edge become attached independently of each other, both the in- and out-degrees of old vertices are larger than those of 'youngsters'. This is the case both for preferential and for indifferent linking. Then in- and out-degrees correlate, and the joint in- and out-degree distribution of vertices does not factorize.

5.13 How to obtain a strong clustering

Another type of correlation in networks is clustering. In Chapter 3, we saw that the clustering coefficients of real networks are usually larger or even much larger than the typical values for classical random graphs, $C \cong \overline{k}/N$, where N is the total number of vertices. Frankly speaking, there exist a thousand different ways to make a highly clustered growing network, and it is easier to obtain high clustering than not to obtain it. (Add, for example, fully connected triples of vertices to a growing graph at each time step.) Trees have zero clustering, since, by definition, they have no loops. Therefore, for example, any citation graph where each new vertex has a single connection, has zero clustering.

We have constructed a simple scale-free growing network with high clustering in Section 5.4 by attaching a new vertex to both ends of a randomly chosen edge (see Fig. 5.6). This produces numerous triangles of edges, which indicates a high clustering. A very similar network can be constructed by attaching each new vertex to a randomly chosen triple of nearest-neighbour vertices of the network (see Fig. 5.15). This graph, like that from Section 5.4, has a power-law degree distribution with an exponent equal to 3. Recall the formula for the clustering coefficient from Section 1.4. To find the clustering coefficient, one should count the total number of triangles of edges in the graph and the total number of triples of connected vertices. The result is $C = 1/2$.

In both these examples, the same idea is realized. New vertices simultaneously become attached to already connected vertices. In other words, a pair or several nearest neighbours 'give birth' to a new vertex. The reader can find examples of such processes in real networks.

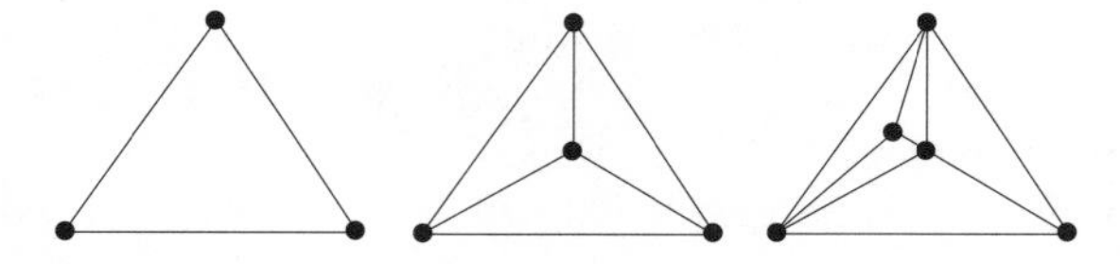

FIG. 5.15. Growing scale-free network with a large clustering coefficient. At each time step, a new vertex with three connections is added. It becomes attached to a randomly chosen triple of the nearest-neighbour vertices.

5.14 Deterministic graphs

Stochastic models of growing scale-free networks[21] are based on three 'whales', that is three leading 'physical' ideas:

(1) The growth is stochastic.
(2) The linking produces 'long-distance' connections and the small-world effect.
(3) The linking is preferential.

Are all these ingredients necessary? In this book we consider only networks with the 'small-world effect', so principle (2) is unavoidable. We search for situations where degree distributions are fat-tailed, which are naturally produced by principle (3). But what about the first 'whale'?

Let us discard the first principle, that is assume that the growth is deterministic, and compare the structure and topology of such deterministic graphs with those of stochastically growing networks. In principle, various deterministic constructions are possible (Barabási, Ravasz, and Vicsek 2001). Here we 'slightly' modify our growing network from Section 5.4 (see Fig. 5.6), so that ingredients (2) and (3) are valid.

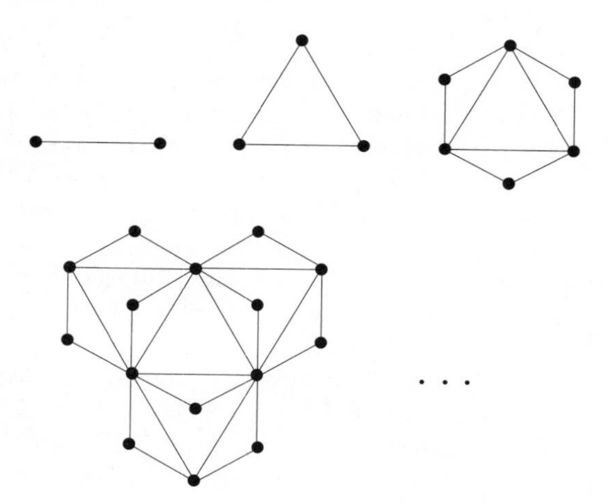

FIG. 5.16. Scheme of the growth of the scale-free pseudofractal graph. The growth starts from a single edge connecting two vertices at $t = -1$. At each time step, every edge generates an additional vertex, which is attached to both end vertices of the edge. Notice that the graph at time step $t + 1$ can be made by connecting together the three t graphs. The total numbers of vertices and edges of the pseudofractal grow with time as $N_t \sim L_t \sim 3^t$. The average degree at long times approaches $\overline{k} = 4$. The average clustering coefficient C tends to $4/5$ as $t \to \infty$.

[21] More generally, with fat-tailed degree distributions.

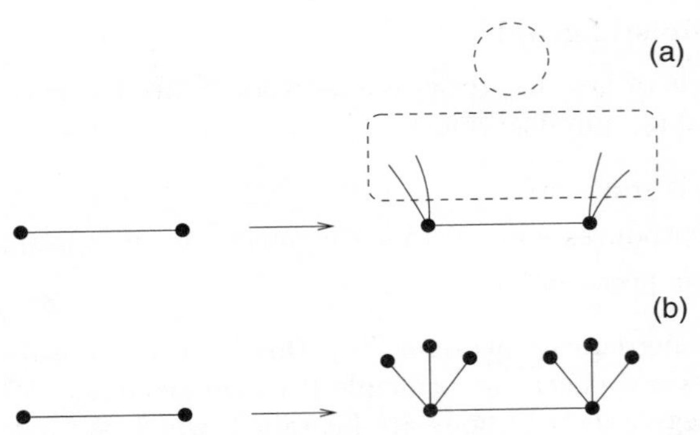

FIG. 5.17. (a) The generalization of the pseudofractal network from Fig. 5.16. This transformation of each edge of a network generates scale-free deterministic graphs (pseudofractals) with various values of exponent γ, which display the small-world effect. The graph structure (inside the dashed curve) that is attached to both end vertices of the 'mother' edge is symmetric. The attached structure may have no path between the parent vertices and may have separate parts. A particular case of this transformation generates trees (b) (Jung, Kim, and Kahng 2002).

Recall that in that network (a citation graph), at each step, a randomly chosen edge 'gives birth' to a new vertex. Contrastingly, in our deterministic construction, at each time step, *each edge* 'gives birth' to a new vertex, which is attached to both the ends of its 'mother' edge. Figure 5.16 demonstrates the growth of this simple graph, which is also a citation graph, by definition. This procedure can be easily generalized to the transformation that is shown in Fig. 5.17. The properties of the graphs generated by the latter transformation are qualitatively similar to the graph in Fig. 5.16, and so we mainly discuss the simpler construction.

The reader who saw pictures of the growth of fractals may conclude that the network in Fig. 5.16 is a fractal. This is incorrect. Indeed, let us look at the scheme of the growth of a perfect fractal, Fig. 5.18. For comparison, we will use the particular case of the Migdal–Kadanoff renormalization group transformations. Each next step in the growth of the Migdal–Kadanoff graph doubles the shortest-path length between a fixed pair of vertices. The total numbers of vertices and edges increase four-fold (asymptotically, at large t). Thus, the average shortest-path length $\overline{\ell}$ grows as a square power of the size of the graph. This is a fractal, and its 'fractal dimension' is 2. Our graph (Fig. 5.16 or Fig. 5.17) is a much more 'compact' structure. The entire network can be set inside a unit triangle. The growth does not increase the shortest-path lengths between old vertices. The structure has no fixed finite fractal dimension. Thus this graph is not a fractal but only a parody of it, and so we call it, for brevity, a *pseudofractal*.

A great advantage of deterministic networks is their simplicity, which allows an exact analysis. For the pseudofractal from Fig. 5.16, the distribution of the

FIG. 5.18. Example of the Migdal–Kadanoff renormalization group transformation (2×2), which generates a fractal with the fractal dimension 2.

shortest-path lengths was calculated exactly (Dorogovtsev, Goltsev, and Mendes 2002a). The asymptote of this exact result for large networks is of the Gaussian form

$$\mathcal{P}(\ell, t) \cong \frac{1}{\sqrt{2\pi(2^2/3^3)t}} \exp\left[-\frac{(\ell - \overline{\ell}(t))^2}{2(2^2/3^3)t}\right], \tag{5.58}$$

which is violated only in narrow regions near the maximum possible shortest-path length and $\ell = 1$.[22] This distribution is determined by the average shortest-path length

$$\overline{\ell}(t) \cong \frac{4}{9}t \cong \frac{4}{9\ln 3}\ln N_t\,. \tag{5.59}$$

The small-world effect in this network is not a great surprise for us. It is determined by our constructions. Notice, however, that the result for the average shortest-path length is surprisingly close to the standard formula from Section 1.5, $\overline{\ell} \sim \ln N/\ln \overline{k}$. The relative difference is really small, $(4/9)(\ln 4/\ln 3) = 0.561\ldots$. But the standard estimate was obtained for specific equilibrium uncorrelated graphs, which differs sharply from the strongly correlated and clustered pseudofractal! Moreover, we will show in Chapter 6 that the standard formula is not valid for equilibrium scale-free networks with a degree distribution exponent less than or equal to 3, and, contrastingly, soon we shall see that the γ exponent of our graph is less than 3. So, we have the classical result in the non-classical situation.[23] Figure 5.19 demonstrates the exact shortest-path-length distribution and its asymptote (5.58) for $t = 7$, that is $N_t = 3282$. For comparison, in the same figure we present the empirical shortest-path-length distribution of the network of autonomous systems in the Internet in November 1997, when this net had nearly the same size and average degree as the pseudofractal at $t = 7$. The reader will notice that the shortest-path-length distributions of these two nets are surprisingly close.

Now it is time to recall the third 'whale': the preferential linking. Look at Fig. 5.16 of the pseudofractal. More connected vertices get a greater fraction of new connections. Then we have a chance to obtain a scale-free structure.

[22] Note that in the limit $t \to \infty$, the length of the shortest path between a pair of vertices is almost surely close to $\overline{\ell}$. We suggest that this is a general feature of infinite-dimensional objects.

[23] For many stochastically growing graphs, the classical formula also turns out to be available even if $\gamma < 3$ (Krzywicki 2001, Szabó, Alava, and Kertész 2002).

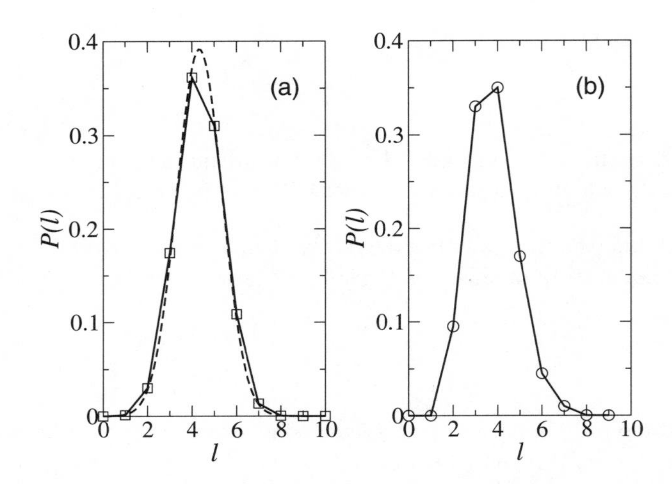

FIG. 5.19. (a) The shortest-path-length distribution of the pseudofractal from Fig. 5.16. $t = 7$, that is $N = 3282$; the average degree is 4.0. The solid curve shows the exact result, the dashed one is the asymptotic formula (5.58). (b) For comparison, the shortest-path-length distribution of the Internet (the network of autonomous systems in November 1997), $N = 3180$; the average degree is equal to 3.5(1) (Vázquez, Pastor-Satorras, and Vespignani 2002).

Stochastic growth produces continuous degree distributions. Deterministic rules of growth produce graphs with a discrete spectrum of degrees. One can see that at time t, the number $m(k,t)$ of vertices of degree $k = 2, 2^2, 2^3, \ldots, 2^{t-1}, 2^t, 2^{t+1}$ is equal to $3^t, 3^{t-1}, 3^{t-2}, \ldots, 3^2, 3, 3$, respectively (check in Fig. 5.16). Then, for large networks, $m(k,t)$ decreases as a power of k, $m(k,t) \propto k^{\ln 3/\ln 2}$, that is the spectrum of degrees has a power-law envelope. Hence these networks can also be called 'scale-free' (Barabási, Ravasz, and Vicsek 2001). We cannot use this exponent $(\ln 3/\ln 2)$ directly for comparison with the exponent of the degree distribution of a random scale-free network. The problem is that the spaces between discrete degrees of the spectrum grow with increasing k. This small problem is overcome by passing to a cumulative distribution $P_{cum}(k) \equiv \sum_{k' \geq k} m(k',t)/N_t \sim k^{1-\gamma}$, for which this difficulty is evidently absent. Here k and k' are points of the discrete degree spectrum. Therefore, exponent

$$\gamma = 1 + \frac{\ln 3}{\ln 2} = 2.585 \ldots \tag{5.60}$$

here plays the role of the γ exponent of random networks.[24] Notice that the maximum degree of a vertex of this graph is equal to $2^{t+1} \sim N_t^{\ln 2/\ln 3} = N_t^{1/(\gamma-1)}$, which precisely coincides with a standard relation for the cut-off of a degree distribution in scale-free networks (see Section 5.6).

[24] Meanwhile, degree spectra of the Migdal–Kadanoff fractals also have power-law envelopes.

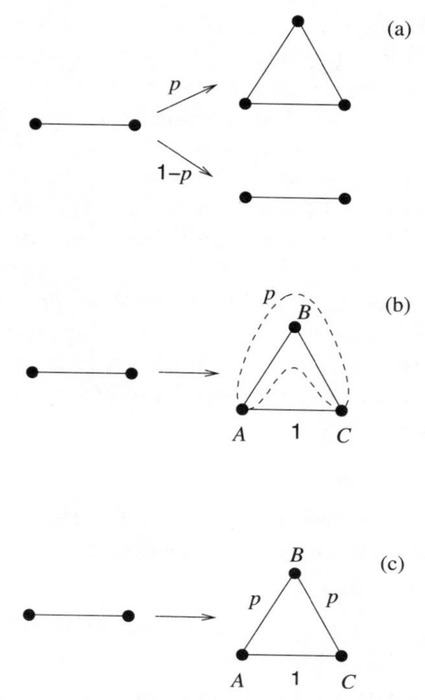

FIG. 5.20. Examples of the transformations of edges, which generate random growing networks. (a) At each time step, each edge of the network is transformed into one configuration of the two shown. p and $1-p$ are the probabilities of these transformations. (b) Another presentation of the same rule for the evolution of edges: the chain ABC is present with the probability p. Particularly, if $p = 1$, the transformation (a) or, equivalently, (b) generates the deterministic scale-free graph shown in Fig. 5.16. For $p \to 0$ but $p \neq 0$, the transformation (a) or (b) generates the stochastically growing scale-free network introduced in Fig. 5.6. (c) A different transformation: at each time step each edge is transformed into the cluster, where edges AB, BC, and AC are present with probabilities p, p, and 1 respectively. If $p = 1$, we again have the pseudofractal from Fig. 5.16. For $p \to 0$ but $p \neq 0$, the transformation (c) generates the Barabási–Albert model.

One can study correlations in the pseudofractal and get the usual formula (5.57) for random scale-free citation graphs. One can find the clustering and distribution of clustering over the network.[25] One can study percolation on this structure and find that it is impossible to eliminate in the infinite pseudofractal

[25]The distribution of clustering is the probability that the clustering coefficient of a vertex is C. For small C, it behaves as $C^{\gamma-2}$. In this case, the clustering coefficient of a vertex is inversely proportional to its degree k. In general, correlations between C and k are more complex.

(we will discuss this phenomenon in detail in Section 6.4). One can obtain the eigenvalue spectrum of the adjacency matrix of the pseudofractal.[26]

In short, one can find practically every structural and topological characteristic of this graph (or the more general pseudofractal graph from Fig. 5.19). Each of them is surprisingly close to the corresponding characteristic of stochastically growing networks (citation graphs) with the same value of the γ exponent.

This closeness indicates that the stochasticity of the growth of networks is not such a crucial factor for their structure and global topology. At least, it is a less crucial factor than one might expect. Of course, real networks are not deterministic graphs. Some differences may be found. For example, the size-dependence of the width of the shortest-path-length distribution in random growing networks may differ from the corresponding dependence $\sim \sqrt{\ln N}$ in the pseudofractal (see eqn (5.58)). However, it is practically impossible to observe this distinction for any reasonably sized network.

How to generate evolving random nets

This section is devoted to deterministic graphs. However, the constructions shown in Figs 5.18 and 5.19 can be easily modified to generate random growing networks. Compare:

(1) The generation of a deterministic graph: at each step, each edge is transformed in the same way with probability 1.
(2) The generation of a stochastically growing network: at each step, an edge can be transformed into distinct configurations of vertices and edges with various probabilities (see Fig. 5.20).

The latter is just one more construction of evolving random nets.

5.15 Accelerated growth of networks

The growing networks we have discussed so far have one very strong restriction. The total number of edges L in them grows proportionally to their size $N \equiv t$ or, at least, linearly. Such linear growth does not change the average degree (at least, $\overline{k}$ approaches a constant in the large-network-size limit). Moreover, the great majority of models of evolving networks assume linear growth, and it is usually supposed to be a natural feature of growing networks. But let us ask ourselves whether this very particular case is so widespread in real networks. Recall the empirical data discussed in Chapter 3 on the growth of the WWW and the Internet, on networks of citations in the scientific literature, on collaboration graphs, on networks of metabolic reactions, and on nets of software components.

[26]This has been done with high precision, and the resulting spectrum has a power-law tail, $G(\lambda) \propto \lambda^{-(\gamma+2)}$. Earlier, power-law tails of the eigenvalue spectra were found for random growing networks (1) empirically (for the Internet: Faloutsos, Faloutsos, and Faloutsos 1999) and (2) numerically (for the Barabási–Albert model: Farkas, Derényi, Barabási, and Vicsek 2001, Goh, Kahng, and Kim 2001a). However, empirical (1) and numerical (2) power-law dependences exist in too narrow a region of eigenvalues λ to draw qualitative conclusions.

In each of these networks, the average degree is not constant but grows, so that the growth is non-linear. We have called it *accelerated growth* (Dorogovtsev and Mendes 2001a), although, in principle, the acceleration may be negative, and the average degree may decrease.

There are a variety of reasons for the acceleration, but the usual reason is simple. Each vertex proceeds to create new connections after its birth, and so the mean degree grows.

General relations

Many things can be found for the accelerated growth without invoking any specific models, but only by using very general considerations. When does this growth produce scale-free networks? Let us assume that the average degree increases as a power of t, $\overline{k} \propto t^a$. Here, $a > 0$ is the growth exponent.[27] Then, the total number of edges is $L(t) \propto t^{a+1}$. The following relations of a general nature are valid for degree, in-degree, and out-degree in directed and undirected networks. We hope that the power-law acceleration provides scale-free distributions of connections. So, we suppose that this is the case and then we check our assumption.

Now, when the average degree grows, the degree distribution does not need to be stationary. Naturally, we take it in the form

$$P(k,t) \sim t^z k^{-\gamma}\,. \tag{5.61}$$

Here a new exponent z is introduced. If $a > 0$, one may think that z cannot be negative. This power-law form is realized only in some range of degrees, $k_0(t) \lesssim k \lesssim k_{cut}(t)$. What are these boundaries? We know all that is necessary to estimate them.

(a) We know the normalization condition $\int_{k_0(t)}^{\infty} dk\, t^z k^{-\gamma} \sim 1$. This integral diverges at the lower boundary, if $\gamma > 1$. This yields (for any $\gamma > 1$) the lower boundary

$$k_0(t) \sim t^{z/(\gamma-1)}\,. \tag{5.62}$$

(b) As usual, the cut-off position can be estimated from the condition $t\int_{k_{cut}(t)}^{\infty} dk\, t^z k^{-\gamma} \sim 1$. Therefore (for any $\gamma > 1$) the position of the cut-off is

$$k_{cut}(t) \sim t^{(z+1)/(\gamma-1)}\,, \tag{5.63}$$

which reduces to a standard estimate (1.8) when $z = 0$.[28]

The resulting degree distribution is schematically shown in Fig. 5.21. Now we can find the γ exponent. For this, recall that $\overline{k} \sim t^a$. The cases $1 < \gamma < 2$ and $\gamma > 2$ are principally different, so we consider them separately.

[27] It is worthwhile to mention the following circumstance. Suppose that the relation $\overline{\ell} \approx \ln t / \ln \overline{k}(t)$ is valid. Then for a power-law dependence $\overline{k}(t)$, the average shortest-path length is independent of the network size in the limit $t \to \infty$. We discussed a similar situation for networks of metabolic reactions in Section 3.4.2.

[28] Recall that we do not account for the mortality of vertices.

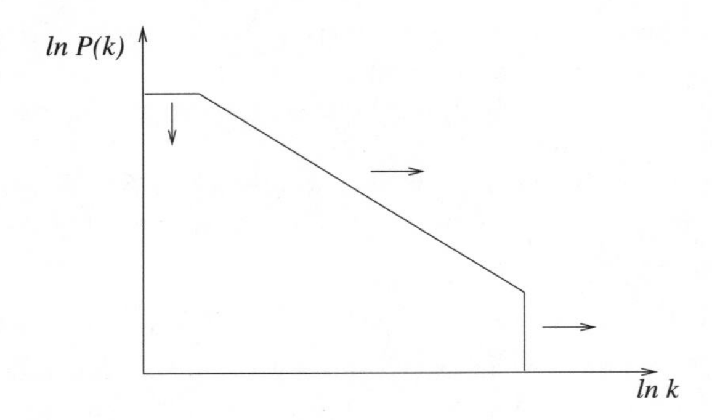

FIG. 5.21. Evolution of the degree distribution of networks that grow in the accelerated mode. Arrows show how the degree distribution evolves in time. If exponent $z = 0$, only the right arrow is present.

(1) $1 < \gamma < 2$. The estimate of the average degree is

$$t^a \sim \int^{t^{(z-1)/(\gamma-1)}} dk\, kt^z k^{-\gamma} \sim t^{-1+(z-1)/(\gamma-1)} \,. \tag{5.64}$$

The value of the integral is determined by its upper limit. Therefore, comparing the powers, we obtain the relation $(z-1)/(\gamma-1) = a+1$. Substituting this into eqn (5.63) shows that in this situation the cut-off is of the order of the total number of edges in the network, $k_{cut}(t) \sim t^{1+a} \sim L(t)$, that is the maximum possible value. Thus we see that the γ exponent is

$$\gamma = 1 + \frac{z+1}{a+1} \,. \tag{5.65}$$

The additional condition $\gamma < 2$ together with eqn (5.65) implies that $z < a$. The lower boundary of the γ exponent, $\gamma = 1 + 1/(a+1)$, is approached when the power-law part of the distribution is stationary, that is $z = 0$.

(2) $\gamma > 2$. In this event, the integral for the average degree is determined by its lower limit

$$t^a \sim \int_{t^{z/(\gamma-1)}} dk\, kt^z k^{-\gamma} \sim t^{z-z(\gamma-2)/(\gamma-1)} \,. \tag{5.66}$$

Comparing the powers readily yields the γ exponent:

$$\gamma = 1 + \frac{z}{a} \,, \tag{5.67}$$

where $z > a$ to keep $\gamma > 2$.[29] In this situation, $z > a > 0$, and so the degree distribution, in principle, cannot be stationary.

[29] Of course, this relation is not valid at $a = 0$.

Scaling

In Section 5.5 we showed how to derive scaling relations for the linear growth. Similarly, one can obtain scaling relations for the accelerated mode. The generalization of the scaling formula (5.29) for average degrees of vertices is of a natural form:

$$\overline{k}(s,t) \propto t^{\delta} \left(\frac{s}{t}\right)^{-\beta} . \tag{5.68}$$

Here δ is a new scaling exponent. The reader may check in the same way as in Section 5.5 that the old relation (5.30) between exponents β and γ again holds, and, in addition, the following relation is valid:

$$z = \delta/\beta . \tag{5.69}$$

The scaling form of the degree distributions of individual vertices is now

$$p(k,s,t) = \frac{s^{1/(\gamma-1)}}{t^{(z+1)/(\gamma-1)}}\, g\left[\frac{s^{1/(\gamma-1)}}{t^{(z+1)/(\gamma-1)}}\right] , \tag{5.70}$$

and the total degree distribution is of the scaling form

$$P(k,t) = t^{z} k^{-\gamma} G(kt^{-(1+z)/(\gamma-1)}) , \tag{5.71}$$

where $g[\;]$ and $G(\;)$ are scaling functions. Compare with eqns (5.36) and (5.37) for linear growth. One can see that it is sufficient to know growth exponent a and only one exponent of γ, β, z, or δ to find all the others.

Expressions (5.65) and (5.67) describe the only two possibilities for scale-free networks growing with acceleration. The values of exponents a and z are model-dependent. Various examples of networks which grow in the accelerated mode are considered by Dorogovtsev and Mendes (2002b). In the next section we describe one of the most intriguing examples.

5.16 Evolution of language

How does language evolve? What is its 'global' structure? This is a major challenge for linguistics and evolutionary biology and an intriguing problem for other sciences (Simon 1955, 1957). In Section 3.4.5 we described the Word Web of human language that was constructed by using texts from the bank of modern British English (the British National Corpus). The Word Web is a very large net which contains 470 000 words and 17 000 000 connections. The empirical degree distribution of this network (see Fig. 3.24 in Section 3.4.5) is of a complex form with two power-law regions.

The weak point of network science is the absence of a convincing comparison of numerous schematic models with reality. But what does a 'convincing comparison' or a convincing description mean? It means that a theory must quantitatively describe sufficiently complex empirical data without fitting. Theoretical curves must be obtained by using only known parameters of the system.

The complex form of the empirical degree distribution of the Word Web hampers any treatment but, on the other hand, makes it possible to propose a convincing concept for the evolution of human language. Indeed, it is hardly possible to describe such a complex form perfectly by coincidence, if we do not use fitting.

The basic ideas of our theory of the evolution of language (Dorogovtsev and Mendes 2001d) are:

- *Language is an evolving network of interacting (collaborating) words.*
- *During its evolution, language self-organizes into a complex structure.*

Words are collaborators in language. Words interact and collaborate when they meet in sentences. The definition and details of the interaction between words are not essential for us. For example, one can naively assume that two distinct words interact if they are the nearest neighbours, at least, in one sentence.

At its birth, a new word already interacts with several old ones. New interactions between old words emerge from time to time, and new undirected edges emerge. All the time a word lives, it enters into new collaborations. Hence, the number of connections grows more rapidly than the total number of words in the language.[30] Thus the growth of the Word Web is accelerated. We use the principle 'popularity is attractive', that is the mechanism of preferential linking.

We suggest the following rules for the Word Web growth (see Fig. 5.22):

(1) At each step, a new vertex (word) is added to the Word Web, and so the total number of words plays the role of 'time' t.
(2) At its birth, a new word connects to several old ones. On average, this number is $m \sim 1$.[31] We use the simplest version of preferential linking for undirected networks. So, the preference is proportional: a new word becomes connected with some old one of degree k with a probability proportional to k, as in the Barabási–Albert model.
(3) Simultaneously, cmt new edges emerge between old words. Here c is a constant. If each vertex of the network makes new connections at a constant rate, this linear dependence on time arises naturally. Each of these new edges connects preferentially chosen pairs of old vertices. The preference is proportional: an edge becomes attached to a pair of words with a probability proportional to the product of their degrees, kk'.

The above rules determine the model of the evolution of language, which includes only three parameters: t, m, and c. At first sight, it seems to be oversimplified: we do not account for the mortality and 'mutability' of words, we assume the simplest form of preference, etc. So, in our theory, words never die, words never change. Of course, this is not true. Nevertheless, we will see that this minimal model is quite sufficient for a quantitative description of the existing empirical data.

[30] Here we forget about the mortality of words.

[31] In principle, m may be non-integer.

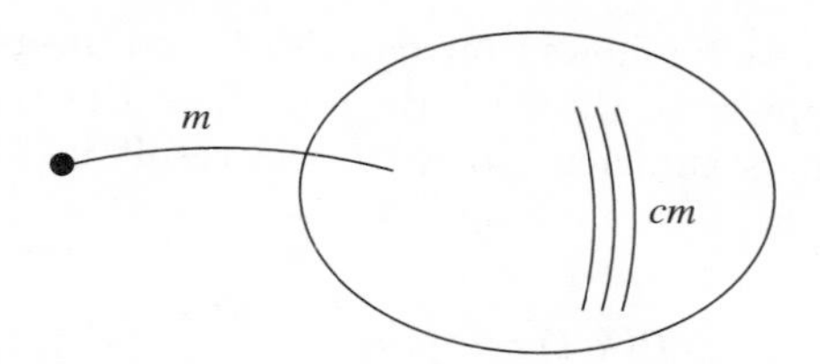

FIG. 5.22. Scheme of the evolution of language, that is the growth of the Word Web (Dorogovtsev and Mendes 2001d). At each time step, a new word emerges. It connects to $m \sim 1$ preferentially chosen old words. Simultaneously, cmt new connections emerge between pairs of preferentially chosen old words. c is a constant.

Even without any calculations one can imagine the form of the resulting degree distribution. In fact we have a combination of two processes: linear and quadratic growth. The first one, which is due to the linking of new words, is the same as in the Barabási–Albert model and naturally produces exponent $\gamma = 3$. The second one, which is the emergence of new connections between already existing words, produces the acceleration and, therefore, a different value of exponent γ. From eqn (5.65) we see that the minimum γ that is possible for such acceleration ($a = 1$) is $3/2$. The question is: how do these two power-law dependences mesh together in the degree distribution of the Word Web?

As usual, to obtain the degree distribution, we use the continuum approach:

$$\frac{\partial \overline{k}(s,t)}{\partial t} = (m + 2cmt)\frac{\overline{k}(s,t)}{2mt + cmt^2}\,, \tag{5.72}$$

with the initial condition $\overline{k}(0,0) = 0$ and the boundary one $\overline{k}(t,t) = m$. To understand the structure of the equation, it is sufficient to look at Fig. 5.22. As a new word emerges, $m + 2cmt$ ends of edges become attached. m ends belong to edges coming from a new word, and the others are the ends of the mct new edges emerging between old words. The total degree of the Word Web is $\int_0^t du\, \overline{k}(u,t) = 2mt + cmt^2$, so that its average degree is equal to $\overline{k}(t) = 2m + cmt$. From eqn (5.72), one can immediately see that exponent β takes two values, $1/2$ and 2, in two limiting regimes. Notice that it is the first time in this book that we find exponent $\beta > 1$. We can easily obtain the solution of this equation:

$$\overline{k}(s,t) = m\left(\frac{cmt}{cms}\right)^{1/2}\left(\frac{2m + cmt}{2m + cms}\right)^{3/2}. \tag{5.73}$$

Evidently, it has two distinct regimes: at small s, $\overline{k}(s,t) \propto s^{-1/2}$, and, at large s, $\overline{k}(s,t) \propto s^{-2}$. Recalling our relation $\gamma = 1 + 1/\beta$, we see that, indeed, the degree distribution must have regions with exponents 3 and $3/2$. These regions are separated by the crossover point

$$k_{cross} \approx m\sqrt{ct}(2 + ct)^{3/2}\,, \tag{5.74}$$

which moves in the direction of large degrees as the Word Web grows.

The degree distribution follows from eqn (5.73) and the general relation (5.27) of the continuum approach. Here we do not show the resulting cumbersome expression. For us, it is sufficient to write the result in terms of the solution $s = s(k,t)$ of eqn (5.73),

$$P(k,t) = \frac{1}{ct}\frac{cs(2+cs)}{1+cs}\frac{1}{k}\,. \tag{5.75}$$

Below the crossover point, this distribution is stationary with $\gamma = 3/2$:

$$P(k) \cong \frac{\sqrt{m}}{2}k^{-3/2}\,. \tag{5.76}$$

Above the crossover, the degree distribution of the Word Web is non-stationary, and $\gamma = 3$:

$$P(k,t) \cong \frac{(2m+cmt)^3}{4}k^{-3}\,. \tag{5.77}$$

There is one more important point in the degree distribution, namely the cut-off that is produced by the finite-size effect. Again we use the condition $t\int_{k_{cut}}^{\infty} dkP(k,t) \sim 1$ and obtain the cut-off

$$k_{cut} \sim \sqrt{\frac{t}{8}}(2m+cmt)^{3/2}\,. \tag{5.78}$$

Now we know all that we need in order to make a comparison with the empirical data. At this point, the meticulous reader may be puzzled. How can it be? Only *two* parameters of the real Word Web are known: the total number of vertices, $t = 0.470 \times 10^6$, and the average degree $\overline{k} = 72$ (Ferrer i Cancho and Solé 2001a). On the other hand, there are *three* parameters in our theory: t, c, and m. Then, how can we hope to describe the empirical data without fitting?

The deciding factor is that, in the log–log scale, the distribution weakly depends on the average number of connections of new words m. Then, m is an inessential parameter of the theory. For m we know only that it is of the order of 1. So, we set its value to 1. The reader has to believe that if the real value is several times greater, the deviations in the log–log plot of the degree distribution are practically unnoticeable. Then we estimate $c \approx cmt/\overline{k}(t) \sim ct/\overline{k}(t)$. With these parameters we can calculate the theoretical degree distribution and compare it with the empirical one. The comparison is presented in Fig. 3.24 from Section 3.4.5. One can see that the agreement is excellent.[32] Furthermore,

[32]Notice that for a better comparison, the theoretical curve is slightly displaced upwards. This multiplication of the degree distribution by a constant cannot be called 'fitting'. Indeed, it is impossible to describe the part of the degree distribution at small degrees. It depends on the peculiarities of the construction of the Word Web, on the grammatical structure of the language, etc. On the other hand, small variations of the degree distribution at small degrees produce a noticeable multiplication of the other part of the distribution by a constant. Here we study only the global organization of language and have not penetrated into this region.

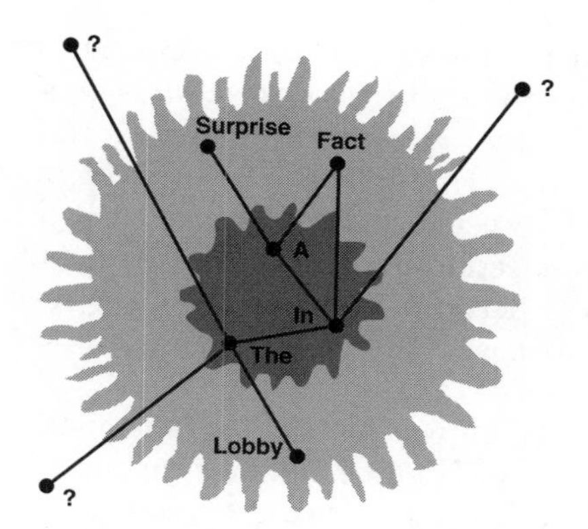

FIG. 5.23. Scheme of the structure of language. The kernel lexicon of the language contains most connected words, which have a different structure of connections than other words. Connections of some words are shown. New words are denoted by '?'.

in Fig. 3.24, we show the position of the crossover point and the cut-off, which was estimated from eqns (5.74) and (5.78). The reader can see the quality of the agreement.

Core of language

Thus, the agreement is convincing. Moreover, attempts to improve the model by accounting for some additional factors would be meaningless: the existing empirical data does not allow a better agreement. Then, the consequences of this theory can be taken seriously. Here we discuss one of them. Few words are in the region above the crossover point $k_{cross} \approx 5 \times 10^3$. These most connected words have a different structure of connections than others. They form a core of the language, its *kernel lexicon* (Ferrer i Cancho and Solé 2001c). This core is much smaller than the lexicon that we use and that is collected in serious dictionaries. Nevertheless, this small part of the language is of primary importance: without it, language does not exist. In Fig. 5.23, we schematically show the complex global structure of language. From eqns (5.74) and (5.77), the number of words in this core is estimated as

$$N_c \sim \frac{m}{8c} \approx \frac{m^2 t}{8\overline{k}} \,. \tag{5.79}$$

This estimate demonstrates that *the number of words in the kernel lexicon does not increase as language grows.* Language develops but its core does not change. Then, the numbers of words in this core in various languages, including 'primitive' ones, must not differ much.[33]

[33]It is hard to claim this more definitely: in principle, parameters m and c must depend on a particular language.

Let us be a little more precise than above, and account for the fact that the average number of connections of a new word, m, is greater than 1. One can see from eqn (5.79) that, indeed, it is worthwhile to do this. What is a reasonable value of m? Even if we allow connections only between the nearest neighbours in sentences, m must be between 1 and 2. However, it is more natural to account for the emergence of connections inside larger 'clusters' of words, and so a more reasonable value of m is between 2 and 3. Then, the resulting estimate for the number of words in this basic part of language is $N_c \sim 5 \times 10^3$ ($m = 1$ yields $N_c \sim 10^3$).

This number seems to be reasonable. There is another indication that language has a separated core. Using the data from the British National Corpus, Ferrer i Cancho and Solé (2001c) studied the frequency of occurrence of distinct words in modern British English. They observed that the functional form of the distribution of occurrence for about 5×10^3 most frequent words differs sharply from that for the other words. One can see that these two (rough) numbers coincide. But do these two sets of words from two distinct distributions coincide?

The model that we use to describe the evolution of language was discussed earlier in the context of collaboration networks (nets of coauthorships) (Barabási, Jeong, Néda, Ravasz, Schubert, and Vicsek 2002). Recall that we considered words as collaborators in language and the evolution of language as the evolution of such collaborations. In principle, formulae and results from this section can be directly applied for some networks of collaborations. However, there are two problems with this way. First, we do not know of any collaboration networks with such 'convenient' parameters as the Word Web, and secondly, there is a 'physical' restriction for the maximum number of collaborations, which produces smaller cut-offs. Thankfully, one cannot have 100 000 coauthors.

5.17 Partial copying and duplication

We speak all the time about fat-tailed degree distributions but up to now have only considered one of the possibilities, namely power-law distributions (one may say, 'scale-free' degree distributions or 'scale-free' networks). Then, the reader may reasonably ask: what do you use this exotic term for? Maybe, the division of networks into scale-free and non-scale-free would be sufficient(?). Maybe, all these attempts to note some supposed generalization of the well-known scale-free distributions are only an indication of an imaginary possibility, which has no relation to reality(?). The aim of this section is to discredit such doubts.

We have explained in Section 1.3 that the power-law degree distribution is of fractal type. This means that the size dependence of the n-th moment of the distribution is of the form $\sim t^{\tau(n)}$ with a *linear* function $\tau(n)$. It is routine to study the property of fractality and so forth by considering the behaviour of the size dependence of higher moments.

Of course, the fractality of the degree distribution must be seen already in the master equation. Let us demonstrate what it comes from. Recall the form of the master equation from Section 5.2 for the time-dependent degree distribution. We wrote it down for the Barabási–Albert model, but it can be naturally generalized. Without going into specific details, the master equation that provides power-law degree distributions usually looks like this:

$$t\frac{\partial P(k,t)}{\partial t} + P(k,t) = \frac{m}{\overline{k}(t)}[(k-1)P(k-1,t) - kP(k,t)] + \delta_{k,k_0} \tag{5.80}$$

(compare with the master equation (5.17) for the Barabási–Albert model). Here m is a number of edges that is preferentially distributed per time step. For us, the last term on the right-hand side is important, that is the Kronecker symbol δ_{k,k_0}. It shows that each new vertex has exactly k_0 edges.

Equations for the moments of the degree distribution, $M_n(t) \equiv \sum_k k^n P(k,t)$, follow from eqn (5.80) if we multiply both its sides by k^n and sum over k:

$$\frac{\partial M_n(t)}{\partial \ln t} + M_n(t) = \frac{m}{\overline{k}(t)} \sum_{k=0}^{\infty} [(k+1)^n - k^n] k P(k,t) + k_0^n \,. \tag{5.81}$$

In particular, for the first moment, we see that

$$\frac{\partial M_1(t)}{\partial \ln t} + M_1(t) = m + k_0 \,, \tag{5.82}$$

and so $M_1 = \overline{k} \to m + k_0$ as $t \to \infty$. For higher moments, saving the leading terms yields

$$\frac{\partial M_n(t)}{\partial \ln t} = \left(n\frac{m}{m+k_0} - 1\right) M_n(t) + \ldots + k_0^n \,. \tag{5.83}$$

We write the negligible term k_0^n above to show what is the role here of the 'birth' (Kronecker symbol) term from eqn (5.80). From this equation, we immediately see that the resulting moments have a power-law form at large sizes of the network, $t \to \infty$:

$$M_n(t) \propto t^{n[m/(m+k_0)]-1} \,. \tag{5.84}$$

Thus, the degree distribution is fractal, which we knew from the very beginning.

When we wrote down the master equation (5.80) we made a very restrictive assumption. We assumed that a new vertex 'knows' nothing about the existing network. It has exactly k_0 connections independently of the state of the network at the moment of birth of the vertex. So, the system is non-adaptive. It is this assumption that leads to the Kronecker symbol in eqn (5.80).

But why must new vertices be so ignorant about the current state of the network? The reader can propose a list of situations where this is certainly not true. How can we account for the influence of the network on the properties on a newborn vertex? This, our baby, has only one characteristic, namely the number of its connections. Then, let this number not be fixed by God but be distributed with some distribution function which depends on the current state of the network. In our case, this state is described by the degree distribution

$P(k,t)$. Hence, let the degree distribution of the newborn vertex be dependent on the degree distribution of the network. In other words, a new vertex is adapted to the current state of the network.

There is an infinite number of ways to introduce such a dependence. We choose one of the simplest (Dorogovtsev, Mendes and Samukhin 2000d, 2002a). Let a new vertex copy a fraction $c < 1$ of the connections of a randomly chosen old vertex.[34] One can say that a new vertex ***partially copies*** the connections of an old one. We also used the term *inheritance*. Then, the degree distribution of a newborn vertex may be taken as $P(ck,t)/c$.[35] Then, instead of the k_0^n term in eqn (5.83), there is now the term $\sim \int dk\, k^n P(ck,t)/c = M_n/c^{n+2}$. Hence, the equation for the moments takes the form

$$\frac{\partial M_n(t)}{\partial \ln t} = \left(n\frac{m}{m+k_0} - 1 + \frac{1}{c^{n+2}} \right) M_n(t) + \dots, \tag{5.85}$$

whence the resulting size dependence of the moments is

$$M_n(t) \propto t^{n[m/(m+k_0)]+1/c^{n+2}-1}. \tag{5.86}$$

But then the function $\tau(n) = n[m/(m+k_0)] + 1/c^{n+2} - 1$ is non-linear. By definition, this shows that the degree distribution is multifractal; that is, it cannot be described by a single exponent. Suppose we try to measure the slope of the multifractal degree distribution in the log–log scale. Then, the result will depend on the point (degree) where we do the measurement. As a rule, the fitting of an empirical distribution by power laws is possible only in narrow regions of degrees. Therefore, it is very possible that what is often treated as a power law is, in fact, a multifractal degree distribution.

The divergence of higher moments is the definitive feature of fat-tailed distributions, or, as we call this, of the modern architecture of random networks. We see that in large scale-free networks, where degree distributions are power laws, the higher moments diverge in a very particular way.

We presented a particular demonstrative example. This situation is in no way unique. Similar processes were used in models (Vázquez, Flammini, Maritan, and Vespignani 2001, Solé, Pastor-Satorras, Smith, and Kepler 2001, Pastor-Satorras, Smith, and Solé 2002) of protein networks. From time to time, vertices in these model networks duplicate themselves with all their connections. After this, part of these connections disappear. Such rules also may produce multifractal distributions.

To finish, we note that models discussed in this section and their variations often provide non-linear, accelerated growth. For example, recall the case of

[34] From the derivation below, one can see why $c = 1$ cannot be used.

[35] Here we are rather impudent. This form, as well as the next equation, is presented in the continuum approximation. Thus, we actually mix discrete and continuum approaches. In principle, this is awkward. However, the final result for moments will be correct, and in this demonstration we cannot be too rigorous. The alert reader must look at the original papers.

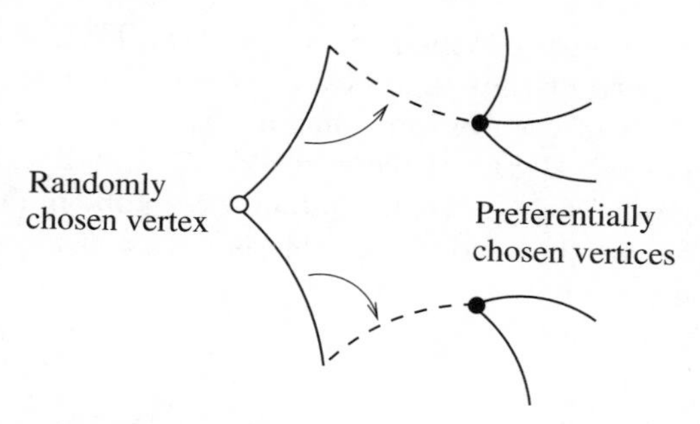

FIG. 5.24. Processes in the scale-free non-equilibrium non-growing network. At each time step, all connections from a randomly chosen vertex are rewired to preferentially chosen vertices.

$c = 1$; that is, a newborn vertex copies *all* the connections of a randomly chosen old vertex. This yields the equation for the first moment: $\partial M_1(t)/\partial \ln t = m$. Hence, the average degree of the network grows at long times as $M_1(t) \cong m \ln t$. This is one of the possible mechanisms of the accelerated growth of networks.

5.18 Non-equilibrium non-growing networks

Can a 'static' network be a non-equilibrium one? Up to now we have discussed only growing networks, and the reader might conclude that only growing networks are non-equilibrium ones. Here we describe a simple scale-free network that has a fixed number of vertices and a fixed number of edges, but, nevertheless, is a non-equilibrium network.

The network has a total number of vertices, N, and a total number of undirected edges, L, so the mean degree is $\overline{k} = 2L/N$. The rules for the evolution of the network are as follows (see Fig. 5.24):

(1) At each time step, a randomly chosen vertex loses all its connections.
(2) These connections become rewired to preferentially chosen vertices. The preference is linear: the probability that such an edge become rewired to a vertex of degree k is proportional to $k + A$, where A is a constant (additional attractiveness).

We can look at this in another way. At each increment of time, one vertex dies, one new bare vertex emerges, and a number of connections are preferentially distributed among the vertices. On average, the number of the ends of edges to be attached is, of course, the average degree of the network, $\overline{k}$. If we are not concerned with the very tail of the degree distribution, this network is equivalent to the basic growing net from Section 5.3 (see Fig. 5.3),[36] where the number m of new attachments per time step is substituted by $\overline{k}$. Then the degree distribution

[36]We mean the equivalence in respect of the degree distributions. Note also that the network from Section 5.3 is directed, unlike the network under discussion. One can easily check that this difference is not essential here.

is of a power-law form with exponent $\gamma = 2 + A/\overline{k}$. This result can be easily derived, but it is important that the network is equivalent to the growing one. One can introduce ages of vertices, and then a vertex acquires connections with increasing age. One can call this a 'current' of edges in the direction of older vertices. The presence of such currents characterizes non-equilibrium networks.

As usual, we remind the reader that this is only a demonstrative example and only one of the possibilities.

6

GLOBAL TOPOLOGY OF NETWORKS

How are the 'distant' vertices connected on a network? How can these 'long-distance' connections be used? What is the 'linear size' of a network? What is the size of the 'core' of a network, that is of its giant connected component? When does the giant component exist? How are finite connected components distributed?

These questions together form a general problem: what is the topology, that is the global structure, of networks? Main theoretical results about global topological properties of random networks were obtained from the analysis of equilibrium graphs. In this respect, growing networks are far less studied. Only the last section of this chapter contains theory for non-equilibrium nets. Nevertheless, many intriguing results for equilibrium graphs are also valid for growing ones.

6.1 Topology of undirected equilibrium networks

Here we use the *basic construction* of equilibrium random graphs. We mean undirected uncorrelated graphs with a given degree distribution, which are maximally random. Mathematicians call them the 'labelled random graphs with a given degree sequence' or 'configuration model'. These networks are constructed by the 'geometrical' construction procedure (see Sections 1.3 and 4.3). In Section 4.3, we explained that such a network, actually, means a microcanonical ensemble of networks.

What the network looks like

Let us try to draw an equilibrium uncorrelated network (see Fig. 6.1), assuming that the net is large. We start from some vertex. All the vertices in the network are statistically equivalent,[1] so we can start from any vertex. Of course, we could equally start from an edge, but to be concrete, we begin our picture from a randomly chosen vertex. Then, we draw all the nearest neighbours of this vertex with all their connections to it. The next step is crucial. We *do not* draw any connections between the nearest neighbours of the starting vertex.

In Chapter 4 we have mentioned that the probability of such connections is negligible when the network is large. To show this more rigorously, one may calculate the average probability that two fixed neighbours of the starting vertex are connected. It is the clustering coefficient of the network, C.

To perform this calculation, we have to recall some claims from Chapter 4:

[1]From the statistical equivalence of vertices in this net, it follows, in particular, that a notion of 'boundaries' cannot be introduced.

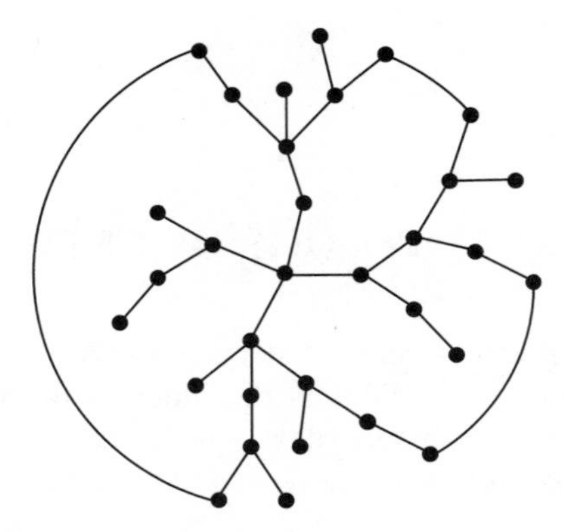

FIG. 6.1. How an uncorrelated equilibrium network looks from the 'point of view' of a randomly chosen vertex. We start from a vertex and successively add to it the first neighbours, the second, ..., n-th with all their connections. In the 'local environment' of this vertex (sufficiently small 'shells') before some value of n, the network has a tree-like structure. Here, each shell contains neighbours (of the starting vertex) of the same order. However, the following shells with remote neighbours of the starting vertex (neighbours of higher orders) already contain loops. Actually, they are connections between some of the vertices of the same shell. Then, if the connected component with the starting vertex is not too small, on a larger scale, it does not look like a tree.

- When we are interested in the properties of a single *vertex*, the basic object of interest is, naturally, a *vertex*, and the main characteristic of the network for us is the *degree distribution* $P(k)$ of vertices.
- But when the point of interest is how connections 'go' from vertex to vertex, how vertices 'interact', how some 'distribution' spreads on a network, the basic object of interest is an *edge*, and the main characteristic of the network is the *distribution of the number of connections of a randomly chosen end vertex of a randomly chosen edge.* One can say, equally, that this is the *distribution of the number of connections of the nearest neighbour of a vertex.* For uncorrelated networks, this probability is equal to $kP(k)/\overline{k}$. Furthermore, the probability that an end vertex of an edge has k extra connections, besides this 'mother' edge (the multiplication number), $(k+1)P(k+1)/\overline{k}$, is equally important.

Using the last probability, one can easily calculate the clustering coefficient (Newman 2002a). Let us consider two nearest neighbours, *1* and *2*, of the starting vertex. Suppose that their degrees are k_1 and k_2. Recall Fig. 1.8 from Section 1.3 (the set of hedgehogs). One should take into account that the connections between the spines of the hedgehogs are equally probable. In the large uncorrelated graph that we are considering, the probability that they are connected is asymptotically equal to $(k_1-1)(k_2-1)/(N\overline{k})$. Indeed, we have two 'hedgehogs' with

$k_1 - 1$ and $k_2 - 1$ free spines (one of the edges of each vertex comes to the starting vertex), and we have nearly $N\overline{k}$ spines of the other vertices (hedgehogs) of the network. One can easily average out this probability, since correlations between vertices are absent. Then, the clustering coefficient is

$$C = \sum_{k_1,k_2} \frac{k_1 P(k_1)}{\overline{k}} \frac{k_2 P(k_2)}{\overline{k}} \frac{(k_1-1)(k_2-1)}{N\overline{k}} = \frac{\overline{k}}{N} \left(\frac{\langle k^2 \rangle - \overline{k}}{\overline{k}^2} \right)^2 . \tag{6.1}$$

In Appendix F, one can find the following relation for the Poisson distributions: $\langle k^2 \rangle - \overline{k} = \overline{k}^2$. Hence, for the classical random graphs, the clustering coefficient is $C = \overline{k}/N$, which is a standard relation. The second equality in eqn (6.1) demonstrates that the probability of connections between the nearest neighbours of a vertex decreases with the size of the network if the second moment of the degree distribution is finite or slowly diverges.

Then, we can be sure there is no mistake: the edges between the nearest neighbours are practically absent. Now we must draw the connections from the first nearest neighbours of the first vertex to all its second-nearest neighbours and must draw the second-nearest neighbours themselves. In a similar way as above, we see that the edges between the second-nearest neighbours of the vertex are practically absent in the large-network limit. We can proceed with our drawing and see that *the network, locally, has a tree-like structure.*

But what is the average number of the second-nearest neighbours of a vertex? We denote this quantity by z_2, and the average number of the nearest neighbours by $z_1 \equiv \overline{k}$. This is not an idle question: z_2 is a very important characteristic of a network. The calculation is very simple because here, as well as nearly everywhere else in this chapter, we can use the facts (1) that these networks, locally, are trees and (2) that their vertices are independent.

The average degree of a randomly chosen end vertex of a randomly chosen edge (we may also say, of any end of any edge) is

$$\sum_k k \frac{kP(k)}{\overline{k}} = \frac{\langle k^2 \rangle}{\overline{k}} . \tag{6.2}$$

So, the average number of edges of this vertex, other than the 'mother edge' is $(\langle k^2 \rangle - \overline{k})/\overline{k}$.

In this event, we have:

- with probability $P(1)$, the first vertex has a single nearest neighbour with, on average, $(\langle k^2 \rangle - \overline{k})/\overline{k}$ edges, other than the connection with the parent vertex;
- with probability $P(2)$, the first vertex has two nearest neighbours with, on average, $2(\langle k^2 \rangle - \overline{k})/\overline{k}$ edges (here we use the fact that the average of the sum of independent quantities is the sum of their averages), and so on.

From this, z_2 follows immediately. These calculations can be made a little more rigorously,[2] but the result is the same: the average number of the second-nearest neighbours of a vertex is

$$z_2 = \langle k^2 \rangle - \overline{k} \,. \tag{6.3}$$

For the classical random graph, where the distribution of connections is Poisson, this yields $z_2 = z_1^2$.

From this equation, we see that the average number of the second-nearest neighbours diverges if the second moment diverges. This is, of course, simple but important, since the divergence of z_2 reflects quite a different type of topology of the network as compared with that of a network where z_2 is finite. In a finite net, the degree distribution has a cut-off (see Section 1.3), and so, instead of the divergence, one gets a size dependence. In the specific case of the power-law degree distribution of the form $P(k > k_0) \propto k^{-\gamma}$ and $P(k < k_0) = 0$,[3] the cut-off point is $k_{cut} \sim k_0 N^{1/(\gamma-1)}$, and thus we easily estimate $\langle k^2 \rangle$ and z_2 for $\gamma < 3$:

$$z_2 \cong \langle k^2 \rangle \sim k_0^2 N^{(3-\gamma)/(\gamma-1)} \sim \overline{k}^2 N^{(3-\gamma)/(\gamma-1)} \,, \tag{6.4}$$

and, for $\gamma = 3$,

$$z_2 \cong \langle k^2 \rangle \approx k_0^2 \ln N \,. \tag{6.5}$$

So, these quantities grow with N for $\gamma \leq 3$.

With z_2 many formulae take on a more elegant form. For example, the average multiplication factor of any end of any edge is, quite naturally, z_2/z_1, and the (average) clustering coefficient may be written in the form

$$C = \frac{z_2^2}{N z_1^3} \,. \tag{6.6}$$

Expressions (6.1) and (6.6) contain a correcting factor z_2^2/z_1^4 to the formula for the clustering coefficient of a classical random graph, z_1/N. This factor may be very big. The relation (6.1) was compared with data for several real growing networks, for which clustering coefficients and the moments of a degree distribution are known. In particular, the e-mail network (Ebel, Mielsch, and Bornholdt 2002) and a Web domain without accounting for the directness of Web links (Newman 2002a) were used. The comparison showed much better results than for the classical formula, although real networks are strongly correlated non-equilibrium objects.

[2] Venture to look at the chain

$$z_2 = \sum_k P(k) \prod_{i=1}^{k} \sum_{q_i} \frac{(q_i+1)P(q_i+1)}{\overline{k}} \sum_{i=1}^{k} q_i = \sum_k P(k) \frac{\overline{k}(\langle k^2 \rangle - \overline{k})}{\overline{k}} = \langle k^2 \rangle - \overline{k} \,.$$

[3] The last assumption is too strong. Actually, $P(k < k_0) \sim$ const or even some 'modest' increase with decreasing k is sufficient. Meanwhile, note the relation $\overline{k} \approx k_0(\gamma-1)/(\gamma-2)$ for such a distribution.

However, let us look attentively at the expressions for the clustering coefficient. If the network is scale-free, substituting eqn (6.4) into eqns (6.1) or (6.6) yields, for $\gamma < 3$,

$$C \sim k_0 N^{(7-3\gamma)/(\gamma-1)} . \tag{6.7}$$

So, the clustering coefficient of such networks *increases* with N when $\gamma < 2+1/3$. In this case, the structure of the network is certainly far from a tree. Moreover, the growth of z_2 with N for $\gamma \leq 3$ actually indicates that the whole region $\gamma \leq 3$ is 'dangerous'.

The tree-like local structure of the network is extremely important, but there is another, no less important circumstance. Look at the expression (6.2) for the average number of connections of an end vertex of a randomly chosen edge. One can see that it is always larger than the average degree of the network: $\langle k^2\rangle/\overline{k} > \overline{k}$.[4] Let us consider a subgraph that contains a randomly chosen vertex, its nearest-neighbour vertices, second neighbours, ..., n-th neighbours with all their connections. Here n must not be 'too large' (see below). The average degree of the starting vertex is $\overline{k}$, but the average degrees of the others vertices of this tree are $\langle k^2\rangle/\overline{k} > \overline{k}$, and their degree distribution is not $P(k)$ but $kP(k)/\overline{k}$. So, on average, the network is locally more 'dense' than the entire network itself. Furthermore, the average degree of vertices in a 'local' environment of a randomly chosen vertex or edge is *much higher* than the average degree of the entire network when the second moment of the degree distribution is large. Such vertices are more highly connected than a vertex selected at random. Many features of the network and numerous processes on the network depend on its 'dense local structure'.[5] Most of results in the present Chapter are obtained by using these two ideas:

(1) *'Locally', the net is tree.*

(2) *This 'local' environment of a (randomly chosen) vertex has a different structure of connections than the network as a whole; vertices from such an environment are more connected than the vertices in the network as a whole.*

Let us explain the meaning of this term 'locally'. The alert reader will already have noticed some contradiction in our claims. A tree, in principle, cannot be highly connected. Indeed, the total numbers of edges and vertices in any tree are rigidly related to each other, $L = N - 1$, and so an infinite tree always has the average degree $\overline{k} = 2$. Finite trees have even smaller average degree. Therefore, a tree cannot be a highly connected object. So, how can we be right? The explanation is obvious. Suppose that the starting vertex belongs to a giant connected

[4] This follows from the trivial inequality $\langle k^2\rangle > \overline{k}^2$.

[5] This text contains too many inverted commas, but we do not want to mix the term 'local' here with 'real' local properties such as degree of a vertex.

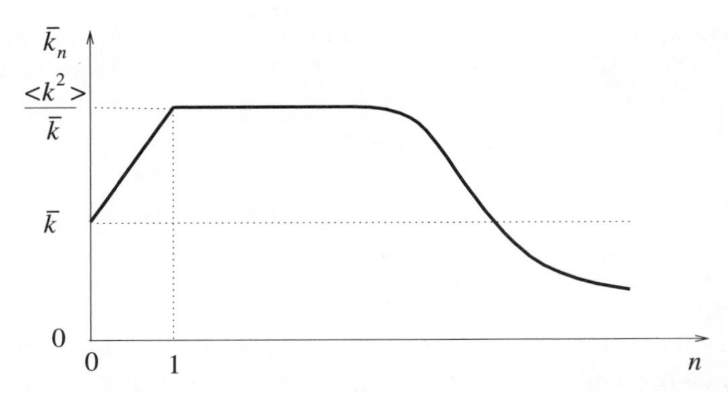

FIG. 6.2. Schematic plot of average degrees of neighbours of n-th order of a randomly chosen vertex versus n (see Fig. 6.1). The average degree of a vertex in the network is $\overline{k}$, but the average degrees of vertices in the 'local' environment of the starting vertex may be much higher.

component.[6] If we proceed with our picture and add to it vertices sufficiently distant from the starting one with all their connections, then, beginning from some shell, the resulting graph contains loops (see Fig. 6.1). In other words, in these 'distant' shells, some of vertices from the same shell are connected. Hence, at large scale, our 'tree' picture is broken by loops. Thus, the 'local' and global properties of such networks are very different. 'Locally', they are trees, but, globally, they are not.

In Fig. 6.2 we schematically show how the average degree of vertices in shells varies from shell to shell. One can see that remote vertices have smaller numbers of connections than the average degree $\overline{k}$ of the entire network. The highly connected 'local' environment of the starting vertex and its 'poor' distant neighbours together yield the proper average degree.

Furthermore, vertices in these networks are statistically equivalent, and so a 'boundary' is absent.[7] Therefore, the structure of a network shown in Fig. 6.1 and the presence of loops at a long scale are natural.

Is it possible to ignore the distant region with loops? There is no complete answer. Sometimes, it is sufficient to account for only the tree-like region and sometimes it is not. This depends on which characteristics of the network we consider and, of course, on the fatness of the tail of the degree distribution. Situations where the loops are important are poorly studied.

Search in networks

A standard seach problem for networks can be formulated as follows.

[6]If such a network is infinite, its finite connected components are almost surely trees. Therefore, to find something more interesting than trees, we must consider a giant connected component. Roughly speaking, in a finite network, this one (if it exists) is the connected component much larger than any of the other connected components.

[7]We thank A.V. Goltsev for indicating this circumstance.

- Suppose that each vertex of a network 'knows' only about its nearest neighbours, and so we can use only this 'local information' of vertices. Suppose that in our search, we do not check the same vertex twice. How many steps do we need to take to find some target vertex if we start from a randomly chosen vertex?[8]

The necessary number of steps depends on a search strategy. Evidently, the average required number of steps, t_s, exceeds the average shortest-path length of the network, $\overline{\ell}(N)$. In particular, in the experiment of Milgram, the average shortest-path length of a network of acquaintances was estimated by the measured t_s. Let us estimate t_s for a rather poor strategy. This is a random search, where at each step we jump to a randomly chosen nearest neighbour of a vertex.

In an undirected uncorrelated network, t_s for this search strategy can be roughly estimated as the ratio of the total number of vertices and the average degree of a randomly chosen nearest neighbour of a vertex. Using the basic relation (6.2) gives $t_s \sim N/(\langle k^2 \rangle / \overline{k})$. Substituting eqn (6.4) for scale-free networks with $\gamma < 3$ yields the estimate

$$t_s(N) \sim \frac{1}{\overline{k}} N^{2(\gamma-2)/(\gamma-1)} . \tag{6.8}$$

So, even the random search strategy leads to $t_s(N) \ll N/\overline{k}$ for $\gamma < 3$.[9] However, one can essentially reduce $t_s(N)$ by choosing a more effective stategy. For example, the target will be approached much faster if at each step, one jumps to the most connected nearest-neighbour vertex (Adamic, Lukose, Puniyani, and Huberman 2001).

Average shortest-path length

Now we know enough to estimate the average shortest-path length of the network. In the same way as eqn (6.3) one can obtain $z_n = z_{n-1}(z_2/z_1)$, so the average number of neighbours of n-th order of a randomly chosen vertex is equal to

$$z_n = \left(\frac{z_2}{z_1}\right)^{n-1} z_1 . \tag{6.9}$$

This formula is valid only in the range of n where the network is still tree-like. Suppose that the starting vertex belongs to the giant connected component (we

[8] A similar search was realized in the experiment of Milgram (1967) which we discussed in Chapter 0. Another example is a search in peer-to-peer networks, which are file-sharing systems without a central server (e.g. GNUTELLA).

[9] These small search times $t_s(N)$ are not an exception. For example, such small expected delivery times for a search in a small-world network were found by Kleinberg (1999b, 2000). Kleinberg studied a small-world network based on a d-dimensional 'mother-lattice'. Vertices of this lattice are connected by shortcuts with probability proportional to a power of the Euclidean distance between a pair of vertices, $r^{-\epsilon}$. When $\epsilon < \infty$, $t_s(N)$ grows slower than N. Moreover, at the specific point $\epsilon = d$, the average seach time increases slower than any power of N(!)

hope that this exists in our network). Therefore, it must be $z_2 > z_1$, otherwise z_n exponentially decreases with growing n. Let us take the average number of the second-nearest neighbours z_2 essentially greater than z_1 to have a sufficiently large giant component (that is, comparable with the size of the network). Then, one can very roughly estimate the average shortest-path length $\overline{\ell}$ of the giant component and the network from the following condition: the total number of vertices at the distance $\overline{\ell}$ or less from a randomly selected vertex is of the order of the size of the giant component N_G (and so N). Very approximately, this is $z_1 + z_1(z_2/z_1) + z_1(z_2/z_1)^2 + \ldots + z_1(z_2/z_1)^{\overline{\ell}-1} \sim N_G \sim N$. Then (Newman, Strogatz, and Watts 2001)

$$\overline{\ell} \approx \frac{\ln(N/z_1) + \ln[(z_2 - z_1)/z_1]}{\ln(z_2/z_1)} . \tag{6.10}$$

Of course, it would be more fair to write roughly $\overline{\ell} \approx \ln N/\ln(z_2/z_1) = \ln N/\ln[(\langle k^2\rangle/\overline{k}) - 1]$. Note that here we ignore the distant non-tree part of Fig. 6.1, substitute the average by the sum of averages, etc. This estimate is *very rough* but it is better than nothing. Besides, if we recall the relation for the classical random graphs, that is $z_2 = z_1^2$ (see Appendix F), the estimate 6.10 yields $\overline{\ell} \approx \ln N/\ln z_1$ in this case, which is the standard expression.

One more remark: on the one hand, the formula (6.10) can be used only when z_2 is essentially larger than z_1, that is $\langle k^2\rangle$ is essentially larger than $\overline{k}$. This is the case when the giant component is large. But, on the other hand, $\langle k^2\rangle$ must not be *too* large. Indeed, eqn (6.10) is certainly not valid if the second moment of the degree distribution (together with z_2) diverges in the limit of infinite networks. We know that for finite scale-free networks, $z_2 \cong \langle k^2\rangle \sim \overline{k}^2 N^{(3-\gamma)/(\gamma-1)}$ for $\gamma < 3$ (see eqn (6.4)). Substituting this estimate into eqn (6.10) gives a constant average shortest-path length, which is surprising. To obtain the correct average shortest-path length in this region, one has to account for the presence of loops in the network. The resulting average shortest-path length grows with N slower than $\ln N$ for $\gamma \leq 3$ (Cohen and Havlin 2002). This can be called a 'supersmall-world effect'.

However, despite all these comments, the formula (6.10) usually gives reasonable values which are in agreement with empirical data for many real networks.

Birth of the giant component

We have explained that the giant connected component is present in the *infinite* network[10] when the average number of the second-nearest neighbours of a randomly chosen vertex exceeds the average number of the nearest neighbours:

$$z_2 > z_1 . \tag{6.11}$$

In other words, the giant connected component is present when

[10]... where we can forget about the non-tree part of Fig. 6.1, at least while speaking about the birth of the giant component, its size, etc.

$$\langle k^2 \rangle - 2\overline{k} = \sum_k k(k-2)P(k) > 0\,, \tag{6.12}$$

and the 'phase transition of the birth of the giant component of the network' or, which is the same, the percolation threshold, occurs at the point where $\sum_k k(k-2)P(k) = 0$. This result is due to Molloy and Reed (1995) (*the Molloy–Reed criterion*). For the classical random graphs this yields $z_1 > 1$.

One can see that isolated vertices and vertices with two connections do not influence the position of the percolation threshold. The first half of this statement is obvious. As for the vertices with two connections, they are only elements of chains of edges. The number of such chains in a network is not essential for its topological properties.

But what is important in this respect is the number of dead ends in the network (the vertices with one edge attached). Indeed, the Molloy–Reed criterion can be written in the form $\sum_{k>2} k(k-2)P(k) > P(1)$. Let us forget about isolated vertices. Then only the dead ends obstruct the existence of the giant component. If the dead ends are absent, and $P(2) < 1$, that is, at least, some of the vertices have more than two connections, the giant component is present. For example, let the degree distribution be exactly of a power-law form with minimum degree 1. It is a simple exercise to substitute this distribution into the Molloy–Reed criterion and find that, in such a network, the giant connected component is present only if $\gamma < 3.479\ldots$ (Aiello, Chung, and Lu 2000). Of course, when the minimum degree is greater than 1, the giant component in the network exists for any γ.

Moreover, it follows from the Molloy–Reed criterion that *if the second moment of the degree distribution diverges or, equivalently, the mean number of the second-nearest neighbours of a vertex is infinite, the giant connected component is always present in the network.* We can make an even stronger claim: the divergence of z_2 guarantees the presence of the giant component in an arbitrary network.[11] In the particular case of scale-free networks, if $\gamma \leq 3$, the giant component exists. Note that practically all real networks are just in this region.

The reader will find more rigorous arguments below, but for the moment let us remain at an intuitive level. If we try to mimic the random networks under consideration by regular structures, what is the best choice? Of course we must choose a tree (look at Fig. 6.1). By definition, a regular tree (the *Bethe lattice*)[12]

[11] Equivalently, the divergence of the average number of the nearest-neighbour vertices of the ends of an edge guarantees the presence of the giant component. In a network with given pair degree–degree correlations $P(k,k')$, this number is equal to $\sum_{k,k'}(k-1+k'-1)P(k,k') = 2\langle k^2 \rangle/\overline{k} - 2$ (see eqn (4.26) in Section 4.6), and so again, if $\langle k^2 \rangle$ diverges, the giant component is present.

[12] Also, it is often called the *Cayley tree*. Rigorously speaking (Baxter 1982), there is a distinction between these two objects. By definition, the Cayley tree has a central vertex and a boundary (vertices with a single connection). The effect of the boundary may be crucial. Contrastingly, all vertices of the infinite, by definition, Bethe lattice are equivalent, all have the same coordination number, and boundaries are absent. In the networks that we consider

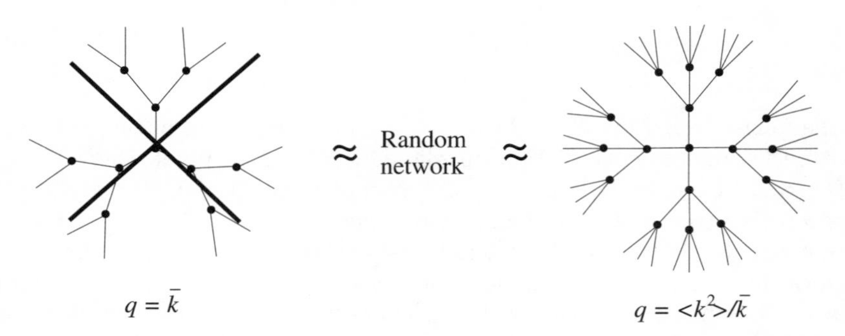

FIG. 6.3. Naive approximation of the local structure of the random networks by a regular tree (the Bethe lattice). It is a mistake to choose the coordination number of such a tree, $q = \overline{k}$. The best choice is $\langle k^2 \rangle / \overline{k}$ (see Figs 6.1 and 6.2). Note that remote vertices are not shown.

has a fixed coordination number, q. A non-alert reader will immediately decide that the best choice for q is the average degree of the vertex in the network, $\overline{k}$, and will make a big mistake. The attentive reader who remembers our conclusions about the 'dense local structure of the networks' and Fig. 6.2 will have certainly guessed that the best coordination number is $\langle k^2 \rangle / \overline{k}$ (see Fig. 6.3). At this point we do not even try to think what the non-integer coordination number means. Instead, we only notice that this Bethe lattice mimicking the networks exists when $q = \langle k^2 \rangle / \overline{k} \geq 2$. This coincides exactly with the Molloy–Reed criterion.

Core of a net

The giant connected component is present when $z_2 > z_1$, but what is its size? Let us denote the relative size of the giant connected component by W. The important quantity in this problem is the probability x that an edge attached to a vertex leads to a finite connected component on its other end. In other words, suppose that we select an edge at random and go in any of two possible directions, go and do not return. The probability that this way leads only to a finite connected component is x.

If we know x, we easily obtain the relative size of the giant component. Indeed, the probability that a randomly chosen vertex belongs to a finite connected component is, naturally, $1 - W$. But this is also the sum of the probabilities that this vertex is isolated, $P(0)$, *or* it has a single connection, which leads to a finite component, $P(1)x^1$, *or* it has two connections, each of which leads to finite components, $P(2)x^2$, *or* it has three such connections, $P(3)x^3$, and so on (see Fig. 6.4). Then, $1 - W = \sum_{k=0}^{\infty} P(k)x^k$. Equivalently, we can write $W = \sum_{k=0}^{\infty} P(k)(1 - x^k)$. This means that, if a vertex has several connections, then at least one of them leads to the giant component. Indeed, suppose that the vertex has k connections. The probability that all of them lead to finite

here, all vertices are statistically equivalent, boundaries are absent, and so the Bethe lattice is more relevant.

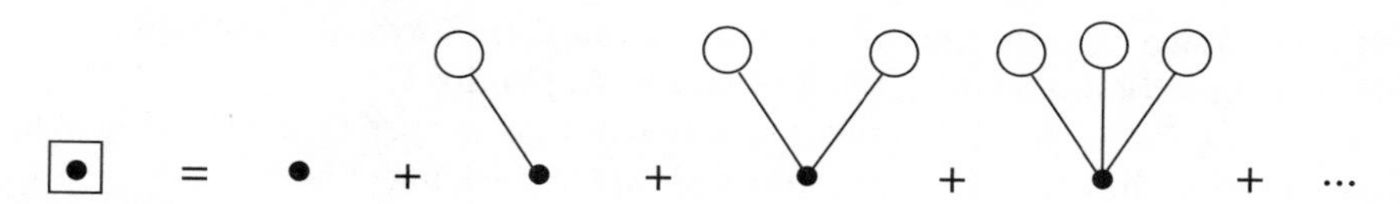

FIG. 6.4. The diagrammatic representation of the probability $1 - W$ that a randomly selected vertex of the network belongs to a finite component. The vertex in the box (in a jug) on the left-hand side denotes just this probability. The circle at the end of an edge shows that this edge leads to a finite component with probability x. The general form of the terms on the right-hand side is $P(k)x^k$. $P(k)$ is the degree distribution of a vertex.

components is x^k. The complementary probability, $1 - x^k$, is the probability that at least one of them leads to the giant connected component.

One can easily find the probability x by using the tree-like structure of the network. See the diagrammatic representation of the self-consistency equation for this probability in Fig. 6.5. Again we use the fact that the degree of the end vertex of an edge is $kP(k)/\overline{k}$. Therefore, $x = \sum_{k=1}^{\infty}[kP(k)/\overline{k}]x^{k-1} = \sum_{k=0}^{\infty}[(k+1)P(k+1)/\overline{k}]x^k$. Finally, one obtains a pair of simple relations (Molloy and Reed 1995, 1998, Newman, Strogatz, and Watts 2001)

$$W = 1 - \sum_{k=0}^{\infty} P(k)x^k \,, \tag{6.13}$$

$$x = \sum_{k=0}^{\infty} \frac{kP(k)}{\overline{k}} x^{k-1} \,, \tag{6.14}$$

from which we can find W for a given degree distribution. If the giant component is absent, x is, obviously, equal to 1. When the giant component is present, eqn (6.14) has a solution $x < 1$. To find the threshold, we linearize eqn (6.14) about $\Delta = 1 - x$. From the result of this linearization, we immediately obtain the Molloy–Reed criterion (6.12).

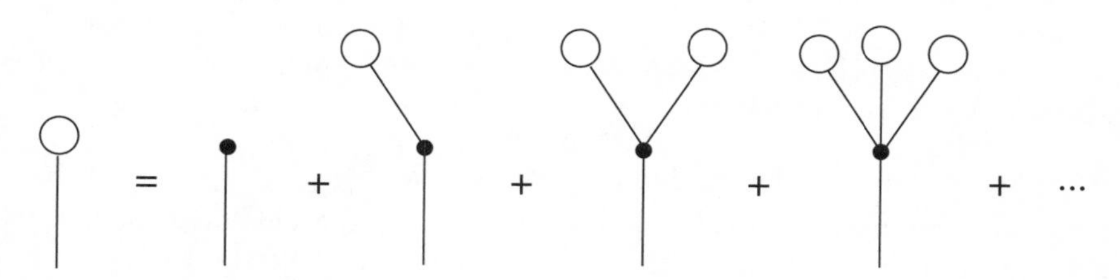

FIG. 6.5. The diagrammatic representation of the self-consistency equation for the probability x that an edge leads to a finite component. This probability is shown by a circle at the end of an edge. The degree distribution of the vertex on the right-hand side of the equation is $kP(k)/\overline{k}$.

The advanced reader may prefer the Z-transform form of these equations.[13] Then use must be made of the definitions in Appendix F.

Note that in the infinite network, finite components are certainly trees. Only the statistics of finite components were used when deriving eqns (6.13) and (6.14).

The results are especially simple when the degree distribution is Poisson. Then one can easily check that $1 - W = \exp(-\overline{k}W)$. This is a well-known result for classical random graphs (see Bollobás 1985). Therefore, in this case, near the phase transition of the birth of the giant connected component, $W(\overline{k}) \cong 2(\overline{k}-1)$, which is proportional to the deviation of the average degree from the critical value, $\overline{k} = 1$. This is the same as in standard percolation on highly dimensional regular lattices (of dimension higher than 6) where mean-field theory is exact.

At first sight this seems to be simple and uninteresting. However, the standard behaviour is realized only when a degree distribution has a rapidly decreasing tail. Let us look at eqns (6.13) and (6.14) once again. $x = 1 - \Delta$, so near the giant component birth point, $W \cong \overline{k}\Delta$, and the size of the giant component is determined by the behaviour of Δ. In principle, one obtains the latter by expanding eqn (6.14) in powers of Δ. When $\langle k^3 \rangle < \infty$, eqn (6.14) takes the form $\Delta = \ldots\Delta - \ldots\Delta^2 + \ldots$ (we do not write down all the coefficients), and so

$$W \cong 2\overline{k}\,\frac{\langle k(k-2)\rangle}{\langle k(k-1)(k-2)\rangle}\,. \tag{6.15}$$

At the birth point, $\langle k(k-2)\rangle = 0$, and the denominator is non-zero. Therefore, when we consider an arbitrary degree distribution, the value of $\langle k(k-2)\rangle = z_2 - z_1$ shows how far or close the network is to the percolation threshold. When $\langle k^3 \rangle < \infty$, the dependence on $\langle k(k-2)\rangle$ is proportional as in standard mean-field theory of percolation. But certainly something non-standard happens when the third moment diverges. In this case, the denominator on the right-hand side of eqn (6.15) diverges, and, moreover, the expansion of eqn (6.14) in Δ fails beginning from the quadratic term. Thus, in scale-free networks with $3 < \gamma \leq 4$, the giant component emerges in an unusual way. We leave the detailed discussion of this problem for a later section (see Section 6.4).

At this point we should emphasize that all these results on giant components, percolation, etc., are obtained only for equilibrium uncorrelated networks.[14] However, the general results of this theory are often applicable even to growing networks, where correlations between degrees of vertices are strong.

Distribution of finite connected components

Other, more complex statistical characteristics of these networks may be obtained in a similar way as above (Newman, Strogatz, and Watts 2001). For such

[13] These relations have a compact form in the Z-transform representation (see Appendix F). If $\Phi(z)$ is the Z-transform of the degree distribution, and $\Phi_1(z) \equiv \Phi'(z)/\Phi(1) = \Phi'(z)/z_1$, then $1 - W = \Phi(x)$, where x is the smallest, real, non-negative solution of $x = \Phi_1(x)$.

[14] We can also add 'non-clustered', although this is rather superfluous. Indeed, a high clustering means the presence of correlations.

calculations, the Z-transform technique is necessary and natural. This powerful technique can be used because of two circumstances: (1) a tree-like structure and (2) the absence of correlations between degrees of the nearest neighbours. For example, one can easily find the distributions of the numbers of the second nearest-neighbours of a vertex, of the third-nearest neighbours, and so on.[15]

To avoid technical details, in this book we discuss only one result of the application of this technique for such networks, namely the distribution of finite connected components. Indeed, we already know how the giant component behaves. Now we discuss how finite connected components are distributed when the network is near its specific point, namely in the neighbourhood of the phase transition of the birth of the giant connected component. We show our cards immediately: the situation is the same as in the standard mean-field percolation. Nevertheless, this result deserves more detailed discussion.

The standard quantity that is considered in percolation theory is the probability $\mathcal{P}(w)$ that a randomly chosen vertex belongs to a connected component of size w. One can also say that this is the probability of belonging to a finite component of w vertices. It is the quantity that is usually calculated in percolation theory, but it is not the distribution of the sizes of connected components, $\mathcal{P}_s(w)$. These two distributions are different, although they are, of course, related to each other:

$$\mathcal{P}_s(w) = \frac{\mathcal{P}(w)/w}{\sum_w \mathcal{P}(w)/w}, \qquad \mathcal{P}(w) = \frac{w\mathcal{P}_s(w)}{\sum_w w\mathcal{P}_s(w)}. \tag{6.16}$$

Near the percolation threshold, both above and below it,

$$\mathcal{P}(w) \propto w^{-3/2} e^{-w/w^*}, \tag{6.17}$$

where w^* approaches infinity at the threshold. Therefore, the power-law form $\mathcal{P}(w) \sim w^{-3/2}$ of the probability of belonging to a finite component of w vertices at the percolation threshold corresponds to the form $\mathcal{P}_s(w) \sim w^{-5/2}$ of the size distribution of connected components.

Recall that precisely the same power law for the size distributions of connected components was observed in the WWW (see Section 3.2.1). An exponent equal to 2.5 was obtained for both the distribution of weakly connected components and the distribution of strongly connected components. The WWW, however, is certainly far from the points of birth of giant connected components: all its giant components are large. So, how could these nice power-law distributions be observed? Why is the exponential cut-off of the empirical size distributions invisible? We cannot answer these questions. Instead, we recall that the WWW is a growing network and so is very distinct from equilibrium nets.

[15]If $\Phi(z)$ is the Z-transform of the degree distribution (see Appendix F), then $\Phi_1(z) \equiv \Phi'(z)/\Phi(1) = \Phi(z)/z_1$ is the degree-minus-one distribution of an end vertex of an edge. The Z-transform of the number of the first-nearest neighbours of a vertex is $\Phi(\Phi_1(z))$, the Z-transform of the number of the second-nearest neighbours is $\Phi(\Phi_1(\Phi_1(z)))$, and so on.

One can see that the first moment of $\mathcal{P}_s(w)$, that is *the average size of a finite connected component, is finite at the threshold.* However, the first moment of $\mathcal{P}(w)$ and the second moment of $\mathcal{P}_s(w)$ diverge at this point. This divergent quantity is not the average size of finite components, as is written in many papers, but the average size of the finite component to which a (randomly chosen) vertex belongs. We repeat: *it is the average size of the finite component to which a vertex belongs that diverges at the giant component birth point.* In percolation theory this quantity has the meaning of susceptibility. It also shows the value of the fluctuations of the sizes of finite components.

We showed above that the deviation from the threshold is characterized by $\langle k^2 \rangle - 2\overline{k} = z_2 - z_1$. As in standard 'mean-field' percolation,

$$\sum_w w\mathcal{P}(w) \propto \sum_w w^2 \mathcal{P}_s(w) \propto \frac{1}{|\langle k(k-2)\rangle|} \,. \tag{6.18}$$

This yields a usual mean-field value 1 for the exponent of the 'susceptibility'.

The power-law critical behaviour of $\mathcal{P}(w)$ has a number of consequences. Using simple scaling arguments as in standard percolation (Stauffer and Aharony 1991), one can estimate the sizes of the largest connected components near the percolation threshold:

(1) When the giant connected component is present in the network, the largest *finite* connected component has a size of the order of $N^{2/3}$. Actually, this is the second-largest connected component in the network.

(2) If the giant component is absent, the largest connected component has a size of the order of $\ln N$. The second-largest component is also of the order of $N^{2/3}$.

6.2 Topology of directed equilibrium networks

Recall the exciting Fig. 1.12 from Section 1.6, which schematically shows the general topological structure of directed networks. One can see a complex combination of giant in- and out-connected components, a giant strongly connected component, tendrils, disconnected components. We recommend that the reader rereads the definitions of these objects in Section 1.6. At first sight, the situation seems much more difficult than in undirected networks (compare Figs 1.11 and 1.12). However, directed networks are so important! Recall the WWW, which has the structure that is shown in Fig. 1.12. Can we really describe this topology quantitatively?

Let us again consider an equilibrium uncorrelated network but let now this network be directed. Speaking more rigorously, we consider an infinite labelled directed graph, maximally random under the constraint that its joint in-, out-degree distribution is equal to a given distribution $P(k_i, k_o)$. We already know how such graphs can be constructed (see Fig. 1.7 from Section 1.3, where a similar construction procedure is described for undirected graphs).

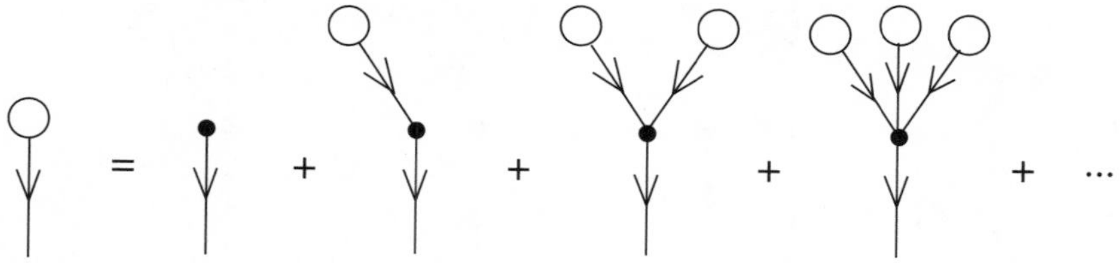

FIG. 6.6. The diagrammatic representation of the self-consistency equation for the probability x that an edge has a finite in-component. This probability is represented by a circle at the source end of an edge. The joint in- and out-degree distribution of the vertex on the right-hand side of the equation is $k_o P(k_i, k_o)/\overline{k}_{io}$.

Thus, the joint in-, out-degree distribution $P(k_i, k_o)$ of a vertex completely determines these networks. If all connections are inside a network, then the average in- and out-degrees of such a network are equal: $\overline{k}_i = \overline{k}_o \equiv \overline{k}_{io} = \overline{k}/2$. The degree distribution, which is obtained by ignoring the directness of edges, is $P(k) = \sum_{k_i} P(k_i, k - k_i)$. So, we can substitute this $P(k)$ into the formulae from the previous section and readily obtain the size of the weakly connected component, W.

Again we stress that while the topology or 'cooperative' properties of a network are considered, and the way vertices are connected with each other is important, the basic quantity of interest is not a degree distribution or a joint in- and out-degree distribution. For directed networks, it is rather a joint in- and out-degree distribution of the nearest neighbour of a vertex. Now the basic object is a directed edge. One can see that the joint in- and out-degree distribution of the target end vertex of a (randomly chosen) edge is $k_i P(k_i, k_o)/\overline{k}_{io}$. The joint in- and out-degree distribution of the source end vertex of an edge is $k_o P(k_i, k_o)/\overline{k}_{io}$.

Let us define two probabilities, x and y. x is the probability that if we start from a randomly chosen edge and move *only against* the directions of all possible edges, we cover a finite connected component. In other words, x is the probability that an edge has a finite in-component. We can write a simple equation for x, which is presented in a diagrammatic form in Fig. 6.6:

$$x = \sum_{k_i, k_o} \frac{k_o P(k_i, k_o)}{\overline{k}_{io}} x^{k_i} . \tag{6.19}$$

Analogously, y is defined as the probability that if we start from a randomly chosen edge and move *only along* the directions of all possible edges, we cover a finite connected component. In other words, y is the probability that an edge has a finite out-component. The equation for y is of the form (see Fig. 6.7)

$$y = \sum_{k_i, k_o} \frac{k_i P(k_i, k_o)}{\overline{k}_{io}} y^{k_o} , \tag{6.20}$$

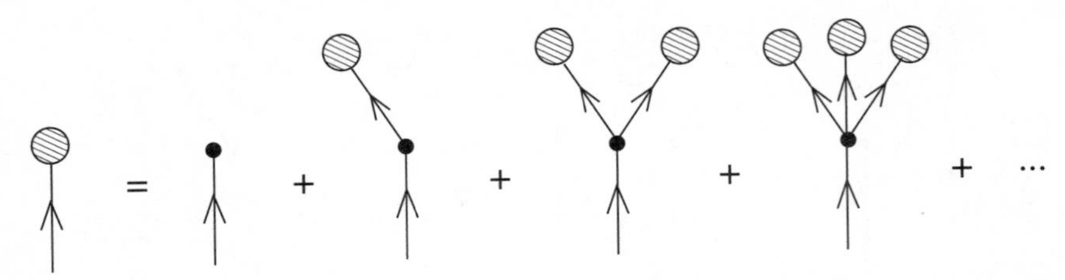

FIG. 6.7. The diagrammatic representation of the self-consistency equation for the probability y that an edge has a finite out-component. This probability is shown by the shaded circle at the source end of an edge. The joint in- and out-degree distribution of the vertex on the right-hand side of the equation is $k_i P(k_i, k_o)/\overline{k}_{io}$.

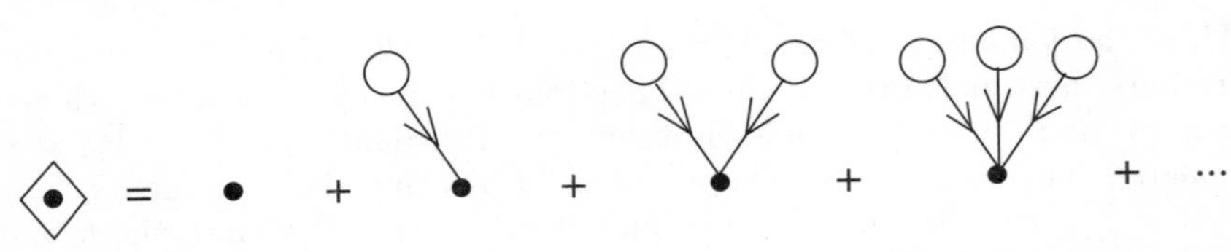

FIG. 6.8. The diagrammatic representation of the probability $1 - I$ that a randomly selected vertex of the network does not belong to the giant in-component; that is, it belongs to a finite in-component. The vertex in the diamond on the left-hand side denotes this probability. The circle at the source end of an edge shows that this edge goes out of the finite in-component with probability x. The general form of the terms on the right-hand side is $\sum_{k_o} P(k_i, k_o) x^{k_i}$.

If the only solution of eqn (6.19) is $x = 1$, the giant in-component is absent. When there exists a solution $x < 1$, the giant in-component exists. If the only solution of eqn (6.20) is $y = 1$, the giant out-component is absent. If there exists a solution $y < 1$, the giant out-component is present. One linearizes these equations about the deviations of x and y from 1 and finds that the giant in- and out-components are present, when[16]

$$\sum_{k_i, k_o} k_o k_i P(k_i, k_o) > \overline{k}_{io} \,. \tag{6.21}$$

This condition is the generalization of the Molloy–Reed criterion to the case of directed networks (Newman, Strogatz, and Watts 2001). In particular, when $\sum_{k_i, k_o} k_o k_i P(k_i, k_o)$ diverges, the giant components exist.

From x and y one can easily find the relative sizes I and O of the giant in- and out-components (see the diagrammatic representations of the expressions in Figs 6.8 and 6.9):

[16]Notice that the giant in- and out-components emerge simultaneously. We will see below that the giant strongly connected component emerges at the same point.

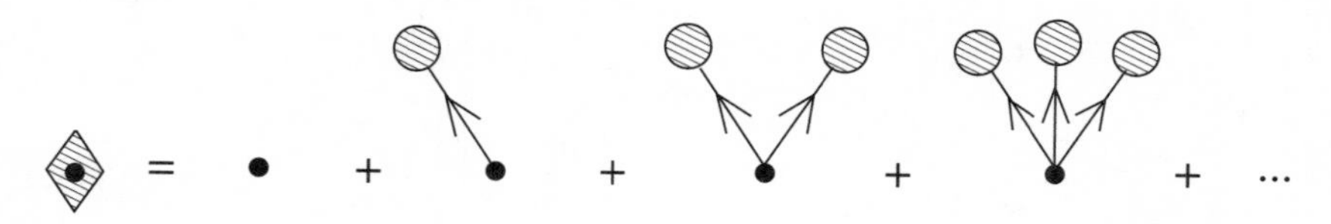

FIG. 6.9. The diagrammatic representation of the probability $1 - O$ that a randomly selected vertex of the network does not belong to the giant out-component; that is, it belongs to a finite out-component. The vertex in the shaded diamond on the left-hand side denotes this probability. The shaded circle at the target end of an edge shows that this edge comes into a finite out-component with probability y. The general form of the terms on the right-hand side is $\sum_{k_i} P(k_i, k_o)y^{k_o}$.

$$1 - I = \sum_{k_i,k_o} P(k_i, k_o)x^{k_i} , \tag{6.22}$$

$$1 - O = \sum_{k_i,k_o} P(k_i, k_o)y^{k_o} . \tag{6.23}$$

The expression for the relative size S of the giant strongly connected component is also very simple. A vertex belongs to the giant connected component if its in- and out-components are both infinite. This means that at least one of its incoming edges goes out of the giant in-component. When the in-, out-degrees of the vertex are (k_i, k_o), the probability of this event is $1 - x^{k_i}$. Also, at least one of the outgoing edges of the vertex must come into the giant out-component: the probability is $1 - y^{k_o}$. Therefore, the relative size of the giant strongly connected component takes the form (Dorogovtsev, Mendes, and Samukhin, 2001a)

$$S = \sum_{k_i,k_o} P(k_i, k_o)(1 - x^{k_i})(1 - y^{k_o}) . \tag{6.24}$$

We see that the giant strongly connected component emerges simultaneously with the giant in- and out-components. One can vary the structure of the network by changing the form of $P(k_i, k_o)$. It is convenient to use the combination $\sum_{k_i,k_o} k_o(k_i - 1)P(k_i, k_o)$ as a control parameter to vary the 'distance' from the point of emergence of these components.

So, the general structure of the directed network changes with the growing control parameter as schematically shown in Fig. 6.10 (for comparison, the evolution of the global structure of an undirected network is also shown). We stress that the point of birth of the giant weakly connected component and the point of birth of the giant in-, out-, and strongly connected components are distinct. There exists then a range of values of the control parameter, where only the giant weakly connected component is present in the network.

Now we know the relative sizes of all the giant components of the directed network. Hence, we also obtain the relative size T of the *tendrils* (see Fig. 1.12):

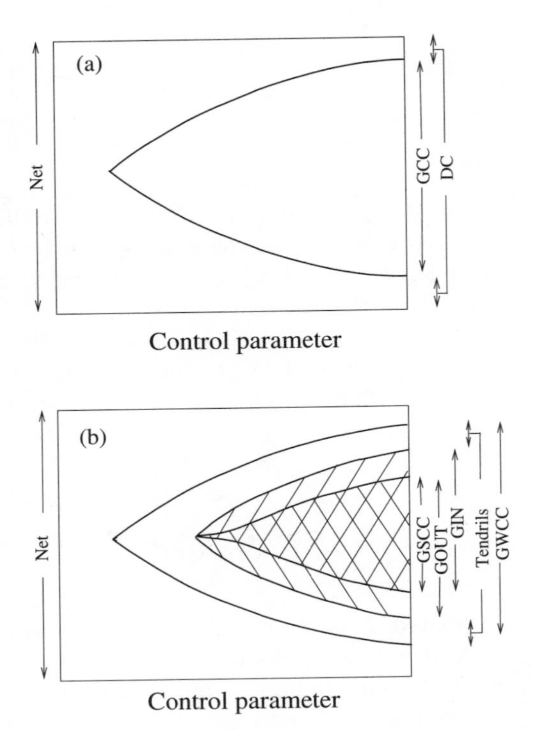

FIG. 6.10. Schematic plots of the variations of all the giant connected components versus a control parameter for the undirected network (a) and for the directed net (b). In the undirected graph, the meanings of the giant connected component (GCC), that is its percolation cluster, and the giant weakly connected component (GWCC) coincide. GIN is the giant in-component. GOUT is the giant out-component. To avoid extra discussions, we assume here that the joint in- and out-degree distribution has a sufficiently rapidly decreasing tail.

$$T = W + S - I - O\,. \tag{6.25}$$

Thus, we have described the global topological structure of these directed networks. Unfortunately, the joint in- and out-distribution of real directed networks, in particular the WWW, is still unknown. Nevertheless, something can be extracted from these formulae. The relative size (6.24) of the giant strongly connected component can be presented in the following form:[17]

$$S = IO + \sum_{k_i,k_o} P(k_i,k_o)x^{k_i}y^{k_o} - \sum_{k_i,k_o} P(k_i,k_o)x^{k_i} \sum_{k_i,k_o} P(k_i,k_o)y^{k_o}\,. \tag{6.26}$$

[17]The advanced reader can write the relations from this section in much more compelling form by using Z-transforms.

Suppose that the joint in- and out-distribution factorizes: $P(k_i, k_o) = P_i(k_i)P_o(k_i)$. Then we have $S = IO$. Such a factorization is certainly absent in real networks. Let us, however, use the last relation for a rough estimate of the size of the giant strongly connected component of the WWW. The relative sizes of the giant in- and out-components of the WWW are known: $I \approx 0.490$ and $O \approx 0.489$ (see Section 3.2.1). The we have $IO \approx 0.240$, which is a little less than the measured value $S \approx 0.277$ but is not far from it.

6.3 Failures and attacks

Let us return, for simplicity, to undirected networks. What happens with networks when somebody tries to spoil them? What is the effect of such a diversion? What is the best way to attack a net?

The integrity of a network is characterized by the presence of its giant connected component. Roughly speaking, if the giant component is absent, we do not have a net but a set of disconnected clusters. Then, to answer the above questions, first of all, one must study the variation of the giant component. The damage is not serious if it does not crucially diminish the giant component, but the damage is fatal if it eliminates the giant component.

The present chapter contains mainly general considerations of the principles of the topological organization of networks. However, at this point, we set aside all these theories and discuss the important results of simulations.

The effect of random damage and attack on communications networks (the WWW and the Internet) was simulated by Albert, Jeong and Barabási (2000). In their simulations they used:

(1) a real sample of the WWW containing 325 729 vertices and 1 498 353 links;
(2) a map of the Internet with 6209 vertices and 24 401 links;
(3) the model of a growing scale-free network with the γ exponent equal to 3 (the Barabási–Albert model); and, for comparison,
(4) the exponential growing network (see Section 5.1).

We emphasize that all the networks from this list are growing and, therefore, correlated. Recall that the networks (1) and (2) have their γ exponents in the range between 2 and 3. In these simulations, networks were treated as undirected.

Failures (random damage) were modelled by the instant removal of a fraction of randomly chosen vertices. *Intentional damage (attack!)* was described by the instant deletion of a fraction of the most connected vertices. The networks were grown and then were *instantly* damaged. The following quantities were measured as functions of the fraction f of deleted vertices:

(a) the average shortest-path length $\overline{\ell}$ of the largest connected component in a network;

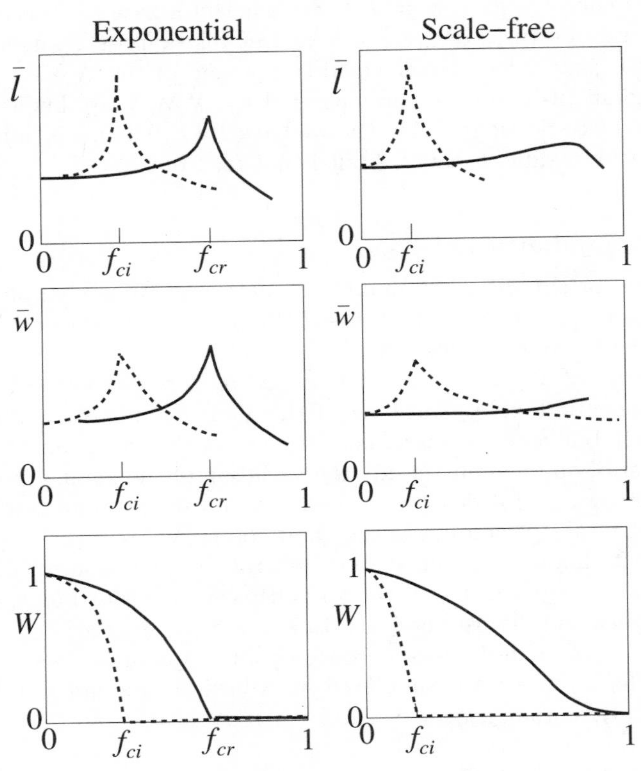

FIG. 6.11. Schematic plots of the effect of intentional and random damage, that is attack and failures, on the characteristics of the exponential network and scale-free ones with exponents $\gamma \leq 3$ (Albert, Jeong and Barabási 2000). The average shortest-path length $\overline{\ell}$ in the largest component, the size of the largest ('giant') connected component, W, and the average size of 'finite' connected components, $\overline{w}(f)$, are plotted versus the fraction of removed vertices $f \equiv 1-p$. The networks are large. The solid lines show the effect of the random damage. The effect of the intentional damage is shown by the dashed lines. f_{ci} and f_{cr} are 'percolation thresholds' for the intentional and random damage, respectively. For the exponential networks, both kinds of damage produce the same dependences, which are typical of standard 'mean-field' percolation. The effect of the intentional damage on the scale-free networks is very similar to that on exponential ones, with the difference that the giant component is destroyed at lower f. However, the random damage cannot completely eliminate a giant connected component in the scale-free networks with $\gamma \leq 3$. In this event, the percolation threshold is at the point $f \to 1$.

(b) the relative size of the largest connected component (in fact, it corresponds to the relative size W of the giant component);[18] and
(c) the average size $\overline{w}$ of connected components (excluding the 'giant' one).

The results of these simulations are schematically shown in Fig. 6.11. A striking difference between the scale-free networks and the exponential one was observed. Whereas the exponential network produces similar dependences $\overline{\ell}(f)$, $W(f)$, and $\overline{w}(f)$ for both kinds of damage,[19] for each of the scale-free nets, (1), (2), and (3), these curves are quite distinct for different kinds of damage.

The qualitative effect of the intentional damage was very similar for all the four networks. The average shortest-path length grows with increasing f as one approaches the point of disappearance of the giant component, f_c: $W(f_c) = 0$. Near this point, $S(f) \propto (f_c - f)$, as in the mean-field theory of standard percolation. At the point f_c, the average size of finite components, $\overline{w}$, has a peak of finite height, as in standard percolation (see the discussion in Section 6.1). In this respect, the behaviour is quite typical of percolation on infinitely dimensional regular lattices, or, in other words, for standard mean-field percolation. However, a generic distinct feature of the scale-free networks (1), (2), and (3) should be emphasized. The giant component in them is destroyed at a very low level of intentional damage. The values of the percolation threshold, f_c, in these nets are very low: only several percentage points, which is much less than for exponential networks. One may say that even a weak attack destroys them.[20]

Figure 6.11 demonstrates that the random damage has a far less pronounced effect on the scale-free nets than the intentional (planned) damage. The simulations show that the elimination of a giant component of the scale-free networks in such a way is not an easy task. So, these scale-free nets are extremely resilient to random damage (robust to failures). To destroy them by random ('silly') wracking, that is to disintegrate them to a set of uncoupled clusters, one has to delete practically all their vertices! One can call such effects *super-resilience, superstability, super-robustness.*

All this is certainly important and partially explains why real networks with fat-tailed degree distributions are so widespread. The next sections are devoted to the discussion of these exciting phenomena.

6.4 Resilience against random breakdowns

The reader will have noticed that the discussion of random damage is actually reduced to the problem of percolation on networks. The deletion of the randomly chosen vertices is, in principle, a classical site percolation problem.[21] Although,

[18] Here this means the relative size of the giant component in the damaged net with respect to the total number of vertices in the undamaged network.

[19] The difference is only quantitative: the removal of more connected vertices, of course, destroys the giant component at a smaller level of damage.

[20] ...weak but smart...

[21] One can easily show that the random removal of edges leads to the same result (Dorogovtsev and Mendes 2001c) but here we discuss only the site problem.

in the previous section, we discussed the effect of damage on growing (that is, correlated) networks, here we consider equilibrium uncorrelated graphs. In Section 6.1, we have mentioned that when one is interested in percolation properties, such a substitution is not criminal.

It is not strange at all that a giant connected component in fat-tailed networks is robust against the random deletion of vertices. If the degree distribution of a network is *very* fat-tailed, then just the tail determines the topology of the net. We can hope that when we remove vertices at random, we do not crucially change the tail of the degree distribution, and so the giant component is not crucially damaged. However, it is surprising that the effect is so strong.

Even these simple considerations allow us to find situations in which the network is super-resilient. Recall the Molloy–Reed criterion (6.12) for the existence of the giant connected component. If the second moment of the degree distribution, $\langle k^2 \rangle$, diverges (in the infinite network), the giant component is always present. This divergence is due to the fat tail of the degree distribution. If the damage does not change the character of this fat tail, the divergence of $\langle k^2 \rangle$ does not disappear in the damaged network. Consequently, *the giant connected component is present in the infinite, randomly damaged network if the degree distribution of the undamaged net has the divergent second moment.*

The first arguments of this kind can be attributed to Cohen, Erez, ben-Avraham and Havlin (2000). Here we use a more strict approach in the spirit of that of Callaway, Newman, Strogatz, and Watts (2000). Let us formulate the problem. Equations (6.13) and (6.14) yield the relative size of the giant component in the net with a given degree distribution, $P(k)$. What will the relative size of the giant component be in this network after the removal at random of a fraction of vertices if before the failure its degree distribution is $P(k)$? In this case, the 'relative' size W of the giant component in the damaged net is usually defined with respect to the total number of vertices in the undamaged network. We follow this tradition. Let a vertex be retained with probability p. What then should we use instead of eqns (6.13) and (6.14) after damaging the net?

In principle, from $P(k)$ we can find what the degree distribution of the damaged net is and substitute it directly into these two equations. However, we can use a simpler way. We want to stay as close as possible to the derivation of this pair of equations for a virgin network. So we introduce the probability x that an edge, randomly chosen in the undamaged network, leads to a finite connected component in the damaged network. But in the damaged network the vertex on the end of this edge, and so the edge itself, may disappear. This must be accounted for as shown at Fig. 6.12, and the consistency equation takes the form

$$x = 1 - p + p \sum_{k=0}^{\infty} \frac{kP(k)}{\overline{k}} x^{k-1} . \tag{6.27}$$

This equation can be considered in the same way as that for an undamaged network (see eqn (6.14) from Section 6.1). It has a non-trivial solution $x < 1$ when the giant component is present. This is the case if

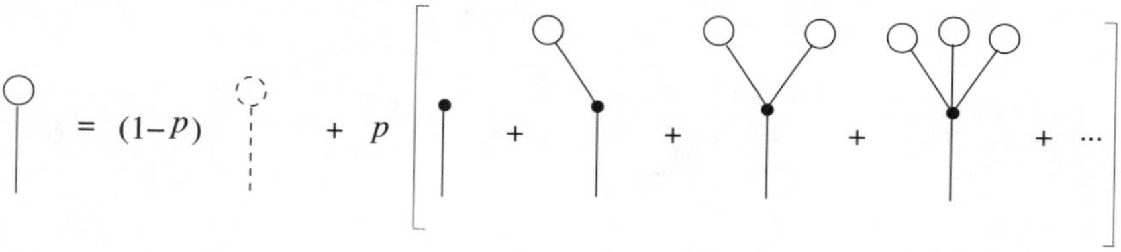

= (1−p) + p [+ + + + ⋯]

FIG. 6.12. Randomly chosen vertices (and edges that are attached to them) are eliminated with probability $1-p$. The diagrammatic representation of the self-consistency equation (6.27) for the probability x that an edge, randomly selected in the undamaged network, does not lead to the giant connected component in the damaged net. This probability is shown by a circle at the end of an edge. The end vertex is absent with probability $1-p$ (the term shown dashed). If it is present (with probability p), then the diagrammatic series are the same as for the virgin network (compare to Fig. 6.5 for $p=1$). In the last case, the degree distribution of the vertex on the right-hand side of the equation is $kP(k)/\overline{k}$.

$$\langle k^2\rangle - \frac{1+p}{p}\overline{k} > 0\,, \tag{6.28}$$

that is the generalization of the Molloy–Reed criterion to randomly damaged networks (the site percolation problem). Interestingly, the form of this criterion is the same for bond percolation. We emphasize that relation (6.28) contains the first and the second moments of the degree distribution of the *undamaged* network.[22] With the moments of the degree distribution of the *damaged* net, we must use the original Molloy–Reed criterion (6.12).

The criterion (6.28) for the existence of the giant component in the damaged net can be rewritten in the form

$$pz_2 > z_1\,, \tag{6.29}$$

where z_1 and z_2 are the average numbers of the first- and second-nearest neighbours of a vertex in the *undamaged* network. Hence the percolation thresholds for both site and bond problems are at the same point,

$$p_c = \frac{1}{(\langle k^2\rangle/\overline{k}\,) - 1} = \frac{z_1}{z_2}\,. \tag{6.30}$$

Notice the beauty of this simple formula. It is eqn (6.30) that shows that when the second moment of the degree distribution of the infinite virgin network diverges, then the percolation threshold of the damages network is zero: the giant component is super-resilient against such damage.

[22]The average degree of the damaged network is $\overline{k}' = p\overline{k}$.

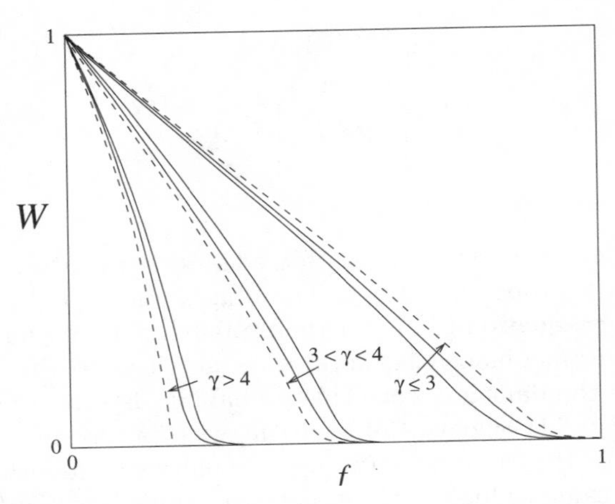

FIG. 6.13. Schematic plot of the relative size W of the largest connected component in the randomly damaged finite scale-free network versus the fraction of removed vertices, $f \equiv 1 - p$ (according to Cohen, Erez, ben-Avraham, and Havlin 2000, Cohen, ben-Avraham, and Havlin 2002a). Distinct curves correspond to different network sizes N and three values of the γ exponent: $2 < \gamma \leq 3$, $3 < \gamma \leq 4$, and $\gamma > 4$. Arrows show displacement of the curves with increasing N. Dashed lines depict the limits $N \to \infty$. Note three distinct regimes in the problem. Notice the strong size effect, which is crucial, when $2 < \gamma \leq 3$ (Dorogovtsev and Mendes 2002a). Also notice that, when $2 < \gamma < 4$, the derivative $(dW/df)(f = f_c, N \to \infty)$ is zero.

In scale-free networks, $\langle k^2 \rangle$ diverges and the the super-resilience (super-robustness, superstability) is realized for $\gamma \leq 3$. Remember, however, that scale-free distributions are only a very particular case of the fat-tailed ones. Furthermore, deterministic graphs with a divergent second moment of the discrete spectrum of degrees are also super-resilient against random damage (Dorogovtsev, Goltsev, and Mendes 2002a).

Elimination of the giant component

Thus, the position of the percolation threshold is clear. Now, how does the random damage diminish the size of the giant component? In Fig. 6.13 we schematically plot these variations. One can see that the behaviour is complex. Three distinct regimes are visible. To obtain W (recall that this is the relative size of the giant component in the damaged network with respect to the total number of vertices *in the undamaged net*) one should find the probability $1 - W$ that a vertex selected at random in the undamaged net does not belong to a giant component in the damaged network in terms of x. See the diagrammatic representation of this equation in Fig. 6.14.

Thus the equation has the form

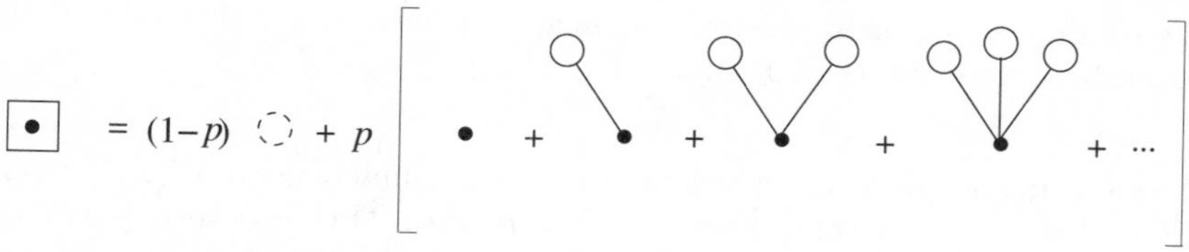

FIG. 6.14. The diagrammatic representation of the expression (6.31) of the probability $1-W$ that a vertex selected at random from the undamaged net does not belong to a giant component in the damaged network in terms of x. Here, x is the probability that an edge chosen at random in the undamaged network does not lead to the giant component in the damaged network. The vertex is absent in the damaged net with probability $1-p$ (the dashed vertex). If the vertex is present, the diagrammatic series are the same as for the virgin network (compare with Fig. 6.4 for $p=1$). In this last case, the degree distribution of the vertex on the right-hand side of the equation is $P(k)$.

$$1 - W = 1 - p + p\sum_{k=0}^{\infty} P(k)x^k \,, \tag{6.31}$$

where x must be taken from the self-consistency equation (6.27). Near the threshold p_c, where the deviation of x from 1 is small, $x \equiv 1-\Delta$, eqn (6.31) yields $W = p\bar{k}\Delta$. Consequently, the form of W in this region is determined by the behaviour of $\Delta(p)$ near p_c. We know what to do in this situation (see Section 6.1). We must expand eqn (6.27) near the threshold as a power series in Δ. Equating the coefficients of the linear terms of this series, we have already obtained the result for p_c. Now we should account for the next term of the series to find Δ. The reader can check that the first terms of the series are

$$0 = \left[1 - p\sum_{k=0}^{\sim 1/\Delta} \frac{P(k)}{\bar{k}} k(k-1)\right]\Delta + \frac{p}{2}\sum_{k=0}^{\sim 1/\Delta} \frac{P(k)}{\bar{k}} k(k-1)(k-2)\,\Delta^2 + \ldots \,. \tag{6.32}$$

We have already met a similar equation for the neighbourhood of the point of birth of the giant connected component in Section 6.1. The upper limits of the sums in eqn (6.32) are $\sim 1/\Delta$. Why? The reason is as follows. Note that the expansion of $(1-\Delta)^k$ as a power series is valid only until $k \sim 1/\Delta$. The region $k \gg 1/\Delta$, where the expansion fails and $(1-\Delta)^k \cong 1$, yields a negligible contribution to the sum $\sum_{k \gtrsim 1/\Delta} P(k)(1-\Delta)^k \cong \sum_{k \gtrsim 1/\Delta} P(k) \to 0$.

Thus, the behaviour of the giant connected component near the percolation threshold is actually described by eqn (6.32). One can see that the results depend on whether the second and third moments of the degree distribution are finite or not. For brevity, let us consider power-law degree distributions, $P(k) \propto k^{-\gamma}$. Then, in the infinite network, the third moment is finite for $\gamma > 4$ and diverges for $\gamma \leq 4$, and the second moment is infinite for $\gamma \leq 3$.

The following regimes are realized near the percolation threshold (Cohen, ben-Avraham and Havlin 2002a):

- $\langle k^3 \rangle < \infty$.
  The sums in eqn (6.32) are convergent and so are independent of Δ. Therefore, the size of the giant component is $W \propto p - p_c$. This regime is particularly realized in classical random graphs.
- $\gamma = 4$.
  Equation (6.32) takes the form $0 = \ldots\Delta + \ldots\ln(1/\Delta)\,\Delta^2$. Here we omit the p-dependent coefficients. So, $W \propto (p-p_c)/\ln[1/(p-p_c)]$.
- $3 < \gamma < 4$.
  The form of the equation is[23] $0 = \ldots\Delta + \ldots\Delta^{\gamma-4}\Delta^2$. So, $W \propto (p-p_c)^{1/(\gamma-3)}$.
- $\gamma = 3$.
  The second moment becomes divergent. We have explained that $p_c = 0$ in this situation. Now the leading terms of the equation are $0 = [1 - p\ldots\ln(1/\Delta)]\Delta$. So, $W \propto p\exp[-2/(p\bar{k})]$. To simplify the coefficient in the exponential a little, we used the continuum approximation for the degree distribution.
- $2 < \gamma < 3$.
  The leading terms are $0 = [1 - p\ldots\Delta^{\gamma-3}]\Delta$, so $W \propto p^{1/(3-\gamma)}$.

Figure 6.13 illustrates these behaviours.[24] As the tail of the degree grows fatter, p_c decreases, and the phase transition becomes more and more continuous. When $\gamma = 3$, all the derivatives of W at the threshold are infinite, that is the phase transition is of infinite order.

Since the critical behaviour of the size of the giant component differs sharply from standard percolation dependences, the reader might conclude that the other basic characteristic, the average size of the finite connected component to which a vertex belongs[25] also behaves non-standardly. This is not quite the case. Cohen, ben-Avraham, and Havlin (2002a) showed that for $\gamma > 3$, this quantity has exactly the same divergence $\propto |p-p_c|^{-1}$, as in standard mean-field percolation. So this dependence is present even in the region $3 < \gamma \le 4$ where the size of the giant component already behaves unusually. Only when $\gamma \le 3$ and the percolation threshold is absent ($p_c = 0$), is the behaviour, indeed, non-standard: the average size of the finite component with a vertex is proportional to p.

Thus, the situation in networks with a fat-tailed distribution of connections is very different from that for percolation on classical random graphs and infinite-dimensional regular lattices. We emphasize that in this section we discuss infinite networks. The reader will learn what happens in finite nets in Section 6.7.

[23] In fact, the situation is a little more complex. The terms omitted in eqn (6.32) also make a contribution $\sim \Delta^{\gamma-2}$, but this only changes the uninteresting constant factor.

[24] While considering these regimes we, naturally, assume that, in the undamaged network, the giant connected component is present.

[25] Recall that this average size plays the role of susceptibility in percolation, and diverges at the percolation threshold, unlike the average size of the finite connected components, which has no divergence at the percolation threshold (see Section 6.1).

6.5 How viruses spread within networks

Computer viruses are a more frequent evil in life than influenza. The WWW and electronic mail are a surprisingly convenient environment for the existence and transmission of, perhaps, the most dangerous diseases of modern times. Why do they spread so easily?

Classical epidemiology could not answer this question, since its main interest focuses on the spread of disease within more traditional systems: mostly within systems without long-range connections, roughly speaking, within lattices. However, real epidemics spread not on lattices but within networks: friendship networks, networks of collaborations, networks of sexual contacts, etc. What are the specifics of networks in the context of the spread of disease? What is the role of the fat-tailed degree distributions?

There are two basic models of the spread of diseases in theoretical epidemiology: the SIS model (*susceptible–infective–susceptible*) and the SIR model (*susceptible–infective–removed*). These models are generically related to the percolation phenomena. But first we discuss the spread of diseases without resorting to the methods from percolation theory.

SIS model

In the SIS model, vertices can be in two distinct states: healthy and ill (see Fig. 6.15). The vertices never die.

(1) Each healthy (susceptible) vertex becomes infected at rate ν when it has at least one infected neighbour.

(2) Infective vertices are cured, that is become susceptible, at rate δ.

Hence, the main parameter of the model is the effective spreading rate $\lambda = \nu/\delta$. We denote the fraction of the susceptible vertices as $S(t)$ and the fraction of infective vertices as $\rho(t)$, so $S(t) + \rho(t) = 1$.

Suppose a vertex falls ill. The standard problem of epidemiology is how the infection spreads. Without going into details, depending on the value of λ, two scenarios are possible: (1) The disease does not spread far and disappears by itself, $\rho \equiv \rho(t \to \infty) = 0$. (2) The disease becomes endemic, and at long times,

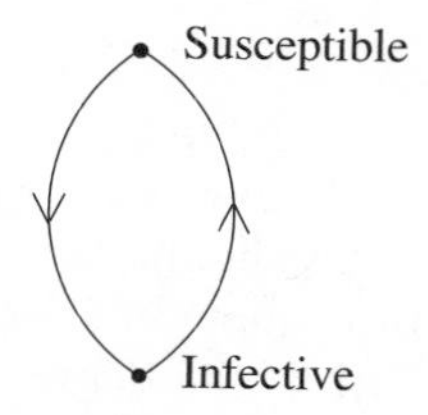

FIG. 6.15. Two states of vertices in the SIS model: susceptible and infective. Susceptible vertices can become infected if they have infected neighbours, but infective vertices recover from time to time and again become susceptible.

a finite fraction of vertices is infected, $\rho > 0$. We are interested only in the final state, and so, to be brief, let us set $\delta = 1$.

The basic notion in epidemiology is an epidemic threshold above which the epidemic spreads and becomes endemic. In the event of the SIS model, this is the value $\lambda = \lambda_c$ above which $\rho > 0$. In the SIS model on regular or inhomogeneous lattices, the epidemic threshold is always present, $\lambda_c > 0$. In networks, the situation is more complex. The arguments of Pastor-Satorras and Vespignani (2000, 2001) for *equilibrium uncorrelated undirected networks* are as follows.

In the naive mean-field approach,[26] the stationary state of a network is described by a balance equation. This simple equation contains the probability $\rho(k)$ that a vertex of degree k is infective in the final state: $\rho = \sum_k P(k)\rho(k)$. The balance condition is obvious: the number of vertices of degree k, which recover per unit time (the left-hand side), is equal to the number of vertices of degree k, which become ill (the right-hand side). The latter rate is proportional (with the constant factor λ) (1) to the degree k of such a vertex, (2) to the probability $S(k) = 1 - \rho(k)$ that it is susceptible, and, in addition, (3) to the probability θ that a randomly chosen nearest neighbour of a randomly chosen vertex is infective, or, in other words, the probability that a randomly selected edge has an infected end vertex. Note that θ is assumed to be independent of k. Consequently, the balance equation is of the form

$$\rho(k) = \lambda k S(k) \theta \,. \tag{6.33}$$

Recalling that the degree distribution of the nearest neighbour of a vertex is $kP(k)/\overline{k}$ yields $\theta = \sum_k [kP(k)/\overline{k}]\rho(k)$. From these relations, the equation for θ follows immediately:

$$\theta = \sum_{k=0}^{\infty} \frac{kP(k)}{\overline{k}} \frac{\lambda k \theta}{1 + \lambda k \theta} \,. \tag{6.34}$$

In fact, this derivation used the tree-like structure of the network and the absence of correlations between vertices.

Above the epidemic threshold, this equation has a non-zero solution. As before, to obtain λ_c, we linearize eqn (6.34) about $\theta = 0$. Finally, we get the epidemic threshold

$$\lambda_c = \frac{\overline{k}}{\langle k^2 \rangle} \,. \tag{6.35}$$

But this mean-field expression is very close to the exact formula (6.30) for the percolation threshold in randomly damaged networks. When $\langle k^2 \rangle \gg \overline{k}$, these relations are of the same form. This should not be surprising, since we have mentioned the similarity between the spread of disease and percolation.

When $\langle k^2 \rangle < \infty$, that is the tail of the degree distribution is not 'too fat', the epidemic threshold is finite, quite similarly to that for the spread of disease within lattices. However, *in an infinite network, when the second moment of the*

[26] In this particular situation, such an approach is sufficient.

degree distribution diverges, the epidemic threshold is absent, $\lambda_c = 0$ (Pastor-Satorras and Vespignani 2000). In particular, in *infinite* scale-free networks, this happens when the γ exponent is less than or equal to 3. Most real scale-free nets have just these values of the γ exponent (see Chapter 3). So, the absence of the epidemic threshold is of primary importance.

We stress that, although the infinite scale-free networks with $\gamma \leq 3$ are extremely robust against random damage (see the previous section), they are, simultaneously, incredibly sensitive to the spread of infections. Both these, at first sight, contrasting phenomena have the same origin, the fat tail of the degree distribution. In other words, the infinite networks with fat-tailed distributions of connections are, simultaneously, super-robust and superweak. A single 'sneeze' can infect the entire WWW. Moreover, infections spread like a bush fire because of the small-world effect. Then, why are we still alive? We shall give the answer in Section 6.7.

Suppose the epidemic threshold is exceeded. How does the final number of 'ill vertices' ρ depend on λ? To find the answer, one must obtain θ from eqn (6.34), substitute it into eqn (6.33), obtain $\rho(k)$, and, finally, $\rho(\lambda)$. The form of the resulting dependences $\rho(\lambda)$ practically coincides with that of $W(p)$ for random damage (see the previous section).[27] All the regimes are the same, and the reader can use the last formulae from Section 6.4 after the change $p \to \lambda$, $p_c \to \lambda_c$. Figure 6.13 can also be used after the change $f \to 1-\lambda$. Again, only for $\langle k^3 \rangle < \infty$ does $\rho(\lambda)$ behave in the same way as in the spread of disease within infinitely dimensional lattices.

Thus, in the SIS model on infinite uncorrelated scale-free networks with $\gamma \leq 3$, the epidemic threshold is absent. Is this valid only for the SIS model?

SIR model

In this model, the vertices may be in three states (see Fig. 6.16): susceptible, infective, and removed (dead or recovered, that is unsusceptible). 'Removed' vertices do not pass the disease on. The rules are as follows:

(1) Each susceptible vertex becomes infected with some probability per time step, if it has at least one infective neighbour.
(2) Infective vertices becomes removed with some probability per time step.

One can see that this model is close to the SIS model. However, notice that infective vertices never become susceptible (see Fig. 6.16). Furthermore, the SIR model is closely related to the bond percolation problem (Grassberger 1983).[28] The epidemic threshold in this model (Newman 2002b) coincides with the threshold for the bond percolation problem on networks. In the previous section we mentioned that the formulae for the bond and site percolation threshold are the

[27] Meanwhile, these results for the spread of disease were obtained earlier than for percolation.

[28] In the final state of the network, only susceptible and removed vertices are present. Indeed, in this model, the infective state of a vertex is only temporal; finally, the vertex becomes removed. If your nearest neighbour dies before infecting you, he or she will never infect you.

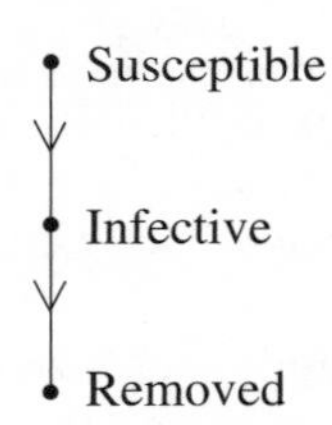

FIG. 6.16. Three states of vertices in the SIR model: susceptible, infective, and removed. Susceptible vertices can become infected if they have infected neighbours; infective vertices become removed from time to time, but removed vertices never change their state—they do not pass the disease on.

same. Then eqn (6.30) also yields the epidemic threshold for the SIR model. Naturally, the results for the relative size of the finite fraction of removed vertices (Moreno, Pastor-Satorras, and Vespignani 2002) are again very similar to those for the percolation problem.

Thus, the absence of the epidemic threshold for networks with the divergent second moment of the degree distribution is a rule. These results were obtained for equilibrium uncorrelated networks, but, as for percolation phenomena, they seem also to be valid for growing nets.

In the present section we have only mentioned the spread of diseases, but evidently, many other processes within networks must have similar features. We mean the transmission of some excitations from vertex to vertex, drug-trafficking, the spreading of rumours, and so on.

The two characteristic features of the 'modern architecture' of networks, namely their super-robustness against failures and the super-vulnerability to diseases, pose an unpleasant dilemma for network designers: should a net be robust against interferences and failures or should it be safe against diseases? Meanwhile, the first network designer, nature, faces the same dilemma. It is very possible that the structure of many natural networks is the result of the resolution of this contradiction.

6.6 The Ising model on a net

The Ising model is one of the immortal themes in physics. It is a traditional starting point for the study of cooperative behaviour and various types of ordering. Suppose that the various vertices interact with each other. What is the effect of this interaction? Let us discuss the specifics of such cooperative phenomena in networks by using, as an example, the Ising model.

We put a spin, $S_i = \pm 1$, on each vertex of the *infinite equilibrium uncorrelated network* with an arbitrary distribution of connections. Suppose that the nearest-neighbour spins interact ferromagnetically. Consider the Ising model with ferromagnetic interactions between nearest neighbours, which is described by the Hamiltonian

$$\mathcal{H} = -J \sum_{\langle ij \rangle} S_i S_j - H \sum_i S_i \,. \tag{6.36}$$

Here, $J > 0$ is the energy of the ferromagnetic interaction between the nearest-neighbour spins, and H is the magnetic field. The first sum is over all the pairs of nearest-neighbour vertices, the second one is over all the vertices in the network.

As usual, below a critical temperature T_c spins are ordered ferromagnetically. Above the critical temperature, the state is paramagnetic. What is the temperature of the phase transition in the Ising model on networks? We again resort to intuitive arguments from Section 6.1 (see Fig. 6.3).

One can use the following circumstance. Everywhere, except at the critical temperature, correlations between spins decay exponentially. Consequently, it is the 'local' structure of the network (see Figs 6.1 and 6.2 from Section 6.1) that determines the cooperative behaviour. So, to estimate roughly the critical temperature, we can use the same trick as in the estimation of the birth point of a giant connected component. We mimic the local structure of the network by the Bethe lattice with a coordination number $q = \langle k^2 \rangle / \overline{k}$ (see Fig. 6.3). Recall that $\langle k^2 \rangle / \overline{k} = 1 + z_2/z_1$ is the average degrees of vertices from the local environment of a randomly chosen vertex.[29] z_1 and z_2 are the average numbers of the first- and second-nearest-neighbours of a vertex. We have found that this reckless approach provides the proper result for the point of birth of the giant component. So, let us try it once again. The exact formula for the critical temperature of the Ising model on the Bethe lattice is $T_c/J = 2/\ln[q/(q-2)]$ (Baxter 1982).[30] Substituting the coordination number q from above into this formula yields

$$\frac{J}{T_c} = \frac{1}{2} \ln \frac{\langle k^2 \rangle}{\langle k^2 \rangle - 2\overline{k}} = \frac{1}{2} \ln \frac{1 + z_1/z_2}{1 - z_1/z_2} \tag{6.37}$$

(see Fig. 6.17). This naive estimate is exact!

Further, the relation between the average numbers of the first- and second-nearest neighbours, z_1 and z_2, and T_c may be written in an even shorter form:

$$\tanh \frac{J}{T_c} = \frac{z_1}{z_2} \,. \tag{6.38}$$

We do not present here the derivation of exact results for the critical behaviour of the Ising model (see Dorogovtsev, Goltsev, and Mendes 2002b and Leone, Vázquez, Vespignani, and Zecchina 2002). However, these results deserve a detailed discussion.

[29] To repeat, interactions are transmitted through edges, from vertex to vertex. Hence, it is the average degree of the nearest neighbour, $\langle k^2 \rangle / \overline{k}$, that is important and not the average degree of a vertex, $\overline{k}$. One can see that the use of the latter would lead to a completely wrong result for networks.

[30] The exact results for critical phenomena on the Bethe lattice coincide with the Bethe-Peierls approximation, so that this approximation can be reasonably used in network theory.

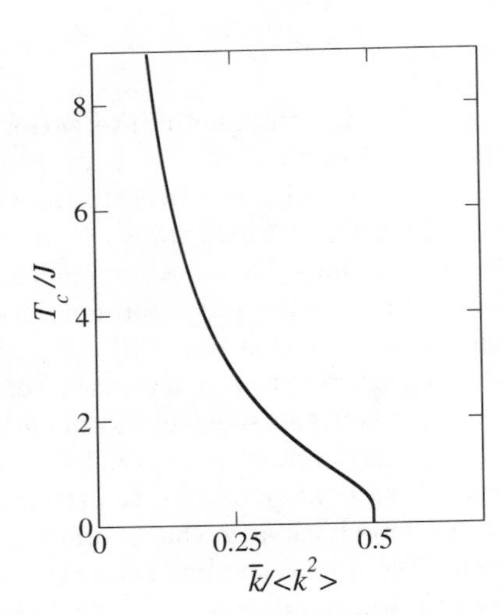

FIG. 6.17. Critical temperature of the Ising model on the infinite equilibrium uncorrelated network with an arbitrary degree distribution versus the ratio of the first and second moments of the degree distribution, $\overline{k}/\langle k^2 \rangle$ (see eqn (6.37)). When $\overline{k}/\langle k^2 \rangle > 1/2$, the giant connected component is absent in the network. When $\langle k^2 \rangle$ diverges, T_c approaches infinity.

First of all, notice that at the birth point of the giant component ($z_2 = z_1$), T_c is, naturally, zero (see Fig. 6.17). Also, notice that, when the second moment of the degree distribution is large, $\langle k^2 \rangle \gg \overline{k}$, the expression for the critical temperature takes the form $T_c/J \cong \langle k^2 \rangle/\overline{k}$. If the second moment of the degree distribution is finite, this leads to a finite T_c, but when the tail of the degree distribution is sufficiently fat, and $\langle k^2 \rangle$ diverges (in the *infinite* network), *the critical temperature turns out to be infinite!* In other words, *when the second moment diverges, any finite temperature does not completely destroy the ferromagnetic ordering!*

How can this be? Is it not surprising that in the system with a finite energy of interaction between the nearest-neighbour spins and a finite average number of connections, the temperature of the phase transition is infinity?

Let us try to recall where the infinite T_c is possible. It is well known that if the ferromagnetic interactions J between each pair of spins in a system are equal, that is the interaction is infinitely ranged, then the mean-field theory is exact, and $T_c \sim JN$. Here $N \to \infty$ is the total number of spins in the system. One can consider this as a ferromagnetic phase transition on a fully connected network. So, we have one more example of the 'infinite critical temperature'. However, this example, where the number of connections of each vertex is infinite, $N \to \infty$, has no relation to our situation. In our case, contrastingly, $\overline{k}$ is finite, and the average energy of a ground state per spin is also finite.

Table 6.1 *Critical behaviour of the magnetization M, the specific heat δC, and the susceptibility χ in the Ising model on networks with a power-law degree distribution $P(k) \sim k^{-\gamma}$ for various values of exponent γ. When the fourth moment is finite, the results are given for an arbitrary degree distribution. Above the critical temperature, $\delta C = 0$. This (and the perfect Curie law) demonstrates that the so-called critical fluctuations are absent. The results for M, δC, and χ in the range of the divergence of the second moment, $\langle k^2 \rangle = \infty$, are given for $T \gg J$. The far right column presents the exact critical temperature in the case $\langle k^2 \rangle < \infty$ and, when $\langle k^2 \rangle = \infty$, the dependence of T_c on the total number N of vertices in a network (see Section 6.7).*

	M	$\delta C(T < T_c)$	χ	T_c
$\gamma > 5$, $\langle k^4 \rangle < \infty$	$\propto \sqrt{T_c - T}$	jump at T_c decreases as $\langle k^4 \rangle$ grows	$\propto \lvert T_c - T \rvert^{-1}$	$2J/\ln \frac{\langle k^2 \rangle}{\langle k^2 \rangle - 2\overline{k}}$
$\gamma = 5$, $\langle k^4 \rangle = \infty$, $\langle k^2 \rangle < \infty$	$\propto \sqrt{(T_c - T)/\ln(T_c - T)^{-1}}$	$\propto 1/\ln(T_c - T)^{-1}$		
$3 < \gamma < 5$, $\langle k^4 \rangle = \infty$, $\langle k^2 \rangle < \infty$	$\propto (T_c - T)^{1/(\gamma-3)}$	$\propto (T_c - T)^{(5-\gamma)/(\gamma-3)}$		
$\gamma = 3$, $\langle k^2 \rangle = \infty$	$\propto e^{-2T/\overline{k}}$	$\propto T^2 e^{-4T/\overline{k}}$	J/T	$\propto \overline{k} \ln N$
$2 < \gamma < 3$, $\langle k^2 \rangle = \infty$	$\propto T^{-1/(3-\gamma)}$	$\propto T^{-(\gamma-1)/(3-\gamma)}$		$\propto \overline{k} N^{(3-\gamma)/(\gamma-1)}$

Thus, our situation is truly non-standard and, moreover, unprecedented. However, since we now know so much about networks, there is no great reason to be so surprised. We have already explained that vertices in the 'local' environment of a vertex have quite a different structure of connections than the network as a whole. The degree distributions of these vertices have fatter tails than the degree distribution of the entire network. In the infinite network, these vertices may have a divergent average number of connections, even when the average number of connections in the entire network is finite. So, the divergence of the average degrees of neighbours of a vertex is routine. These 'local' properties of the network determine the critical behaviour, and so the divergence of T_c is not a mystery.

The behaviour of the main thermodynamic quantities in the Ising model is presented in Table 6.1 and Fig. 6.18. We show the temperature dependences of the magnetization M, the magnetic susceptibility χ, and the anomalous critical contribution to the specific heat δC. The table reveals a sharp difference from the standard Ising model on lattices. Let us look at it attentively.

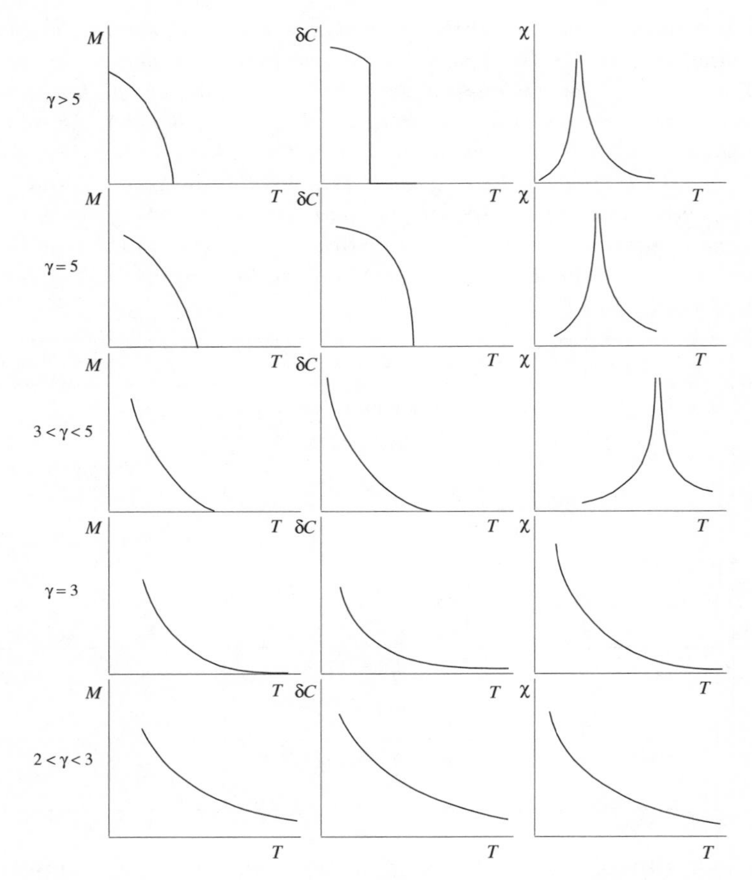

FIG. 6.18. Schematic plots of the evolution of the temperature dependences of the main thermodynamic quantities of the Ising model on infinite scale-free networks with decreasing γ. The temperature dependences of the magnetization M, critical contribution to the specific heat δC, and magnetic susceptibility χ are shown in the critical region. Compare with Table 6.1.

- We see that the so-called critical fluctuation phenomena are absent. Indeed: (1) $\delta C(T > T_c) = 0$, that is any fluctuation corrections are absent, and (2) the Curie law is perfect: when the critical temperature is finite, $\chi \propto |T - T_c|^{-1}$. So, the behaviour is mean-field-like, but these mean-field dependences are very non-standard.
- Notice that there is no rather customary peak of the specific heat at the critical temperature.
- Only when the fourth moment of the degree distribution is finite is the critical behaviour of the thermodynamic quantities the same as in the standard mean-field solution of the Ising model on lattices.

- When the fourth moment diverges, the behaviour changes crucially. One may use power-law degree distributions for convenient parametrization. Then, for $\gamma \leq 5$, the critical behaviour is abnormal. *As γ decreases from 5 to 3, the critical temperature grows, and the phase transition becomes more and more continuous.* Note that the last term is quite rigorous. The reader can see that the exponents of the expressions for the magnetization, $M \propto (T_c - T)^{1/(\gamma-3)}$, and the specific heat, $\delta C \propto (T_c - T)^{(5-\gamma)/(\gamma-3)}$, get higher and higher.[31] Below $\gamma = 4$, these exponents exceed 1. In the traditional Ehrenfest classification, this means that the phase transition is of order higher than second.
- Moreover, at the point $\gamma = 3$, where T_c becomes equal to infinity, the phase transition is of infinite order! All the derivatives of the specific heat and the magnetization are zero at $T_c = \infty$.
- In the whole region of the divergent second moment, that is for $2 < \gamma \leq 3$, the critical temperature is infinite and $\chi \propto 1/T$ as in a paramagnet.[32] The magnetization and specific heat are decreasing functions of temperature.
- In this region, size effects are strong (see the discussion in Section 6.7).

In summary, one may say that fat tails of degree distributions increase the critical temperature and make the phase transition more continuous. Some of the above exact results for $\gamma = 3$ were compared with the data obtained in the simulations of Aleksiejuk, Holyst, and Stauffer (2002) for the Ising model on the Barabási–Albert network, and an excellent agreement was found. And this is despite the fact that, strictly speaking, this theory is for equilibrium uncorrelated networks and not for growing ones![33]

All these features (see the list above, Table 6.1, and Fig. 6.18) resemble those of the percolation phenomena on networks (Section 6.4) and the spread of disease within them (Section 6.5). The reader can consult the corresponding sections and find that these critical behaviours are surprisingly close to each other. What is the reason for such closeness?

Everybody has heard about the Ising model. However, there exists a little less well-known but essentially more general model with discrete symmetry. We mean the n-state Potts model. The Ising model is only a particular (two-state) case of the Potts model. The latter has one more important particular case. The basic result of percolation due to Kasteleyn and Fortuin (1969) and Fortuin

[31] Loosely speaking, the critical exponent of specific heat $C(T)$ is defined as follows, $\delta C(T) \propto |T - T_c|^{-\alpha}$. So, in our case, α is negative, and we actually speak about the growth of the modulus of this critical exponent.

[32] Notice this 'paramagnetic' temperature dependence of the susceptibility in the ferromagnet phase.

[33] Recall that in the Barabási–Albert model, degree–degree correlations are weak, which is an exception for growing networks (see Section 5.12). However, the growth of networks provides a wide spectrum of correlations, and the Barabási–Albert model by no means can be called uncorrelated. In the equilibrium nets that were considered in this chapter, vertices are statistically equivalent. Contrastingly, in the growing networks each vertex has its age, and the resulting non-equivalence (and correlations), in principle, can produce unconventional critical behaviour (see Section 6.10).

and Kasteleyn (1972) is that *the site percolation problem is equivalent to the one-state Potts model.* The rule of the mapping is $p = 1 - e^{-2J/T}$. Here p is the probability of the percolation problem that a site is occupied, and J and T are the coupling in the one-state Potts model and temperature, respectively. Naturally, this equivalence is also valid for percolation on networks. One can see that $p_c = 0$ in the percolation problem corresponds to the infinite critical temperature in the one-state Potts model, and so on. Thus, in all the last sections, we considered the critical behaviour on networks of models from the same class. Hence, the observed similarity is not surprising.

But that is not all. The results that we speak about in these sections were obtained for graphs with a tree-like structure in the environment of a randomly chosen vertex (see Fig. 6.1). But it is commonplace in the physics of phase transitions that *critical fluctuations are impossible in the absence of loops.* So, in such a situation, mean-field approaches are quite adequate. Although the critical phenomena in networks differ dramatically from standard 'mean-field' behaviour in lattices, *all these non-standard phenomena, actually, have a very simple mean-field nature*, without critical fluctuations. We saw that the critical exponents for networks may differ sharply from the standard mean-field ones for phase transitions on lattices. This difference is not because of the critical fluctuations as in low-dimensional lattices. The reason is the fat-tailed distribution of connections, which makes the critical behaviour fascinating even without critical fluctuations.

Thus, for these networks, mean-field-like solutions are valid. But, in fact, mean-field theory provides a narrow set of behaviours. In the case of lattices, the set of mean-field exponents is dull, and it is critical fluctuations that produce such a rich variety of critical exponents for various models. If critical fluctuations are not essential, diversity is practically absent, and the critical behaviour of various models is very similar. Hence, the observed intriguing type of behaviour is not the specificity of the two or three closely related models on equilibrium uncorrelated graphs, but it is typical of general critical phenomena in a wide circle of networks (Goltsev, Dorogovtsev, and Mendes 2002, see also Appendix G).

6.7 Mesoscopics in networks

All real networks are finite. However, up to now we have mostly considered the so-called 'thermodynamic limit', that is infinite networks. The reader may be excited by strong results above, but are they valid for real finite networks? Some of these results were truly exciting. Recall:

- The disintegration of the scale-free network with exponent $\gamma \leq 3$ of the degree distribution by random damaging is practically impossible; to eliminate its giant component, one has to delete all vertices (Sections 6.3 and 6.4).
- There is no epidemic threshold in scale-free networks with exponent $\gamma \leq 3$. Any diseases spread easily within such networks (Section 6.5).
- The critical temperatures of various phase transitions in scale-free networks with exponent $\gamma \leq 3$ are infinite (Section 6.6).

So, the claims are strong. But then, *why are we still alive?* Indeed, most of the observed scale-free networks in nature (see Fig. 3.32 and Table 3.7 from Section 3.11) have a γ exponent between 2 and 3. If the claim about the absence of the epidemic threshold in these networks is true, pandemics would never stop. In short, all these seem, at the very least, strange.

We emphasize that all these effects are obtained only for *infinite* networks. *In finite networks, these three statements are incorrect.* Finite networks are not super-resilient against random damage. The epidemic threshold exists in finite networks. Phase transitions in finite networks have a finite critical temperature.[34]

So, the finiteness of networks is not a fact of secondary importance, and it is utterly incorrect to say: 'The main effects may be obtained for infinite networks, and finite-size effects are only some corrections and subtleties for a subsequent study.' Moreover, in these networks with a 'modern architecture', that is with fat-tailed degree distributions, finite-size effects are crucial. These nets cannot be understood if we forget that they are finite. So, *they are basically mesoscopic objects.*

Finite-size effects

Let us consider the position of the three critical points from the above list in equilibrium uncorrelated scale-free networks with $\gamma \leq 3$ (Dorogovtsev and Mendes 2002a). We mean the percolation threshold p_c for the random damage, the epidemic threshold λ_c, and the critical temperature T_c of the Ising model.[35] When the second moment of the degree distribution is finite but much greater than the average degree, all these three quantities take the same form:

$$J/T_c \cong \lambda_c \cong p_c = \frac{\overline{k}}{\langle k^2 \rangle} \cong \frac{z_1}{z_2}\,. \tag{6.39}$$

Consequently, if the second moment of the degree distribution of the infinite network diverges, then $p_c(N=\infty) = \lambda_c(N=\infty) = J/T_c(N=\infty) = 0$. However, in finite networks, the degree distribution has a size-dependent cut-off, and so the second moment is finite. We have estimated $\langle k^2 \rangle$ by the expressions (6.4) and (6.5) for $\gamma < 3$ and $\gamma = 3$, respectively, in Section 6.1. Rigorously speaking, relations (6.39) were obtained for infinite networks, but let us take the next step: let us estimate these values in finite networks by substituting the estimate for the second moment in these finite networks into eqn (6.39). Of course, this is not rigorous, but this is only an estimate which will be compared with simulations, so we have a chance to check ourselves.

[34] The alert reader has certainly noticed that, rigorously speaking, the notions of the percolation and the epidemic thresholds and the critical temperature are inapplicable to finite systems, since, in finite systems, real phase transitions are absent and a real ordering is impossible. The authors are aware of this absolute fact. Of course, here, we can speak only about large enough samples to observe some parody of phase transitions with slightly smeared anomalies.

[35] Note that the considerations below are valid for a much wider circle of critical phenomena.

Then we have

$$J/T_c(N) \cong \lambda_c(N) \cong p_c(N) \approx \frac{(3-\gamma)(\gamma-1)}{(\gamma-2)^2\,\overline{k}} N^{-(3-\gamma)/(\gamma-1)} \quad \text{for} \quad 2<\gamma<3 \tag{6.40}$$

and

$$J/T_c(N) \cong \lambda_c(N) \cong p_c(N) \approx \frac{4}{\overline{k}\ln N} \quad \text{at} \quad \gamma = 3\,. \tag{6.41}$$

Here we thoroughly wrote out the factors involved, which where obtained by using the continuum approximation for the degree distribution. These factors are necessary for comparison with the data of simulations of scale-free networks.[36]

The readers may not be quite satisfied by the quality of our estimate, so some extra arguments are necessary to convince them. All the critical points, p_c, λ_c, and T_c, can be extracted from a number of simulations of various networks. For example, p_c can be estimated from the data of the simulations of equilibrium uncorrelated scale-free graphs (Cohen, Erez, ben-Avraham, and Havlin 2000a), λ_c can be found from the simulation of the Barabási–Albert model (Dezső and Barabási 2002), and $T_c(N)$ of the Ising model on the Barabási–Albert model was measured by Aleksiejuk, Holyst, and Stauffer (2002). The agreement turns out to be more than convincing. For example, the simulation of Aleksiejuk, *et al.* yields $T_c/J \approx 2.6 \ln N - 3$ over a wide range of sizes for the network with average degree $\overline{k} = 10$. The estimate (6.41) gives $T_c/J \approx 2.5 \ln N$. In fact, this seems to be even 'too close' to the simulation, since the estimate is rough, but what can we do? Furthermore, it is not the first time that we note that the estimate was obtained for equilibrium uncorrelated graphs, but the agreement with the results for growing, that is correlated, networks, is excellent.

This finite-size effect is most pronounced in networks with $\gamma = 3$. For example, if $\gamma = 3$ and and the average degree is $\overline{k} = 6$, then $p_c(N = 10^4) \approx 0.07$, $p_c(N = 10^6) \approx 0.05$, and $p_c(N = 10^9) \approx 0.03$. Nevertheless, for smaller γ, it is also strong. When γ is close to 3, the thermodynamic limit can be approached only for an unthinkably huge net. Even if, say, $\gamma = 2.5$, and the net is very large, we get noticeable values of the threshold (see eqn (6.40)). For example, $p_c(N = 10^6) \sim 10^{-2}$ and $p_c(N = 10^9) \sim 10^{-3}$.

We showed above that the finite-size effect is of primary importance. Are there situations where it may not be essential? Let us return to eqn (6.40). When γ is close to 2, the exponent $-(3-\gamma)/(\gamma-1)$ is close to -1, $p_c(N)$ decreases with growing N rapidly enough, and then the thermodynamic limit behaviour can be observed for quite reasonable sizes of networks. Inspection of Fig. 3.32 and Table 3.7 from Section 3.11 shows that there exist a number of real scale-free networks with such an exponent γ. The Internet and some basic biological networks are in this region, and it is these networks that are robust and super-resilient against random damage but defenseless against 'a single sneeze'.

[36] The epidemiological aspect of the finite-size effect is discussed in more detail by May and Lloyd (2001) and Pastor-Satorras and Vespignani (2002a).

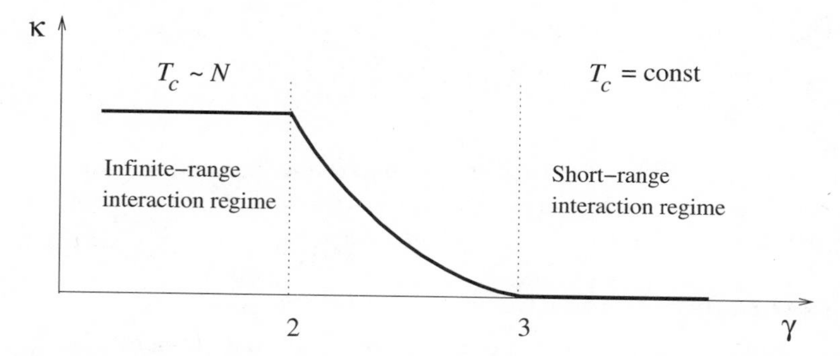

FIG. 6.19. Exponent κ from the size dependence of the critical temperature, $T_c \propto N^\kappa$, versus the γ exponent of the degree distribution of a network. The size dependence of the critical temperature of the ferromagnetic phase transition on scale-free networks evolves from $T_c(N) \propto N$ (the 'fully connected network' regime) to $T_c(N) = \text{const}$ (the standard behaviour of systems with short-range interactions). Nevertheless, the entire crossover takes place in the narrow range of the degree distribution exponent: $2 < \gamma < 3$.

We are discussing finite-size effects in the range $2 < \gamma \leq 3$. There is one more aspect of the problem. While discussing the Ising model in Section 6.6 we mentioned that a ferromagnetic phase transition on a fully connected network has a critical temperature proportional to the total number of vertices of the network. On the other hand, critical temperatures of phase transitions in systems with short-range interactions are independent of N. So there exist these two marginal behaviours. Roughly speaking, when we change γ from 0 to ∞, we cross from a fully connected net to the short-range interaction regime. Nevertheless, one can see that the entire crossover takes place in the 'very narrow' range $2 < \gamma < 3$ (see Fig. 6.19), where $T_c \propto N^{(3-\gamma)/(\gamma-1)}$.

Mesoscopic fluctuations

In this book we discuss mesoscopic objects. Mesoscopic physics is a well-populated discipline. However, in network science, mesoscopic effects are an open field. As we have touched on this topic, let us make a short digression from global topology problems. What is the standard mesoscopic approach? What should be studied in finite networks?

Recall our discussions from Section 4.1. Let a graph g be a particular realization of a statistical ensemble of random networks, each one of size N. Suppose that the number of vertices of degree k in this graph is $N(k, g)$. Then the degree distribution of the ensemble is $P(k) = \langle N(k, g)\rangle / N$. Here the averaging is over all realizations g of the ensemble. Most of the time, researchers in network science study this average $\langle N(k, g)\rangle$ or generically related quantities.

Now, what is the matter of interest of mesoscopic physics in this, actually standard, situation? For each realization g we must find $N(k, g) - \langle N(k, g)\rangle$ and calculate (or measure) the fluctuations $\langle [N(k, g) - \langle N(k, g)\rangle]^2\rangle$. Usually, it is worthwhile to consider the ratio

$$\frac{\langle [N(k,g) - \langle N(k,g)\rangle]^2 \rangle}{\langle N(k,g)\rangle^2} .$$

At the very tail of the degree distribution, these fluctuations are essentially strong, and everybody in network science tries to get rid of them (see Appendix B), but nobody has, as yet, studied them.

6.8 How to destroy a network

The discussions in Sections 6.3 and 6.4 demonstrate that random sabotage (random damage, indifferent damage, and so on) is not an effective way of destroying a network with a fat-tailed degree distribution. Furthermore, it is an *extremely* inefficient way, if the second moment of the degree distribution diverges. The obvious efficient method is to 'attack' a network and eliminate its most connected vertices (see Sections 6.3). Let us discuss such 'attacks' (intentional damage) in more detail.

We saw in Section 6.4 that the random damaging of a network, is, in fact, the site percolation problem for the network. Vertices (with all their connections) are removed at random with probability $f \equiv 1 - p$. Then, if the degree distribution of the undamaged network is $P(k)$, the size of the giant component is given by eqns (6.27) and (6.31), which are represented in graphical form in Figs 6.12 and 6.14.

For percolation on regular lattices, such a formulation (the random deletion of, for example, sites) is the only possibility, since all sites of a regular lattice are equal. Contrastingly, in networks, distinct vertices may have quite different properties. For example, they have various numbers of connections (degrees). Therefore, the following generalization of the percolation problem is natural. Let us delete the vertices (with all their connections) not at random but intentionally: a vertex of degree k is deleted with probability $f(k) \equiv 1 - p(k)$. Again, the degree distribution of the undamaged, equilibrium uncorrelated infinite network is given: $P(k)$.

In the same way as in Section 6.4 (eqns (6.27) and (6.31) and Figs 6.12 and 6.14) one can easily obtain the pair of equations for the size of the giant connected component (Callaway, Newman, Strogatz, and Watts 2000):

$$x = 1 + \sum_{k=0}^{\infty} \frac{kP(k)}{\overline{k}} p(k)(x^{k-1} - 1) , \tag{6.42}$$

$$W = \sum_{k=0}^{\infty} P(k)p(k)(1 - x^k) . \tag{6.43}$$

As previously shown, to find the size of the giant connected component, one must substitute the non-trivial solution $x < 1$ of the first equation from this pair into the second equation. The giant connected component exists only if this solution is present, that is for

$$\sum_{k=0}^{\infty} k(k-1)P(k)p(k) > \sum_{k=0}^{\infty} kP(k)\,. \qquad (6.44)$$

With this criterion, one can study the general effects of the damaging of the networks. We have considered the random (indifferent) damaging: that is, the particular case of the degree-independent probability that a vertex is retained, $p(k) = p = \text{const}$. The other marginal particular case is the deletion of the most connected vertices. This means that the probability $p(k)$ is zero when degree k exceeds some value: $p(k > k_{max}) = 0$, and $p(k \leq k_{max}) = 1$.

Let us attack a network in such a manner and delete a fraction f of its vertices—all its most connected vertices. Then the vertices of degrees greater than $k_{max}(f)$: $f = 1 - \sum_{k=0}^{k_{max}(f)} P(k)$ will be deleted. Naturally, we attack our favourite network: the scale-free net. To be specific, we choose the network with the perfect power-law degree distribution for each degree and without isolated vertices. Using the criterion (6.44) immediately gives the result presented in Fig. 6.20 (Dorogovtsev and Mendes 2001c, Newman 2002a). One can see that the deletion of a very small fraction of vertices already disintegrates the scale-free net, if the most connected vertices are removed. Indeed, the maximum of f_c in Fig. 6.20 is only at several percentage points. This is in contrast to classical random graphs, where there is no great difference between the effects of intentional and random damage (see Section 6.3).

At first sight, one might conclude that the attack on a scale-free net is an easy undertaking. However, this is not quite as straightforward as it sounds. Indeed, we must delete only a percentage of vertices, but one can check that the removal of this percentage means the deletion of each vertex of degree greater than k_{max} of the order, say, 10, which is a very small number. In this sense, the damage should be tremendous. In Section 3.2.1, we saw that even the deletion of all hyperlinks to vertices of in-degree higher than 3 does not destroy the giant weakly connected component of the WWW.

The result of the attack is not universal: the dependence in Fig. 6.20 is determined by the entire degree distribution. Furthermore, the low-degree part of the degree distribution may be crucial. If, for example, dead ends are absent, then the giant component is always present in the undamaged network, in contrast to Fig. 6.20.

Equations (6.43) and (6.44) allow us to find the variation of the size of the giant component near the threshold. In the event of the deletion of the most connected vertices (Callaway, Newman, Strogatz, and Watts 2000), this variation is quite similar to that in standard mean-field percolation: W is proportional to the deviation from the percolation threshold.

Using these equations, one can easily study the effect of any kind of damage on infinite networks, which is determined by the probability $p(k)$. The results may be very different from the two marginal cases that we discussed.

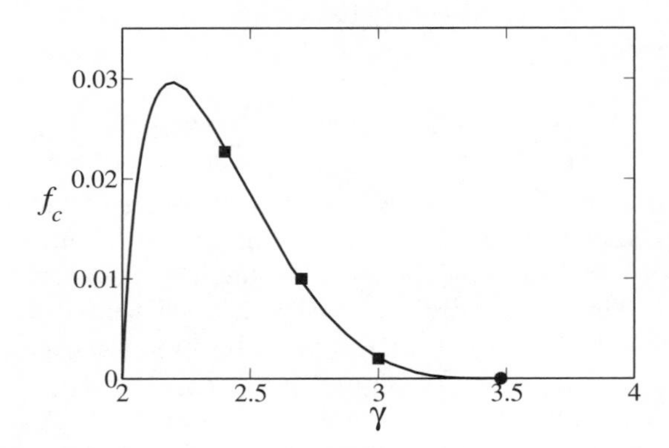

FIG. 6.20. The fraction f_c of the most connected vertices in an infinite scale-free network, which must be deleted to disintegrate the network, versus the γ exponent of the degree distribution (Dorogovtsev and Mendes 2001c, Newman 2002a). The disintegration of the network means the elimination of its giant connected component. Such a dependence was originally obtained in the framework of a continuum approach (Cohen, Erez, ben-Avraham, and Havlin 2001a). Here we present the exact curve. Isolated vertices are absent in the network. The circle indicates the point $\gamma = 3.479\ldots$ above which there is no giant component in the undamaged network with this degree distribution (see Section 6.1). The squares represent the results of calculations and simulation by Callaway, Newman, Strogatz, and Watts (2000).

6.9 How to stop an epidemic

We have explained that the problem of the spread of disease within networks is very close to percolation problems (see the discussion in Section 6.5). So, the considerations from Sections 6.4 and 6.8 can be directly applied to the spread of diseases. We only have to change the terminology from these sections. The deletion of a vertex now means its immunization, so that it cannot pass on the virus. The probability $p(k)$ that the vertex of degree k is still alive after the attack now means the probability that it is not immunized and can pass on the infection. The percolation threshold is now the epidemic threshold. The formulae for these thresholds are practically the same.

From a formal point of view, the fight against epidemics is a package of measures directed at increasing the epidemic threshold. The only measures we can take are immunization and isolation of individuals (vertices). These measures are actually equivalent. Our discussion in Section 6.4 readily shows that if the second moment of the degree distribution of a network diverges, the random immunization of vertices cannot increase the epidemic threshold and is totally ineffective.

The most effective way to stop an epidemic in such a network is *the targeted immunization* of vertices (Pastor-Satorras and Vespignani 2002b, 2002c, Moreno,

Pastor-Satorras and Vespignani 2002, and Dezső and Barabási 2002). Similarly to the previous section, the immunization of highly connected vertices allows one to increase the epidemic threshold without great expenses.

This is indeed very simple and obvious. However, it is hardly obvious for epidemiologists that we have to immunize 100% of vertices to stop an epidemic in these networks if we carry out random immunization, and so targeted immunization seems to be the only effective possibility.

Unfortunately, very often the necessary information for the targeted immunization is absent: the highly connected vertices of a network are unknown. There is a simple way to approach a good result in this situation (Cohen, ben-Avraham, and Havlin 2002b). Recall that the average degree of randomly chosen nearest neighbours of randomly chosen vertices, $\langle k^2 \rangle / \overline{k}$, in uncorrelated networks is greater than the mean degree $\overline{k}$ of a network. Moreover, for $\gamma \leq 3$, $\langle k^2 \rangle / \overline{k}$ is much greater than $\overline{k}$. So, the immunization of the randomly chosen nearest neighbours of randomly selected vertices produces nearly the same effect as the targeted immunization of highly connected vertices.[37]

6.10 BKT percolation transition in growing networks

The reader will certainly have noticed that percolation on networks is a non-standard phenomenon. We emphasize that up to now we have considered percolation on equilibrium uncorrelated networks. In this 'simple' situation, the theory is well developed. But what is the situation for growing nets?

Two distinct formulations of the percolation problem for growing networks are possible:

(a) First, grow the network, then abruptly remove from it a fraction of vertices or edges, and immediately study the effect. To study percolation here means to study the variation of the giant and finite connected components of the resulting network versus the fraction of removed vertices or edges. As a rule, the simulations of such immediate damage do not show a great difference from percolation on equilibrium uncorrelated networks. The reasons for this similarity are not quite clear, but, usually, nothing unexpected happens in this case.

(b) The network grows under permanent damage (constant sabotage), which is, in this case, an intrinsic component of the evolution of the net. Varying this factor provides networks with various 'densities' of connections. In this event, to study percolation means to study the variation of the giant and finite connected components of the network in the long-time asymptotic regime versus the 'intensity' of ceaseless sabotage.[38]

[37] In this case, the absence of correlations between degrees of nearest-neighbour vertices is a significant circumstance. This immunization strategy is effective in uncorrelated networks, but its effect in correlated nets essentially depends on the form of correlations.

[38] There is a case, where these two kinds of damage coincide (see below), and the instant damage also produces non-trivial effect.

The latter case (Callaway, Hopcroft, Kleinberg, Newman, and Strogatz 2001, Dorogovtsev, Mendes, and Samukhin 2001b) is surprising and is crucially different from percolation on equilibrium networks. Let us discuss it.

So, consider the emergence of the giant connected component in a growing network under the variation of growth conditions. As Callaway, Hopcroft, Kleinberg, Newman, and Strogatz (2001), we use the simplest model (see Section 1.1):

(1) At each time step, a new vertex is added to the network.

(2) Simultaneously, b new undirected edges emerge between pairs of randomly chosen vertices (b may be non-integer).

One can see that the main parameter of the problem, b, plays the role of a control parameter, which allows us to vary the structure of the network, change the size of the giant component, approach the point of its birth, and so on.

The attachment of new edges is not preferential. Consequently, the degree distribution is exponential (see Section 5.1). We emphasize that this exponential form is valid for any value of the parameter b. Our aim is to find the distributions of connected components in this net. The matter of interest is the probability $\mathcal{P}(w,t)$ that a randomly chosen vertex belongs to a connected component of w vertices at time t. Let us obtain the master equation for $\mathcal{P}(w,t)$.

Up to now we derived equations only for degree distributions, but the derivation of the master equation for $\mathcal{P}(w,t)$ is very similar. At first, consider the average number $\langle N(w,t)\rangle$ of connected components of w vertices in the network at time t. For the large network, it is easy to write the master equation for this quantity:

$$\langle N(w,t+1)\rangle = \langle N(w,t)\rangle + \delta_{w,1} + b\sum_{u=1}^{w-1}\frac{u}{t}\frac{w-u}{t}\langle N(u,t)\rangle\langle N(w-u,t)\rangle$$
$$- 2b\frac{w}{t}\langle N(w,t)\rangle\,. \tag{6.45}$$

The terms on the right-hand side of this equation have the following nature:

(1st) The first term is obvious.

(2nd) At each time step a connected component of size 1 is added (an extra isolated vertex), which just means the Kronecker symbol.

(3rd) The number of components of size w increases because of the fusion of pairs of connected components into components of w vertices. This occurs when a new edge connects a pair of components with the overall number of vertices equal to w. Note that the probability of a new edge becoming attached to a component is proportional to its size.

(4th) The number of components of size w diminishes by 1 when a new edge becomes attached to one of vertices of such a component. The factor 2 is due to the simple fact of an edge having two ends with each one having to be accounted for.

We do not have to account for the possibility that both the ends of a new edge become attached to vertices of the same finite connected component: in large networks, this probability is negligible. Now, using the evident relation $\mathcal{P}(w,t) = w\langle N(w,t)\rangle/t$ readily yields

$$t\frac{\partial \mathcal{P}(w,t)}{\partial t} + \mathcal{P}(w,t) = \delta_{w,1} + bw\sum_{u=1}^{w-1}\mathcal{P}(u,t)\mathcal{P}(w-u,t) - 2bw\mathcal{P}(w,t)\,. \quad (6.46)$$

This is a basic master equation for the evolution of connected components in growing networks. Compare it with the master equations for degree distributions in such nets (see, for example, eqn (5.17) from Section 5.2). The principal difference is that eqn (6.46) is non-linear. The origin of this non-linearity is the fusion of finite components that we discussed. Of course, non-linear equations are more complex than linear ones, but this equation, in fact, is very simple. Indeed, the non-linearity is present here in the form of a convolution. Such equations are ideal for the treatment by the Z-transform technique. The original analysis was numerical (Callaway, Hopcroft, Kleinberg, Newman, and Strogatz 2001) but one can also obtain all the main results for giant and finite connected components exactly[39] (Dorogovtsev, Mendes, and Samukhin 2001b).

This analysis was carried out in the large-network limit, where the probability $\mathcal{P}(w,t) = \mathcal{P}(w)$ is stationary. The most surprising result is the size of the giant connected component, which is present in the network for $b > b_c = 1/8$:

$$W(b) = 0.590\ldots\exp\left(-\frac{\pi}{2\sqrt{2}}\frac{1}{\sqrt{b-b_c}}\right)\,. \quad (6.47)$$

This dependence is also shown in Fig. 6.21(a). One can see that all the derivatives of $W(b)$ are zero at the critical point. Thus, *the phase transition is of infinite order*. We have met situations with infinite-order phase transitions in equilibrium networks, but those anomalies occurred in a very particular case: at a single point where the second moment of the degree distribution starts to diverge. In scale-free networks, this corresponds to $\gamma = 3$. Now we are observing something even more intriguing. Also, notice the unusual type of singularity at the critical point.

The behaviour of other characteristics is no less interesting. Recall that the role of susceptibility in percolation is played by the average size w_1 of a finite connected component that contains a randomly chosen vertex. This is the first moment of $\mathcal{P}(w)$, $w_1 \equiv \sum_w w\mathcal{P}(w)$, and is very distinct from the average size of a finite component (see the discussion in Section 6.1). In standard percolation and in percolation on equilibrium networks, w_1 diverges. This is standard behaviour for phase transitions. Contrastingly, in the present situation, w_1 *stays finite at the critical point* (see Fig. 6.21(b)).

Now we turn to the distribution $\mathcal{P}(w)$, that is to the distribution of sizes of finite connected components, to which a vertex belongs. In standard percolation

[39] Equation (6.46) is derived for finite components, but the reader will recall that the size of a giant connected component also follows from the equations for finite components.

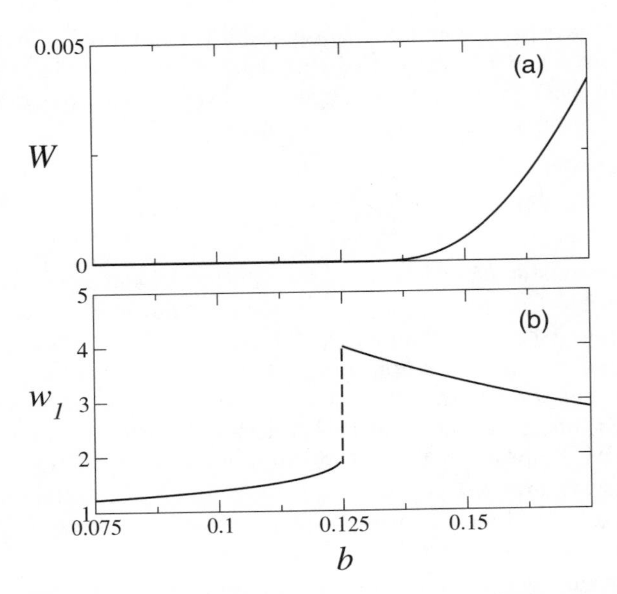

FIG. 6.21. The Berezinskii–Kosterlitz–Thouless percolation transition in the growing network. (a) The size W of the giant connected component versus the rate b of the creation of new edges in the growing network (see eqn (6.47)). The point of birth of the giant component is $b_c = 1/8$. Notice that all the derivatives of $W(b)$ are zero at the critical point. (b) The average size w_1 of a finite connected component that contains a randomly chosen vertex versus b (Callaway, Hopcroft, Kleinberg, Newman, and Strogatz 2001). Notice that $w_1 \equiv \sum_w w\mathcal{P}(w)$ is finite at the critical point, unlike standard percolation and percolation on equilibrium networks.

on lattices and in percolation on equilibrium networks, the situation is classical (see Section 6.1). At the critical point, $\mathcal{P}(w)$ is of a power-law ('critical') form, and on both sides of the threshold this dependence is cut off exponentially ('non-critical behaviour'). This 'critical power law' in standard mean-field percolation and in percolation on equilibrium random graphs is $\mathcal{P}(w) \propto w^{-3/2}$.

In the percolation problem that we are considering now, the distribution at the point of birth of the giant component is of the form

$$\mathcal{P}(w) \sim \frac{1}{w^2 \ln^2 w} \tag{6.48}$$

at large w. At first sight, the difference is not great, but this is an illusion. The reader must notice that the first moment of this distribution is finite, unlike the situation for equilibrium nets.

In the phase with a giant component, the distribution is, roughly speaking, an exponentially decreasing function, $\propto \exp[-w/w_0(b-b_c)]$, as in percolation on equilibrium networks. Here, $w_0(b-b_c)$ turns out to be zero at the critical point, and is proportional to the size of the giant component near b_c. However, the behaviour in the 'normal' phase without a giant component dramatically differs from that one might expect. *In the entire phase without a giant component, the distribution $\mathcal{P}(w)$ is a power-law function.* The exponent of this function gradually grows from 2 to ∞ as b decreases from the critical value to 0.

We emphasize that this behaviour is unprecedented for percolation phenomena. The power-law form of a distribution means criticality, and here this sign of criticality is observed not at an isolated critical point but in the entire phase. Furthermore, in the model that we discuss here the preferential linking is absent: edges become attached without any preference and the degree distribution is exponential. In a similar way, we can consider a network growing under the mechanism of preferential linking, that is with a power-law degree distribution (Dorogovtsev, Mendes, and Samukhin 2001b). The results are very similar to those for the exponential net. The formula (6.48) remains valid. In the expression (6.47) for the size of the giant component, the constant factors and the critical point b_c become dependent on the γ exponent of the degree distribution.[40] We see that such behaviour is of a general nature.

Thus, we face an extremely surprising phenomenon. How can all this be explained? First, let us look for similar critical phenomena in other fields of physics—maybe some parallels will be useful. In fact, a close analogy exists. There is a very important specific case for phase transitions in marginal situations with unusual critical behaviour. We mean the Berezinskii–Kosterlitz–Thouless (BKT) phase transition (Berezinskii 1970, Kosterlitz and Thouless 1973) in the two-dimensional XY model. Generally, this phase transition takes place in systems with the continuous symmetry of a two-component order parameter, when they are at their lower critical dimension. In other words, below this dimension, any long-range order is absent at any temperature, and there is no phase transition. Above the critical dimension, the critical behaviour is quite standard, and so the BKT is present just at the point where the regime changes.

The BKT transition has the following main unusual features:

(1) Above some temperature T_c, the pair correlations between magnetic moments decrease exponentially $\sim \exp[-r/\xi(T)]$ (r is the distance between the moments), and the correlation length near the phase transition behaves as

[40]However, there is a significant difference between these two networks. The net growing under the mechanism of random linking posesses a unique property: the instant and permanent deletion of edges are equivalent. The reason is the fact that the attachment of new edges is independent of the state of the net. The instant removal of a fraction p of randomly chosen edges from the network (with the parameter b) growing under the mechanism of random linking leads to the same result as the growth with the parameter $b' = pb$ and, consequently, to the same phase transition. Furthermore, other cooperative phenomena in networks growing under the mechanism of random linking show the same BKT critical singularities as in eqns (6.47) and (6.48). In particular, this has already been observed for the spread of disease (the SIS model) by Boguñá and Pastor-Satorras (2002). Contrastingly, in networks growing under the mechanism of preferential linking, the instant and permanent damage are not equivalent.

$$\xi(T) \propto \exp\left(-\frac{\text{const}}{\sqrt{T - T_c}}\right) \tag{6.49}$$

(compare with eqn (6.47)).

(2) Below and at the phase transition temperature, the correlations between magnetic moments are power law. Thus, the entire low-temperature phase is in the critical state, and so, very often, the region $T \leq T_c$ is called 'the line of critical points'.

We schematically show this situation in Fig. 6.22(a).

Now we compare this with the 'non-equilibrium percolation' in growing networks. In Fig. 6.22 we schematically present the characteristic features for both

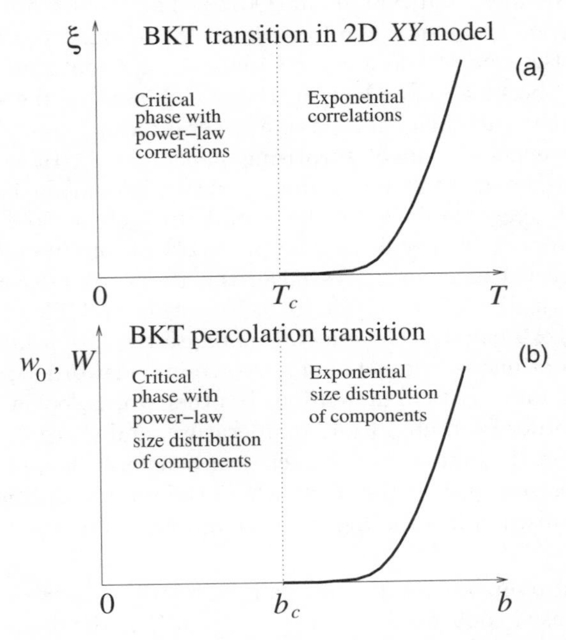

FIG. 6.22. The parallels between the classical BKT phase transition (a) and the non-equilibrium percolation on growing networks (b). (a) For the classical BKT transition, we schematically show the correlation length ξ versus temperature. Note that all the derivatives of ξ at T_c are zero. Below and at the phase transition point, the pair correlations decrease as a power law, and so the entire low-temperature phase is in the 'critical state'. (b) For the non-equilibrium percolation, we show the size of the giant connected component and the scale w_0 of the exponential decay of the distribution $\mathcal{P}(w) \propto \exp[-w/w_0]$ versus the control parameter b. All the derivatives of W and w_0 at b_c are zero. Below and at the phase transition point, the size distribution of connected components decreases as a power law.

the phase transitions that we discuss. The similarity is quite visible. Note that networks have no metric structure, and so correlations between locations of their vertices cannot be described in the way that is used for models on lattices. Nevertheless, these correlations can be naturally characterized by the distribution $\mathcal{P}(w)$, and so by the size distribution of connected components. Thus, the structure and the law of the variation of correlations are practically the same in both cases. In both cases, a 'critical phase' contacts the normal phase. This, evidently, yields such a kind of critical singularities. Once again, since the claim is important: (a) The contact of two phases with exponentially decreasing correlations at a single critical point, where correlations decrease as a power, corresponds to a 'standard' critical singularities. (b) The contact of a critical phase (the power-law decrease of correlations in the whole phase) with a normal phase (the exponential decrease of correlations) yields singularities of the BKT type. Then the non-equilibrium phenomenon that we consider may be called the BKT percolation transition.

However, we have not yet explained the most interesting thing: why is it that the entire phase without the giant component turns out to be in the critical state? We know that preferential linking (preferential attachment of edges to vertices) produces power-law degree distributions in growing networks. This takes place over a wide range of parameters of a network. One may call this a 'self-organized criticality'.

Now, let us return to our present problem where we are interested in the growth of finite connected components, and so in the attachment of new edges not to vertices but rather to components. Obviously, the larger the component, the higher its probability to get a new connection and to grow. Recall eqn (6.45). One can see that in the exponential growing network (that is, the indifferent attachment of edges to vertices), this probability is proportional to the size of a connected component. This 'preference' naturally exists not just in exponential or scale-free growing networks. Here, we only stress that it does not relate to the preferential attachment of edges to vertices.

Thus, the origin of a power-law form of the size distribution of connected components is rather similar to the origin of scale-free degree distributions. For the distribution of components, it is the preferential growth of large components and the preferential fusion of larger components. As for scale-free degree distributions, power-law distributions of connected components exist over a wide range of parameters. This *self-organized critical state* is realized in the growing networks only if the giant connected component is absent, that is when $b < b_c$.

As soon as the giant connected component emerges, the situation changes radically. A new channel for the evolution of connected components comes into play. With a high probability, large connected components do not grow up to even larger ones but join the giant component. This channel decreases the number of large finite components and produces the exponentially decreasing $\mathcal{P}(w)$ in the presence of the giant component.

Thus, the phenomenon can be interpreted quite naturally. It should be present in many growing networks. The BKT percolation transition was also observed in other growing nets (see the models of Kim, Krapivsky, Kahng, and Redner (2002) and Bauer and Bernard (2002) for biological networks).

6.11 When loops and correlations are important

In many cases in this chapter, results which were obtained for equilibrium uncorrelated networks, taking into account the tree-like structure of the environment of a vertex, turned out to be valid for networks with numerous loops and correlations between degrees of their vertices. We mentioned that, as a rule, cooperative phenomena in networks and percolation on them do not depend dramatically on whether or not the correlations are present.[41] Many times in this chapter we used the scheme of the network in Fig. 6.1 without accounting for the far regions with numerous loops.

However, correlations and loops may be important in many other cases.[42] We finish this chapter with a very brief indication of situations where the presence of correlations and loops is crucial.

Distribution of shortest-path lengths

(1) From Section 6.1, we know that the average shortest-path length $\overline{\ell}$ in equilibrium uncorrelated networks with a given degree distribution depends on the total number of vertices as $\sim \ln N/\text{const}$. The constant here is determined by the form of the degree distribution. When the second moment of the degree distribution diverges, the constant approaches infinity and this relation is no longer valid. However, this divergence indicates that the size dependence $\overline{\ell}(N)$ should be a function increasing slower than $\ln N$ (the 'supersmall-world effect').[43] Even if, in scale-free equilibrium nets, the γ exponent exceeds 3, the constant in the expression for the average shortest-path length may be much greater than the average degree $\overline{k}$.

(2) Contrastingly, numerous simulations and calculations for various growing scale-free networks (for example, see Section 5.14) show that $\overline{\ell}$ is close to $\ln N/\overline{k}$, even if $\gamma \leq 3$. Hence, the formula for classical random graphs is valid for these strongly correlated networks.

Distribution of traffic

The betweenness was introduced in Section 1.5. In fact, this quantity characterize traffic through a vertex. Its distribution $\mathcal{D}(\sigma)$ characterizes the distribution of traffic in the network.

(1) This distribution was calculated analytically for scale-free trees by Szabó, Alava, and Kertész (2002) and Goh, Oh, Jeong, Kahng, and Kim (2002). The

[41] In this respect we must exclude the BKT percolation transition, which can only occur in growing networks (see the previous section).

[42] Recall that loops and clustering are specific correlations.

[43] In this situation, the functional form of $\overline{\ell}(N)$ is determined by the dependence of the position of the cut-off of the degree distribution on N.

result is $\mathcal{D}(\sigma) \propto 1/\sigma^2$ for various scale-free trees.[44] So, in a large network of this type, a finite fraction of all the shortest paths passes through one or several vertices (hubs). In other words, one or several vertices control traffic over the network.

(2) Also, this distribution was obtained from the simulations of scale-free non-tree-like networks by Goh, Kahng, and Kim (2001b) and Goh, Oh, Jeong, Kahng, and Kim (2002). In such an event, the distribution $\mathcal{D}(\sigma)$ noticeably deviates from the $1/\sigma^2$ law: $\mathcal{D}(\sigma)$ decreases with σ more rapidly than the $1/\sigma^2$ dependence.[45] In this case, in contrast to scale-free trees, hubs that keep a finite fraction of traffic are absent. Here we again suppose that a network is large.

Both these cases are related to shortest paths, and in both cases, the difference between behaviours (1) and (2) is great.

[44] Goh, Oh, Jeong, Kahng, and Kim (2002) obtained the exact asymptotic behaviour of the distribution $\mathcal{D}(\sigma) \propto 1/\sigma^2$ for growing scale-free trees.

[45] Note, however, that in three real networks (the WWW and Internet and the metabolic networks of the species of Archaea), despite the presence of loops and appreciable clustering, the behaviour close to $\propto 1/\sigma^2$ was observed (Vázquez, Pastor-Satorras, and Vespignani, 2002, Goh, Oh, Jeong, Kahng, and Kim 2002).

7

GROWTH OF NETWORKS AND SELF-ORGANIZED CRITICALITY

Several chapters of this book have been devoted to the problem: how do networks arrive at their complex structure? That is, how does the 'modern architecture' of networks arise? We have given the answer: the origin of this architecture is the self-organization of networks. In this chapter we shall discuss several other processes, generically related to the self-organization of networks.

7.1 Preferential linking and the Simon model

Fat-tailed distributions and the particular case of power-law distributions have been observed in various situations in various systems for many years. And for many years, scientists strove to find their nature. An important advance was made by Simon[1] (1955), who proposed a growth model producing scale-free distributions. This simple model has numerous applications, and is conceptually important.

It is convenient to formulate the Simon model in the following way (Zanette and Manrubia 2001):

(1) At each time step, a new individual is added to the system.

(2) (a) With probability p (Simon used the notation α), this new individual establishes a new family;

(b) with the complementary probability $1-p$, the new individual chooses at random some individual and joins its family.

Rule (2)(b) simply means that new individuals are distributed among families with probability proportional to their sizes. Note that this is very similar to the rules of preferential linking. The number of individuals equals t, and, at long times, the number of families is pt. Let us consider the long-time limit. The statistics of this system are described by a distribution $P(k,t)$, that is the probability that a family has k members at time t. The master equation for this distribution may be obtained in the same way as in Sections 5.1 and 5.2. One introduces the average number $\langle N(k,t)\rangle$ of families with k members at time t. The equation for this quantity is

$$\langle N(k,t+1)\rangle = \langle N(k,t)\rangle + p\delta_{k,1} + (1-p)\left[\frac{k-1}{t}\langle N(k-1,t)\rangle - \frac{k}{t}\langle N(k,t)\rangle\right] . \tag{7.1}$$

[1] Herbert A. Simon (1916–2001) was a Nobel Prize winner in economics (1978).

(Recall the derivations from Sections 5.1 and 5.2.) $P(k,t) = \langle N(k,t)\rangle/(pt)$, since the total number of families approaches pt. Therefore, the master equation for the distribution $P(k,t)$ is of the form

$$\frac{\partial[tP(k,t)]}{\partial t} = \delta_{k,1} + (1-p)[(k-1)P(k-1,t) - kP(k,t)] \tag{7.2}$$

(compare with eqn (5.17) in Section 5.2). This equation has a stationary solution. One can see that it is a power-law distribution with exponent $\gamma = 1 + 1/(1-p)$. The mathematically alert readers can check by direct substitution that the exact stationary distribution is

$$P(k) = (\gamma - 1)B(k,\gamma) \overset{k \gg 1}{\propto} k^{-\gamma}\,. \tag{7.3}$$

If the readers have never heard of beta-functions, $B(\ ,\ )$, they may easily check the asymptote by using the continuum approximation.

In fact, in 1955 Simon proposed a natural mechanism for the self-organization of growing systems into a critical state (we mean power laws). In earlier chapters we called this general mechanism *popularity is attractive*. Originally Simon applied his model to the problem of the frequency of word occurrence in a text: what is the frequency of the occurrence of distinct words? Empirical studies (see Zipf 1949) show that the distribution of the frequency of occurrence of distinct words has a power-law tail with exponent close to 2. How does the Simon model explain this form?

Suppose you are a writer and write your book at the rate of one word per second. With a small probability p, this new word has not yet occurred in the text, but with much greater probability you have already used it once or twice or many times. Suppose the word 'drama' has already been used in your dramatic text 1000 times. What is the probability that the next word in your text will be just this 'drama'? According to the Simon model, this probability is proportional to 1000. If the word 'love' was repeated 100 000 times, it will be the next one with probability 100 times higher than 'drama'. This is Simon's assumption. One can say that distinct words are 'families', and any words are 'individuals'. So, the frequency of occurrence of distinct words in a text is described by eqn (7.3) with exponent $\gamma = 1 + 1/(1-p)$. But our speech is so poor that we mostly use the same words, so p in a long text is very small. Hence γ should indeed be very close to 2.

With the same success as for words and families, the Simon model can be reformulated for networks, in terms of vertices and, for example, directed edges (Bornholdt and Ebel 2000):

(1) At each time step, a new edge is added to the network.

(2) (a) Also, with probability p, a new vertex is added, and the target end of the new edge becomes attached to this vertex;

 (b) with the complementary probability $1-p$, the target end of the new edge becomes attached to the target end of a randomly chosen old edge.

According to rule (2)(b), new edges become attached to vertices with probability proportional to their in-degrees. One can see that this is just preferential linking. In this case, it produces the in-degree distribution of the form (7.3).

Note, however, that the matter of interest to Simon was one-particle distributions, but in networks this is only the tip of the iceberg.

7.2 Econophysics: wealth distribution in evolving societies

In the previous section we demonstrated that the ideas of network science are related to general problems of self-organization. Below we present an example of how concepts from network growth can be applied to other fields.

In Section 4.4 we discussed wealth distribution in an equilibrium situation, where the size of a population is fixed. Let us consider wealth distribution in evolving societies by using a technique for growing networks. For simplicity, we do not account for mortality, redistribution and loss of money, inflation, and many other very important factors. So, our demonstration is very schematic. Furthermore, in contrast to Section 5.11, the society is assumed to be homogeneous: 'strong guys' are absent. Let there be one birth at each increment of time. Hence, at time t, our society consists of t members.

What are the main features of wealth distribution in this society? Ignoring nuances and subtleties, the distribution and society may be *unfair*, *fair*, or *superfair*. For example, if the wealth distribution $P(k,t)$ is a power law, $P(k) \sim k^{-\gamma}$, and $\gamma \leq 2$, that is the second moment of the distribution diverges, the society is *unfair*: few persons own a finite fraction of the total wealth. If $\gamma > 2$, the society is *fair*: the treasure that belongs to several rich individuals is nothing compared with the total wealth. The wealth condensation transition (Bouchaud and Mézard 2000) occurs when the wealth distribution passes over k^{-2} dependence.[2] When the wealth distribution decreases more rapidly than a power law, that is when all the moments are finite, the society is *superfair*.

On the other hand, societies may be *stagnating*, *developing*, or *degrading*, according to the evolution of the mean wealth. This notion does not depend on the rate of the population growth. In all cases the population grows. In *stable societies*, the average wealth of an individual (the average capital, the average amount of money) does not change with time, and the input flow of capital is constant. In *developing societies*, the average wealth and the input flow of capital grow with time. In *degrading societies*, these quantities decrease. Notice that we do not care where new money comes from.

When is fairness possible? Is it possible to approach a fairer distribution in these societies by some special measures? The first such measure that comes to mind is the equal distribution of some money among all the people. Is this way effective?

[2] A different kind of wealth condensation transition was discussed in Section 5.11

To parametrize various societies, we take the input flow of capital in the power-law form, $\propto t^\alpha$, as for the accelerated growth of networks (see Section 5.15). Growth exponent α indicates the type of a society. $\alpha = 0$ corresponds to stagnating societies. Positive and negative α exponents provide developing and degrading societies respectively.

We reformulate the network growth concepts for this wealth distribution problem. Let us assume that money attracts money, which corresponds to preferential attachment. In attempting to diminish inequality, the society permanently distributes some fraction of wealth 'fairly' (equally) among its members. Another way to make life better for all is to provide everyone with starting capital. For simplicity we assume that it is proportional to the average wealth. This choice seems to be natural.

The case of stagnation is trivial. In Chapter 5, this situation was considered many times. The resulting wealth distribution is a power law with exponent $\gamma > 2$, and so the stagnating society is fair.

The degrading and developing societies can be considered in the spirit of the accelerated growth of networks. Let the rules of the evolution of a society be as follows:

(1) At each time step, a new individual is born.

(2) The wealth mt^α is distributed among members of the society at each increment of time.

   (a) The wealth pmt^α is distributed equally.

   (b) The wealth $(1-p)mt^\alpha$ is distributed preferentially (money attracts money). The preference is proportional.

(3) The starting capital of an individual is proportional to the current average wealth of the society. Hence, the starting capital is dmt^α, where d is a constant coefficient.

According to these rules, in the continuum approximation, we have the equation for the average wealth $\overline{k}(s,t)$ of an individual (where s is the time of its birth and t is the time of observation):

$$\frac{\partial \overline{k}(s,t)}{\partial t} = mt^\alpha \frac{p}{t} + (1-p)mt^\alpha \frac{\overline{k}(s,t)}{\int_0^t du\, \overline{k}(u,t)} \,. \tag{7.4}$$

The boundary condition is $\overline{k}(t,t) = dmt^\alpha$. From this equation, one can easily obtain the wealth distribution, which depends on all the parameters of the problem: α, p, and d. In some region of parameters, the wealth distribution is non-stationary. For $\alpha < (1-p)/(p+d)$, the wealth distribution is a power law with exponent $\gamma = 2 + (1+\alpha)(p+d)/[1-p-\alpha(p+d)]$. If $\alpha > (1-p)/(p+d)$, the society is 'superfair'. The resulting phase diagram of evolving societies is shown in Fig. 7.1.

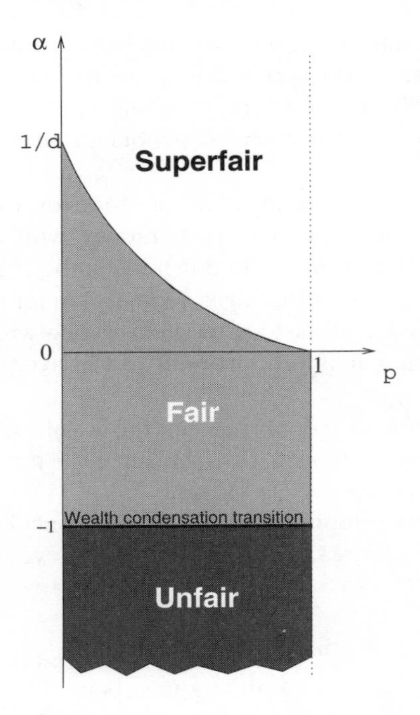

FIG. 7.1. Phase diagram of evolving societies (Dorogovtsev and Mendes 2002b). Exponent α characterizes the 'acceleration' of the development of a society. p is a fraction of the new wealth, which is distributed equally among all members of the society. The complementary part of the money is distributed preferentially (money attracts money). The starting capital of an individual is proportional to the current mean wealth with the coefficient d. In superfair societies, the wealth distribution decreases more rapidly than any power law. In fair and unfair societies the γ exponent of wealth distribution is $\gamma > 2$ and $\gamma \leq 2$ respectively. The wealth condensation transition $\gamma = 2$ is present at $\alpha = -1$. Notice that it is impossible to transfer a degrading society into the 'fair' phase by increasing p.

Our approach is, of course, minimal and not quite serious. Nevertheless, this phase diagram is natural. Notice that the position of the wealth condensation transition does not depend on p and d. Therefore, even if a significant part of new wealth is distributed equally, and even if the starting capital is large, *rapidly degrading societies are necessarily unfair!* So, it is, in principle, impossible to approach any 'fairness' by the 'fair' distribution of any part of new wealth in such a situation. 'Fair' societies are possible only if there is some progress or the rate of degradation is rather modest. 'Fair' distribution of new wealth only produces visible results in fair societies.

7.3 Multiplicative stochastic processes

In this short section our aim is to indicate the place of the preferential linking mechanism amongst the processes of statistical mechanics.

Again, we ask: what is the preferential linking? In brief terms, the answer is: more connected vertices get new connections with higher probability. We saw that, in addition to this process, others may also be present. In particular, more connected vertices may lose their connections with higher probability. In principle numerous variations are possible. Very often they produce fat-tailed distributions. It is important that in each of these variations, the probability of fluctuation of the number of connections of a vertex is higher for more connected vertices.[3] Moreover, we saw that fat-tailed distributions arise when the probability of the fluctuation of the number of connections of a vertex depends linearly on this number (degree).

Thus, the probabilities are very different, but, in standard situations, the amplitudes of these fluctuations are equal. As a rule, at one moment, a vertex can either obtain a new connection or lose one of its links. However, this circumstance is not crucial. We have demonstrated several times that usually discreteness is not a critical factor in determining the resulting self-organized state. So let us discard the discreteness. Then we can take another step. Instead of fluctuations (1) of equal amplitudes (2) with probabilities linearly dependent on degrees of vertices, we pass to fluctuations ($1'$) of amplitudes linearly dependent on degrees of vertices but ($2'$) with degree-independent probabilities. Notice that property ($1'$) can only be introduced for continuum degrees.

Here we do not discuss the origin of these fluctuations, and how they are distributed. Furthermore, for brevity, we shall only discuss a proportional dependence on degrees. Fluctuations ($1'$), ($2'$) determine so-called *multiplicative stochastic processes*. One can see that these fluctuations multiply quantities (note that the factor of the multiplication may be both above and below 1). One of the simplest examples of such a process is a random walk on the logarithmic scale.

Multiplicative stochastic processes are intensively studied in various contexts, including econophysics and evolutionary biology (Sornette and Cont 1997, Solomon and Maslov 2000). There is a wide circle of multiplicative stochastic models, which often provide fat-tailed distributions.[4] This explains the interest behind such models. The most known model of this type with coupled multiplicative fluctuations is described by the *generalized Lotka–Volterra equation*, though this is only one of many possibilities.

From a mathematically rigorous point of view, fluctuations (1), (2) and fluc-

[3] Note that here we do not separate equilibrium and non-equilibrium networks, since, in principle, fat-tailed degree distributions are also possible in equilibrium nets (see Chapter 4).

[4] For example, Huberman and Adamic (1999) considered the following way of generating power-law distributions. Suppose that each new vertex is born with the same starting degree, and afterwards its degree grows in the same multiplicative stochastic manner, with the same parameters. All the vertices evolve independently. If the number of vertices grows exponentially with time, the resulting degree distribution is of a power-law form.

tuations (1′), (2′) from above differ seriously from each other, though, to a physicist, this difference is not fundamental.[5] Thus the preferential linking mechanism is, actually, a discrete version of multiplicative stochastic processes.[6]

[5]The reader who is left unsatisfied by these intuitive verbal arguments may refer to papers on multiplicative stochastic models (for example, see Solomon and Levy 1996, Bouchaud and Mézard 2000, Solomon and Richmond 2001).

[6]A mathematician should prefer a more precise formulation: these processes are close relatives.

8

PHILOSOPHY OF A SMALL WORLD

We live in a world of networks, and, moreover, networks are the core of our civilization. Nowadays, in the epoch of economic, political, and cultural globalization, one can see this with perfect clarity. Thus, the world of networks is our inevitable future, for better or worse.

We are in a period of revolution in network science. Communication networks, the WWW, the Internet, peer-to-peer networks, genome, chemical reaction networks, catalytic networks, nets of metabolic reactions, protein networks, idiotypic networks, neural networks, signalling networks, artery nets, transportation networks, river networks, ecological and food webs, social networks, networks of collaborations, terrorist networks, nets of citations in scientific literature, telephone call graphs, mail networks, power grids, relations between enterprises, nets of ownership, networks of influence, the Word Web, electronic circuits, nets of software components, landscape networks, 'geometric' networks ... —what will be next?

The progress is so immediate and astounding that we actually face a new science based on a new concept, and, one may even say, on a new philosophy: *the natural philosophy of a small world.* Old ideas from mathematics, statistical physics, biology, computer science, and so on take quite a new form in application to *real* evolving networks. Let us list some of basic points of this new conception:

- Evolving networks are one of the *fundamental objects* of statistical physics.
- Evolving networks *self-organize* into complex structures.
- The result of this self-organization is nets with *the crucial role of highly connected vertices.* The entire spectrum of connections is important in networks with such an architecture, but just the highly connected vertices determine the topology and basic properties of the networks, stability, transmission of interactions, spread of diseases, and so on.
- These networks have *several levels of structural organization, several distinct scales*: the local structure of connections of a vertex, the structure of connections in its environment, and the long-range structure of a network. Each of these levels determines a distinct set of properties of the networks.
- Networks are *extremely 'compact' objects* without well defined metric structure. Their organization varies between a tree-like form and a highly correlated structure, rich in loops, which are the two faces of a network.

More than 40 years ago, in 1960, Erdős and Rényi wrote their seminal paper 'On the evolution of random graphs', which is one of the starting points of mathematical random graph theory. What Erdős and Rényi called 'random graphs' were simple equilibrium networks with the Poisson distribution of connections. What they called 'evolution' was, actually, their construction procedure for these equilibrium graphs. Major recent achievements in network science are related to the transition to the study of the *evolving, self-organizing networks with fat-tailed, non-Poisson, distributions of connections.*

Most basic networks, both natural and artificial, belong to this class: the Internet, the WWW, networks of protein interactions, and many, many others. As such, the impressive recent progress in this field represents a significant step towards understanding the most exciting networks of our world: the Internet and WWW, and basic biological networks. It is a step towards understanding one of the few fundamental objects of the Universe: *a network.*

APPENDIX A

RELATIONS FOR AN ADJACENCY MATRIX

Here we demonstrate how to obtain several main characteristics of an undirected graph by using its adjacency matrix $\hat{a}$.

For simplicity, suppose that 'tadpoles' and 'melons' are absent. So, the diagonal elements of the adjacency matrix are zeros, $a_{ii} = 0$, and the non-diagonal elements $i \neq j$ are 0 or 1.

(1) The degree of a vertex is

$$k_i = \sum_j a_{ij} \,. \tag{A.1}$$

(2) The total degree of the graph, $K \equiv \sum_i k_i$, that is the doubled total number of edges, L, is

$$K = 2L = \sum_{ij} a_{ij} = \sum_i (\hat{a}^2)_{ii} \equiv \mathrm{Tr}\,\hat{a}^2 \,, \tag{A.2}$$

where $\mathrm{Tr}\,\hat{A} \equiv \sum_i (\hat{A})_{ii}$ is the trace of a matrix. (3) The total number of loops of length 3 in the graph is[1]

$$N_3 = \frac{1}{6} \sum_i (\hat{a}^3)_{ii} \equiv \frac{1}{6} \mathrm{Tr}\,\hat{a}^3 \,. \tag{A.3}$$

(4) The total number of connected triples of vertices in the graph is

$$T = \frac{1}{2} \sum_{i \neq j} (\hat{a}^2)_{ij} = \frac{1}{2} \sum_i k_i (k_i - 1) \,. \tag{A.4}$$

(4) The clustering coefficient of the graph is

$$C = \frac{1}{9} \frac{\sum_i (\hat{a}^3)_{ii}}{\sum_{i \neq j} (\hat{a}^2)_{ij}} \,. \tag{A.5}$$

(5) The length ℓ of the shortest path between vertices i and j is equal to the minimal power of the adjacency matrix with a non-zero $\{ij\}$ element:

$$(\hat{a}^{\ell-1})_{ij} = 0 \,, \quad (\hat{a}^{\ell})_{ij} \neq 0 \,. \tag{A.6}$$

The generalization of eqns (A.2)–(A.3) is strightforward.

[1] We thank A.N. Samukhin for indicating this convenient relation.

APPENDIX B

HOW TO MEASURE A DISTRIBUTION

Most empirical data on networks are degree (in-degree, out-degree) distributions. Usually, because network size is restricted and the statistics are poor, it is rather difficult to obtain the functional form of a distribution from these data. Here we describe the standard procedure for the fitting of the empirical distribution.[1]

In a log–log scale (see Fig. B.1(a)) we plot the typical empirical degree distribution of a finite scale-free network. One can expect that its form is close to a power-law dependence, $P(k) \propto k^{-\gamma}$, so the first thing we can do is to take a ruler, draw a fitting line by eye, and measure its slope. Many researchers use another option. They select by eye a region with a power-law behaviour and apply the method of least squares to all points in the region. This is a standard option in graphics packages, and most of the data on the degree distribution exponents were obtained precisely in such a way.

This 'method' is certainly not the best one since fluctuations in the large-degree region of the distribution hinder accurate fitting. To improve this situation and make the fluctuations less pronounced, one should consider the corresponding cumulative distribution $P_{cum}(k) = \sum_{k'=k}^{\infty} P(k')$ (see Fig. B.1(b)). Naturally, for the power-law distribution, the corresponding cumulative distribution is $P_{cum}(k) \propto k^{-(\gamma-1)}$. One can see that in the cumulative distribution, the fluctuations are nearly invisible. The picture looks better although the cumulative distribution still has a cut-off due to the finite-size effect, and nothing more can be done. Nevertheless, there is another difficulty which can be overcome.

For our fitting, $P_{cum}(k)$ is naturally presented in log–log scale, so many more points are in the large degree region than in the area of small k. If we use all the points, then the large degrees will carry more weight in the determination of the exponent value, and the resulting error may be large. Therefore, it is necessary to divide the region of $\log k$ chosen (by eye) for fitting to equal-size windows and to replace all the points in each window by a single point. This can be done in several ways, for example by simple averaging.

The pitfalls of such a procedure are obvious but, unfortunately, there is nothing more reliable than this.

[1] We do not consider moment analysis.

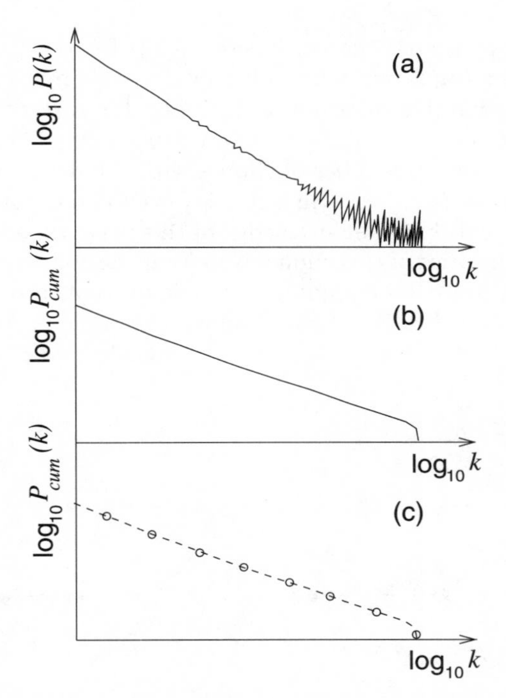

FIG. B.1. (a) The log–log plot of the empirical degree distribution for a finite scale-free network. Fluctuations are strong in the large-degree region of the distribution. (b) The log–log plot of the cumulative degree distribution corresponding to the degree distribution (a). The fluctuations are far less pronounced. (c) The same cumulative degree distribution as (b) but prepared for fitting by a power-law dependence. The points are obtained by averaging over equal-size windows in the $\log k$ axis. The fitting line is the result of the application of the least squares method to the points in the indicated region.

APPENDIX C

STATISTICS OF CLIQUES

In Section 5.4 we presented the expression (5.32) for the number-of-hubs distribution for cliques (bipartite subgraphs) in the scale-free growing network in Fig. 5.6. Let us derive this distribution by using the directed variation of this network (see Fig. C.1).

Recall that the average number of cliques with h hubs at time t was denoted as $N_{\text{cliq}}(h, a = 2, t) \equiv N_{\text{cliq}}(h, t)$ (in our case, each clique can include only two authorities). A clique is based on each edge of this network, and two end vertices of the edge are authorities of this clique. Note that, formally speaking, there may be cliques without hubs. For example, the 'clique' that is based on vertices A and B in the graph in Fig. C.2 has no hubs. The same is true for the 'clique' based on vertices A and C.

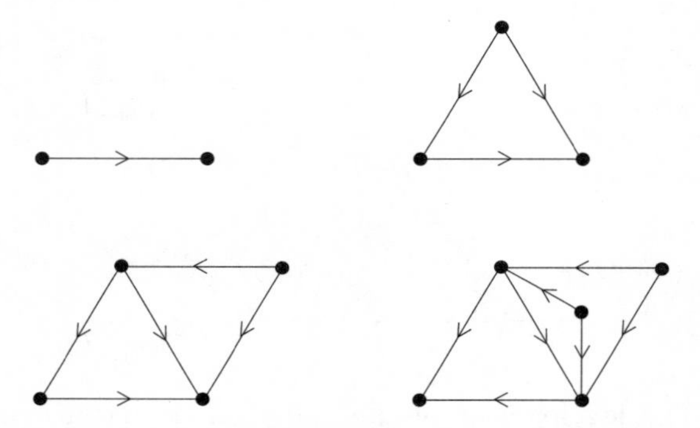

FIG. C.1. The directed variation of the network introduced in Fig. 5.6 from Section 5.4. Each new vertex has two outgoing connections to the end vertices of a randomly chosen edge.

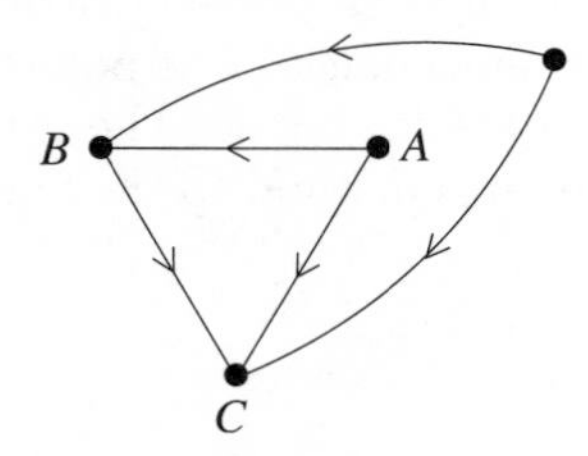

FIG. C.2. The clique which is based on the vertices B and C (authorities) in this graph has two hubs. The 'clique' which is based on the authorities A and B has no hubs.

The equation for $N_{\text{cliq}}(h, t)$ has the form

$$N_{\text{cliq}}(h, t+1) = N_{\text{cliq}}(h, t) + 2\delta_{h,0} + \frac{1}{t}N_{\text{cliq}}(h-1, t) - \frac{1}{t}N_{\text{cliq}}(h, t)\,. \qquad \text{(C.1)}$$

The first term on the right-hand side of the equation is obvious. The second term is due to two new 'cliques' without authorities (the clique based on edge AB and the one based on edge AC). A new vertex in our network is, in fact, a new hub in a randomly chosen clique. This yields the third (gain) and the fourth (loss) terms.

By direct substitution, one can check that the stationary solution of eqn (C.1) is $N_{\text{cliq}}(h, t) = 2^{-h}t$. This is the expression (5.32) from Section 5.4.

APPENDIX D

POWER-LAW PREFERENCE

In Section 5.9 we described the structure of networks growing under the mechanism of preferential linking, when the preference is of the power-law form $f(k) = k^y$; $y = \text{const}$. Here we shall derive the results for these networks in a simple way.

Recall that, for simplicity, we consider an undirected citation graph, where each new vertex has a single connection to a preferentially selected vertex.

At first, we consider the region $0 < y < 1$. The continuum approach equation is similar to eqn (5.23):

$$\frac{\partial \overline{k}(s,t)}{\partial t} = \frac{\overline{k}^y(s,t)}{\int_0^t du\, \overline{k}^y(u,t)}, \tag{D.1}$$

$k(0,0) = 0$, $k(t,t) = 1$. Here, the continuum approach is quite sufficient to describe the essential features of the degree distribution. The equation is consistent. Indeed, applying $\int_0^t ds$ to both the sides of the equation yields the correct result for the total degree of the network, $\int_0^t du\, \overline{k}(u,t) = 2t$.

One can easily check that eqn (D.1) has a scaling solution: $\overline{k}(s,t) = \kappa(s/t)$. In the scaling variables, eqn (D.1) takes the form

$$-\frac{\partial \kappa(\xi)}{\partial \ln \xi} = \mu^{-1} \kappa^y(\xi), \tag{D.2}$$

where we introduced a new constant μ,

$$\int_0^1 d\zeta\, \kappa^y(\zeta) = \mu. \tag{D.3}$$

Also, the relation $\int_0^1 d\zeta\, \kappa(\zeta) = 2$ is valid.

The solution of eqn (D.2) with as yet unknown constant μ is

$$\kappa(\xi) = \left(1 - \frac{1-y}{\mu} \ln \xi\right)^{1/(1-y)}. \tag{D.4}$$

To obtain μ, we substitute this solution into eqn (D.3) or, equivalently, into the expression for the total degree. We can solve each of the two resulting transcendental equations for μ and, in each case, obtain the same $\mu(y)$ dependence. We plotted it in Fig. 5.11 from Section 5.9. Substituting the scaling solution (D.4) into the continuum approach expression (5.27) for a degree distribution readily yields $P(k) \propto k^{-y}\exp[-\mu k^{1-y}/(1-y)]$, that is the result (5.44). The corresponding exact expression is slightly more complex (see Krapivsky, Redner, and Leyvraz 2000, Krapivsky and Redner 2001) but any principal difference is absent.

Now we shall discuss the case $y > 1$. We shall follow Krapivsky, Redner, and Leyvraz (2000). Here we have to use the master equation for this network (recall that $\langle N(k,t)\rangle$ is the average number of vertices of degree k at time t):

$$\langle N(k,t+1)\rangle = \langle N(k,t)\rangle + \frac{(k-1)^y}{\sum_n n^y \langle N(n,t)\rangle}\langle N(k-1,t)\rangle - \frac{k^y}{\sum_n n^y \langle N(n,t)\rangle}\langle N(k,t)\rangle + \delta_{k,1} \tag{D.5}$$

(compare with eqn (5.16) for the Barabási–Albert model). We saw that $\mathcal{M}(t) \equiv \sum_n n^y \langle N(n,t)\rangle \propto t$ for $y \leq 1$. In particular, for the evolution of the average number of vertices with a single connection, eqn (D.5) yields

$$\langle N(1,t+1)\rangle = \langle N(1,t)\rangle - \frac{1}{\mathcal{M}(t)}\langle N(1,t)\rangle + 1\,. \tag{D.6}$$

For example, when $y = 1$, and so $\mathcal{M}(t) = 2t$, this yields $\langle N(1,t)\rangle \cong 2t/3 < t$. How many vertices of degree 1 are there in the network where $y > 1$? To find the answer, we must know the behaviour of $\mathcal{M}(t)$.

First, we estimate $\mathcal{M}(t)$ from below. Recall that the total number of connections in the network is t. In this case there are no vertices of degree greater than t at time t. Therefore, from eqn (D.5), we have

$$\frac{\langle N(k,k)\rangle}{\langle N(k-1,k-1)\rangle} = \frac{(k-1)^y}{\mathcal{M}(k-1)}\,. \tag{D.7}$$

But $\langle N(t,t)\rangle$ must decrease with time. Indeed only the oldest vertex can attach all t connections, and each next attachment can only diminish this quantity. Hence $\langle N(k,k)\rangle/\langle N(k-1,k-1)\rangle < 1$, and so $\mathcal{M}(t) > t^y$ at long times.

Moreover, $\mathcal{M}(t) \sim t^y$, since one can also estimate $\mathcal{M}(t)$ from above:

$$\mathcal{M}(t) \leq \sum_n t^{y-1} n \langle N(n,t)\rangle = t^{y-1} 2t = 2t^y\,. \tag{D.8}$$

Substituting $\mathcal{M}(t) \sim t^y$ into eqn (D.6) shows that the second term on the right-hand side of the equation is negligible. So, $\langle N(1,t)\rangle \cong t$ at long times. This means that the number of vertices of degrees greater than 1 is negligible. Therefore, a small number of the oldest vertices get all the connections in the network, that is t connections.

APPENDIX E

INHOMOGENEOUS GROWING NET

In Section 5.10, we discussed the possible effects of the inhomogeneity of growing networks. We spoke about 'weak effects' of the inhomogeneity of fitness, which do not crucially change degree distributions. Let us show how one can obtain a degree distribution in such a situation (Bianconi and Barabási 2001a).

To demonstrate, we choose the directed growing network:

(1) At each step a new vertex is added. It has n incoming connections.
(2) Simultaneously, m new edges emerge. Their target ends become preferentially attached to vertices.
(3) Each vertex s has its own value of $G = G(s)$, which is time-independent. G is a random variable that is distributed according to a function $P(G)$. The probability that a new edge become attached to a vertex s of in-degree $q(s) \equiv k_i(s)$ is proportional to $G(s)q(s)$.

What is the in-degree distribution of this net? We write, as usual, the equation

$$\frac{\partial \overline{q}(s,t)}{\partial t} = m \frac{G(s)\overline{q}(s,t)}{\int_0^t du G(u)\overline{q}(u,t)} \tag{E.1}$$

with the boundary condition $\overline{q}(t,t) = n$. The distribution of 'fitness' $P(G)$ does not change with time, and $\int_0^t ds\overline{q}(s,t) = (m+n)t$, so that we can expect that at long times $\int_0^t du G(u)\overline{q}(u,t)$ grows proportionally to time. Therefore, we can write

$$\frac{\partial \overline{q}(s,t)}{\partial t} = \beta(G(s))\frac{\overline{q}(s,t)}{t}\,. \tag{E.2}$$

Here we introduced an analogue of our old exponent β, which now depends (only) on G. In a standard manner, we obtain the solution of eqn (E.2), $\overline{q}(s,t) = n(s/t)^{-\beta(G(s))}$. Substituting this solution into the definition $\beta(G) \overset{t\to\infty}{\to} mGt/\int_0^t du G(u)\overline{q}(u,t)$ yields

$$\frac{mGt}{\beta(G)} = n\int dG G P(G)\frac{t}{1-\beta(G)} \equiv \frac{nt}{c}\,. \tag{E.3}$$

Here we introduced a new positive constant c. Then we see that $\beta(G)$ is a linear function: $\beta(G) = cGm/n$.

Thus we have the equation for the constant c:

$$\int dG P(G) \frac{G}{1 - cGm/n} = \frac{1}{c} \,. \tag{E.4}$$

If this transcendental equation has a solution, the degree distribution can be easily found.

As usual, to obtain the in-degree distribution $P(q)$ in the continuum approximation we use eqn (5.27), taking into account that $s = tq^{-1/\beta}$. However, now we must average it over G. The result is the relation

$$P(q) \propto \int dG P(G)(1 + 1/\beta(G)) q^{-1-1/\beta(G)} \,, \tag{E.5}$$

which coincides with eqn (5.47) from Section 5.10.

If $P(G) = \delta(G - G_0)$, we return to the ordinary scale-free homogeneous network. If G homogeneously distributed in the range $(0, 1)$ and $n = m$, $q = k$ as in the undirected Barabási–Albert model, one can easily obtain $P(k) \propto k^{-2.255\ldots}/\ln k$, that is eqn (5.48) from Section 5.10.

A stricter approach to this problem was developed by Ergun and Rodgers (2002) but the same results were obtained.

Whether eqn (E.4) has a solution or not depends on the specific form of the distribution $P(G)$, especially on the form of the tail $P(G)$ at large degrees. The absence of the solution indicates that there is condensation of edges in the network.

APPENDIX F

Z-TRANSFORM

The Z-transform (generating function) technique is the main approach for the solution of discrete difference equations. In graph theory, the Z-transform is ideally suited for tree-like uncorrelated graphs. Here we list the basic definitions, theorems, and formulae for the Z-transform.

The definition is[1]

$$\Phi(z) \equiv \sum_{k=0} P(k) z^k \,. \tag{F.1}$$

The inverse Z-transform, that is the coefficients of the Taylor series for $\Phi(z)$, is

$$P(k) = \frac{1}{k!} \frac{d^k \Phi(z)}{dy^k}\bigg|_{z=0} = \frac{1}{2\pi i} \oint_C dz \, \frac{\Phi(z)}{z^{k+1}} \,. \tag{F.2}$$

Here C is a contour around 0, which does not enclose the singularities of $\Phi(z)$.

The limiting theorems are

$$\Phi(0) = P(0) \tag{F.3}$$

and, if the finite limit $P(k \to \infty)$ exists,

$$\lim_{z \to 1} (1 - z)\Phi(z) = P(k \to \infty) \,. \tag{F.4}$$

The formula for moments is

$$\langle k^n \rangle \equiv \sum_{k=0} k^n P(k) z^k = \left[\left(z \frac{d}{dz} \right)^n \Phi(z) \right]_{z=1} \,, \tag{F.5}$$

so $\overline{k} = \Phi'(1)$, $\langle k^2 \rangle = \Phi'(1) + \Phi''(1)$, etc.

The Z-transform of the Poisson distribution is

$$P(k) = e^{-\overline{k}}\, \overline{k}^k / k! \;\to\; \Phi(z) = e^{\overline{k}(z-1)} \,, \tag{F.6}$$

and so, for the Poisson distribution, it holds that $\langle k^2 \rangle = \overline{k} + \overline{k}^2$. The average number of the nearest neighbours is $z_1 \equiv \overline{k}$. The average number of the second-nearest neighbours of a vertex in equilibrium uncorrelated graphs is $z_2 = \langle k^2 \rangle - \overline{k}$ (see Section 6.1). Hence in such graphs with the Poisson degree distribution, $z_2 = z_1^2$.

[1]Often, the Z-transform is introduced with the changed sign of the power in the sum. We use the Z-transform version that is equivalent to generating functions.

Useful relations are

$$\begin{aligned}
\sum_{k=0} P(k-1)z^k &= z\Phi(z)\,, \\
\sum_{k=0} P(k+1)z^k &= \frac{1}{z}[\Phi(z) - P(0)]\,, \\
\sum_{k=0} kP(k)z^k &= z\Phi'(z)\,, \\
\sum_{k=0} (k-1)P(k-1)z^k &= z^2\Phi'(z)\,, \\
\sum_{k=0} (k+1)P(k+1)z^k &= \Phi'(z)\,,
\end{aligned} \tag{F.7}$$

The Z-transform of the convolution is: if $P(k) \to \Phi(z)$ and $Q(k) \to \Psi(z)$, then

$$\sum_{j=0}^{k} P(j)Q(k-j) \;\to\; \Phi(z)\Psi(z)\,. \tag{F.8}$$

As for singularities: for a non-integer b, it holds that

$$P(k) \overset{k\to\infty}{\sim} k^{-b} \;\to\; \Phi(z) \overset{1-z\to 0}{\cong} c_1 + c_2(1-z)^{b-1} + \text{analytical terms}\,, \tag{F.9}$$

where c_1 and c_2 are constants (one may include c_1 in the last term). The relations may be slightly more complex when b is integer. For example, asymptotically $P(k) \cong k^{-3} \;\to\; \Phi(z) \cong \frac{1}{2}(1-z)^2 \ln(1-z)^{-1}$ + analytical terms.

APPENDIX G

CRITICAL PHENOMENA IN NETWORKS

Here we outline a unified approach to critical phenomena in networks with an arbitrary degree distribution (Goltsev, Dorogovtsev, and Mendes 2002).

Because of the local tree-like structure of equilibrium uncorrelated networks, critical fluctuations are absent (see Section 6.6). In such a situation, mean-field approaches are sufficient, and the general phenomenology can be developed in the spirit of the Landau theory of phase transitions. Furthermore, in more complex networks with correlations between vertices and loops, the mean-field theory must be valid outside of the fluctuation region around a phase transition (if the region of critical fluctuations exists at all). In this 'non-fluctuation' range, such a phenomenological approach is also applicable.

Conjectures

(1) Assume that the free energy of an interacting system near the phase transition is a linear functional of the degree function of the network:

$$F(\Delta, E) = \sum_{k=0}^{\infty} P(k)G(\Delta, k) - E\Delta\,. \tag{G.1}$$

Here, Δ is the order parameter, and E is its conjugated field. For brevity, we only consider the one-component scalar order parameter and write only the leading term in E.

(2) Assume that the function $G(\Delta, k)$ is an analytical function of the order parameter Δ and of product $k\Delta$.[1] Then, when Δ and $k\Delta$ are sufficiently small, $G(\Delta, k)$ can be written in the form of a series:

$$G(\Delta, k) = \sum_{n=2}^{\infty}\sum_{l=1}^{n} b_{nl}\, k^l \Delta^n\,. \tag{G.2}$$

The coefficients in the expansion are analytical functions of, for example, temperature or any other control parameter. Note that the fact that the sums in eqn (G.2) start from $n = 2$ and $l = 1$ means some additional assumptions which are, in principle, not essential.

(3) Assume that the free energy is finite at any finite Δ and E.

Ground

In principle, the form of the free energy (G.1), (G.2) is not obvious *a priori*. The arguments supporting this form follow.

[1] We acknowledge A.N. Samukhin who indicated the importance of this conjecture.

The equation of state, from which one can find the order parameter, is

$$\frac{\partial F(\Delta, E)}{\partial \Delta} = 0\,. \tag{G.3}$$

For several models on the networks with a local tree-like structure, where mean-field theory is exact, the equations of state are known, so let us check whether they correspond to the form (G.1), (G.2).

Recall, for example, the 'equation of state' (6.14) for the phase transition of the birth of a giant component. Introducing $\Delta \equiv 1 - x$, that is the probability that an edge leads to the giant connected component, yields the equation

$$1 - \Delta = \sum_{k=0}^{\infty} \frac{kP(k)}{\overline{k}}(1-\Delta)^{k-1}\,. \tag{G.4}$$

It can be seen that the form of this equation just corresponds to that of the expansions (6.14) with the difference that here $E = 0$.

Also, in a very naive form, a mean-field consideration for spin models on networks leads to equations of the form $\Delta = \sum_k [kP(k)/\overline{k}] \tanh(Jk\Delta/T)$.[2] This and other known equations of state for spin models have similar structures corresponding to the free energy (G.1), (G.2).

Thus, assumptions (1) and (2) are fulfilled in the particular case of graphs with a local tree-like structure. Assumption (3) is natural.

Consequences

If all the moments of the degree distribution are finite, the free energy turns out to be an analytic function of Δ. It is simply a series in the moments of the degree distribution and powers of Δ,[3]

$$F(\Delta, E) = \sum_{n=2}^{\infty} \sum_{l=1}^{n} b_{nl}\, \langle k^l \rangle \Delta^n - E\Delta\,. \tag{G.5}$$

So, the standard Landau theory of phase transitions is valid in this event with the difference that the coefficient of the series in the order parameter depends on the moments. In particular, the coefficient of Δ^2 is a linear combination $a\overline{k} + b\langle k^2 \rangle$, where a and b depend on temperature or any other relavant control parameter. At the critical temperature T_c, the coefficient of Δ^2 is equal to zero, and is an analytical function of temperature near T_c. As usual, the critical behaviour is determined by the linear dependence of this coefficient on temperature in the neighbourhood of T_c. Note that from the form of the coefficient of Δ^2, it readily

[2] We stress that here the order parameter Δ does not mean magnetization, although the latter can be expressed in terms of Δ.

[3] This expansion can also be used as the starting point of the theory (Goltsev, Dorogovtsev, and Mendes 2002).

follows that the critical temperature is determined by the ratio of the second and first moments, $T_c = T_c(\langle k^2 \rangle / \overline{k})$.

When the degree distribution is fat-tailed, and higher moments are infinite, the series (G.1) in powers of Δ is only formal: its coefficients diverge. This indicates that the free energy, unlike $G(\Delta, k)$, is a non-analytical function of Δ. In this case, one must extract a singular contribution to the free energy at small Δ. In principle, one can see from the above state equations for percolation and spin models that at small Δ the essential contribution to free energy is only

$$F(\Delta, E) = \sum_{k=0}^{\text{const}/\Delta} P(k) G(\Delta, k) - E\Delta \,. \tag{G.6}$$

Here, the constant in the upper limit of the sum over k is actually temperature-dependent. Its particular value is of secondary importance and does not influence the critical exponents and functional form of critical singularities.

We were not quite strict in these arguments. One can see that if the first moment of the degree distribution is finite, eqn (G.6) directly follows from assumptions (1)–(3) from above. If $F(\Delta) = \sum_k P(k) G(\Delta, k) < \infty$, at large k, $G(\Delta, k)$ should grow slower than k grows. From this and assumption (2), eqn (G.6) follows immediately.[4] So, the idea of this approach is very simple: *although G is an analytic function, the fat tail of the degree distribution leads to a non-analytic contribution to the free energy $F(\Delta)$*. One can see a distinction from the Landau theory of phase transitions. While the Landau theory supposes an analyticity of the free energy as a function of the order parameter, analytic properties of the free energy are not assumed here. Instead, the analytic properties of the function G are postulated. In principle, one may present analytic and non-analytic terms in the free energy separately. The former are junior powers of Δ, whose coefficients contain only finite moments.

Using eqn (G.6), in a similar way as in Section 6.4, one may analyse all possible cases: various degree distributions and various symmetries of the order parameter. For example, for the phase transition in an interacting system with the symmetry of the Ising model, $F(\Delta, E) = A\Delta^2 + B\Delta^{\gamma-1} - E\Delta$ when $3 < \gamma < 5$. Here, $A \propto (T - T_c)$ and B is a constant.

The scenarios of the critial behaviour for different power-law degree distributions $P(k) \propto k^{-\gamma}$ are presented in Fig. G.1. One can see that there are few scenarios.[5] The type of the critical behaviour of a system at large γ is the same as in the 'standard' Landau theory, or, equivalently, in mean-field theory for the corresponding model. Suppose we know this 'standard' mean-field critical behaviour of some interacting system. Then, looking at Fig. G.1, we can readily describe the critical behaviour of this system on networks with *any* exponent γ.

[4] For brevity, one may consider a power-law degree distribution in the continuum approximation.

[5] We do not discuss situations with several coupled order parameters, which may be slightly more complex.

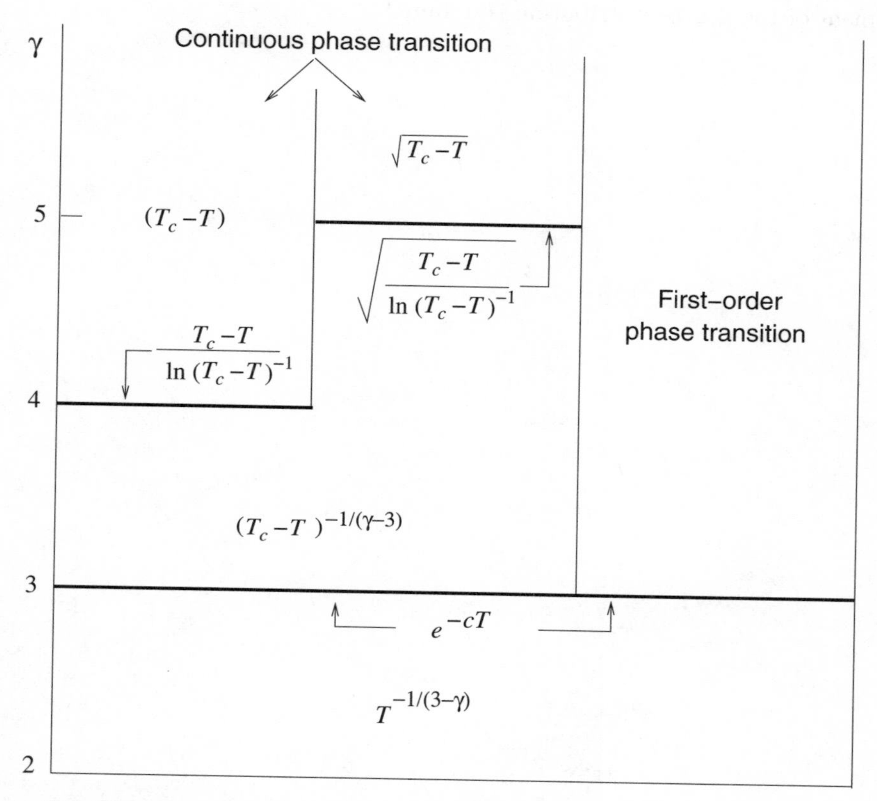

FIG. G.1. Scenarios of critical behaviour for scale-free networks. The order parameter is one-component. The temperature dependences of the order parameter Δ are shown for various regions of critical exponent γ. The role of temperature can play any other relevant control parameter. A specific scenario of critical behaviour is realized depending on what the order of this phase transition is in the 'normal' Landau theory for lattices, or, equivalently, in the mean-field theory for the corresponding model. The latter yields the behaviour at large γ. Note, that, for $\gamma \leq 3$, the behaviour is the same in all the cases. Depending on the symmetry of the order parameter, phase transitions in 'standard' Landau theory are continuous or of first order. If the 'standard' Landau theory provides a phase transition of first order, the phase transition in such networks is of first order when $\langle k^2 \rangle$ is finite. Note that for a continuous phase transition, there exist two distinct scenarios depending on the symmetry. The constant c in the exponential $\Delta \propto e^{-cT}$ at $\gamma = 3$ is determined by the complete form of the degree distribution.

Note that the critical behaviour is universal for $\gamma \leq 3$; that is, when the second moment of the degree distribution is infinite.[6]

[6]Rigorously speaking, the phenomenological approach cannot be directly applicable when the critical temperature approaches infinity. In this case, the temperature dependences of all the coefficients of the free energy expansion are essential. Then, these dependences may be taken from the mean-field equations of states of particular models. For all known models, for $\gamma \leq 3$, this provides the $\Delta(T)$ that is shown in Fig. G.1.

APPENDIX H

A GUIDE TO THE NETWORK LITERATURE

The list of references in the Bibliography is long, and many works were not cited in the main chapters. Moreover, a number of interesting problems were not even touched upon. To help the reader we present a brief guide to numerous valuable sources.[1]

Books about networks

Introductory and popular science books: Barabási (2002), Huberman (2001), Buchanan (2002).
A book on small-world networks: Watts (1999).
Mathematical books on graph theory and its applications: Bollobás (1985, 1998), Chung (1997), Cvetcović, Doob, and Sachs (1979), Janson, Luczak, and Rucinski (2000), Knuth (1977).
Some books on the Internet and the WWW: Huberman (2001), Huiterna (2000), Belew (2000), Dodge and Kitchin (2000, 2001).
A book on ecological webs: Cohen, Briand, and Newman (1990).
Books touching on the network topic (biology): Kauffman (1993, 1995, 2000).

Books on related problems

Some books on related problems in statistical physics: Bak (1997), Barabási and Stanley (1995), Harris (1989), Mandelbrot (1983), Stauffer and Aharony (1991), Bunde and Havlin (1994).
Power laws in economics and finance, and econophysics: Pareto (1897), Gibrat (1931), Zipf (1949), Simon (1957), Mantegna and Stanley (1999), Bouchaud and Potters (2000).
Science indexing: Garfield (1979), Egghe and Rousseau (1990).
Books on social analysis: Scott (1979), Wasserman and Faust (1994).

Recent reviews on networks

Reviews on complex evolving networks: Strogatz (2001), Albert and Barabási (2002), Dorogovtsev and Mendes (2002), and the collection of contributions, including brief reviews, in Bornholdt and Schuster (2002).
A brief review on biological networks: Bose (2002).

[1] In many cases the sources already cited in the main chapters of the book are omitted from this guide.

Graph theory papers

Bauer and Golinelli (2001), Bollobás and Riordan (2001), Bollobás, Riordan, Spencer, and Tusnády (2001), Flajolet, Knuth, and Pittel (1989), Janson, Knuth, Luczak, and Pittel (1993).

Papers on the Internet and WWW written by computer scientists

Brin and Page (1998), Butler (2000), Chakrabarti, Dom, Gibson, Kleinberg, Kumar, Raghavan, Rajagopalan and Tomkins (1999), Claffy (1999), Claffy, Monk, and McRobb (1999), Dean and Henzinger (1999), Gibson, Kleinberg, and Raghavan (1998), Gunther (2002), Kumar, Raghavan, Rajagopalan, Sivakumar, Tomkins, and Upfal (2000a), Lawrence (2000), Lawrence and Giles (1998a, 1998b, 1999a, 1999b), Pansiot and Grad (1998), Rafiei and Mendelzon (2000).

Papers on the Internet and WWW written by physicists

Abe and Suzuki (2002), Adamic and Huberman (2000), Barabási, Albert, Jeong, and Bianconi (2000), Barabási, Freeh, Jeong, and Brockman (2001), Caldarelli, Marchetti, and Pietronero (2000), Capocci, Caldarelli, Marchetti, and Pietronero (2001), Goh, Kahng, and Kim (2002), Huberman, Pirolli, Pitkow, and Lukose (1998), Kahng, Park, and Jeong (2001), Maslov and Sneppen (2002b), Maurer and Huberman (2000, 2001).

Empirical and experimental study of cellular networks

Barkai and Leibler (1997), Bastolla, Porto, Roman, and Vendruscolo (2001), Becskei and Serrano (2000), Bhalla and Iyengar (1999), Bilke and Peterson (2001), Elowitz and Leibler (2000), Farkas, Jeong, Vicsek, Barabási, and Oltvai (2002), Ito, Chiba, Ozawa, Yoshida, Hattori, and Sakaki (2001), Ito, Tashiro, Muta, Ozawa, Chiba, Nishizawa, Yamamoto, Kuhara, and Sakaki (2000), Podani, Oltvai, Jeong, Tombor, Barabási, and Szathmáry (2001).

Theory and models of cellular networks

Brede and Behn (2001), Fox and Hill (2001), Fraser and Reidys (1997), Holland (2001), Hörnquist (2001), Kauffman (1969), Maritan, Colaiori, Flammini, Cieplak, and Banavar (1996), Pastor-Satorras, Smith, and Solé (2002), Wagner (2001b, 2001c).

Ecological networks

Camacho, Guimerà, and Amaral (2001, 2002), Dunne, Williams, and Martinez (2002b), Lässig, Bastolla, Manrubia, and Valleriani (2001), Montoya and Solé (2001), Quince, Higgs, and McKane (2001), Williams and Martinez (2000), Williams, Martinez, Berlow, Dunne, and Barabási (2001).

Collaboration and social networks

Abramson and Kuperman (2000), Alberich, Miro-Julia, and Rossello (2002), Asvanund, Clay, Krishnan, and Smith (2001), Calvó-Armengol and Zenou (2001),

Davidsen, Ebel, and Bornholdt (2001), Gilbert (1997), Jin, Girvan, and Newman (2001), Watts, Dodds, and Newman (2002).

Other real networks

Transportation nets: Banavar, Maritan, and Rinaldo (1999).
Networks in economics: Kim, Lee, Kim, and Kahng (2001), Kullmann, Kertész, and Kaski (2001).
Networks of citations: Garfield (1972), Price (1965), Silagadze (1997).
Networks in semantics: Steyvers and Tenenbaum (2001).

Small-world networks

Almaas, Kulkarni, and Stroud (2002), Araujo, Mendes, and Seixas (2002), de Arcangelis and Herrmann (2002), Barbour and Reinert (2001), Barrat (1999), Barthélémy and Amaral (1999), Comellas and Sampels (2000, 2001), Jasch and Blumen (2001), Jespersen and Blumen (2000), Jespersen, Sokolov, and Blumen (2000a, 2000b), Karimipour and Ramzanpour (2002), Kasturirangan (1999), Kulkarni, Almaas, and Stroud (2000), Kuperman and Abramson (2001a), Latora and Marchiori (2001, 2002a), Marchiori and Latora (2000), Mathias and Gopal (2001), de Menezes, Moukarzel, and Penna (2000), Moukarzel (2000), Moukarzel and de Menezes (1999), Newman (2000a, 2000b), Newman, Moore, and Watts (2000), Newman and Watts (1999a, 1999b), Pandit and Amritkar (1999, 2001), Ramzanpour and Karimipour (2002), Souma, Fujiwara, and Aoyama (2001).

Structural properties of various networks

Bagnoli and Bezzi (2001), Dall and Christensen (2002), Dorogovtsev, Mendes, and Samukhin (2002c), Eckmann and Moses (2001), Gleiss, Stadler, Wagner, and Fell (2000), Gudkov and Johnson (2001), Han and Li (2002), Holme and Kim (2001, 2002), Holme, Kim, Yoon, and Han (2002), Kansal and Torquato (2001), Kim, Yoon, Han, and Jeong (2001), Latora and Marchiori (2002b), Manna and Sen (2002), Moreira, de Andrade Jr, and Amaral (2002), Mossa, Barthélémy, Stanley, and Amaral (2002), Newman (2001c, 2002c), Rodgers and Darby-Dowman (2001), Sen and Chakrabarti (2001), Svenson (2001), Vandewalle and Ausloos (1997), White and Newman (2001).

Models of evolving networks

Bianconi (2002b), Bornholdt and Rohlf (2000), Christensen, Donangelo, Koiller, and Sneppen (2000), Eppstein and Wang (2002), Ergun (2002), Ipsen and Mikhailov (2001), Jain and Krishna (1998), Jost and Joy (2002), Klemm and Eguíluz (2002a, 2002b), Kotrla, Slanina, and Steiner (2002), Kullmann and Kertész (2001a, 2001b), Levene, Fenner, Loizou, and Wheeldon (2002), Newman (2001b, 2002c), Rohlf and Bornholdt (2002), Roehl and Bornholdt (2001), Slanina and Kotrla (1999, 2000), Tadić (2001a, 2001d), Vázquez (2001a), Volchenkov and Blanchard (2002), Xulvi-Brunet and Sokolov (2002), Zheng and Ergun (2001).

Cooperative phenomena in networks

Percolation: Cohen, Erez, ben-Avraham, and Havlin (2001b), Lancaster (2002), Moore and Newman (2000a, 2000b), Newman, Jensen, and Ziff (2002), Schwartz, Cohen, ben-Avraham, Barabási, and Havlin (2002), Weigt and Hartmann (2000, 2001a, 2001b).

Spread of deseases: Ancel, Newman, Martin, and Schrag (2001), Kuperman and Abramson (2001b), Newman (2001d, 2002b), Zanette and Kuperman (2001), Zekri and Clerc (2001), *rumour propagation*: Zanette (2001a, 2001b, 2002).

The Ising model: Bianconi (2002a), Gitterman (2000), Janke, Johnston, and Stathakopoulos (2002), Svenson and Johnston (2002).

The XY model: Kim, Hong, Holme, Jeon, Minnhagen, and Choi (2001).

Synchronization phenomena: Barahona and Pecora (2002), Hong, Choi, and Kim (2002a, 2002b), Jalan and Amritkar (2002), Lago-Fernández, Huerta, Corbacho, and Sigüenza (2000), Wang and Chen (2002).

Self-organized criticality models on networks: Kulkarni, Almaas, and Stroud (1999), Moreno, Gómez, and Pacheco (2002), Moreno and Vázquez (2002).

Other cooperative phenomena: Bernardes, Stauffer, and Kertész (2002), Copelli, Zorzenon dos Santos, and Sa Martins (2001), Glass and Hill (1998), Jain and Krishna (2001, 2002), Kuperman and Zanette (2002), Lise and Paczuski (2002), Lux and Marchesi (1999), Maslov and Zhang (2001), Sánchez, López, and Rodríguez (2002).

Communications, congestion, and walks on networks

Adamic, Lukose, and Huberman (2002), Arenas, Díaz-Guilera, and Guimerà (2001), Guardiola, Díaz-Guilera, Perez, Arenas, and Llas (2002), Guimerà, Arenas, and Díaz-Guilera (2001), Fuks and Lawniczak (1999), Kinouchi, Martinez, Lima, Lourenco, and Risau-Gusman (2001), Lahtinen, Kertész, and Kaski (2001, 2002), Lih and Chern (2001), Monasson (1999), Puniyani, Lukose, and Huberman (2001), Sachtjen, Carreras, and Lynch (2000), Tadić (2001b, 2001c), Valverde and Solé (2002).

Electronic archives with papers on networks

Many recent papers on networks can be found in electronic archives:

- http://arXiv.org/ (mostly cond-mat/),
- http://www.unifr.ch/econophysics/ (Econophysics Forum),
- http://www.santafe.edu/sfi/publications/ (Working Papers of Santa Fe Institute).

The history of the Internet

The page of Vint Cerf:
http://www1.worldcom.com/global/resources/cerfs_up/internet_history/.

The history of the WWW

The page of Tim Berners-Lee:
http://www.w3.org/People/Berners-Lee/ShortHistory.html.

REFERENCES

S. Abe and N. Suzuki (2002), Itineration of the Internet over non-equilibrium stationary states in Tsallis statistics, cond-mat/0204336.

G. Abramson and M. Kuperman (2000), Social games in a social network, *Phys. Rev.* E **63**, 030901; nlin.AO/0010015.

L.A. Adamic (1999), The small world web, Proceedings of ECDL'99, LNCS 1696, pp. 443–452 (Springer, Berlin).

L.A. Adamic (2000), Zipf, power-laws, and Pareto — a ranking tutorial, http://www.parc.xerox.com/istl/groups/iea/papers/ranking/.

L.A. Adamic and B.A. Huberman (2000), Power-law distribution of the World Wide Web, *comment*, *Science* **287**, 2115a; cond-mat/0001459.

L.A. Adamic, R.M. Lukose, and B.A. Huberman (2002), Local search in unstructured networks, *Handbook of Graphs and Networks: From the Genome to the Internet*, ed. S. Bornholdt and H.G. Schuster (Wiley-VCH, Berlin); cond-mat/0204181.

L.A. Adamic, R.M. Lukose, A.R. Puniyani, and B.A. Huberman (2001), Search in power-law networks, *Phys. Rev.* E **64**, 046135; cs.NI/0103016.

W. Aiello, F. Chung, and L. Lu (2000), A random graph model for massive graphs, Proceedings of the Thirty-Second Annual ACM Symposium on Theory of Computing, pp. 171–180.

R. Alberich, J. Miro-Julia, and F. Rossello (2002), Marvel Universe looks almost like a real social network, cond-mat/0202174.

R. Albert and A.-L. Barabási (2000a), Topology of evolving networks: Local events and universality, *Phys. Rev. Lett.* **85**, 5234; cond-mat/0005085.

R. Albert and A.-L. Barabási (2000b), Dynamics of complex systems: Scaling laws for the period of Boolean networks, *Phys. Rev. Lett.* **84**, 5660.

R. Albert and A.-L. Barabási (2002), Statistical mechanics of complex networks, *Rev. Mod. Phys.* **47**, 74; cond-mat/0106096.

R. Albert, H. Jeong, and A.-L. Barabási (1999), Diameter of the world-wide web, *Nature* **401**, 130; cond-mat/9907038.

R. Albert, H. Jeong, and A.-L. Barabási (2000), Attack and error tolerance of complex networks, *Nature* **406**, 378; cond-mat/0008064.

A. Aleksiejuk, J.A. Holyst, and D. Stauffer (2002), Ferromagnetic phase transition in Barabási–Albert networks, *Physica* A **310**, 260; cond-mat/0112312.

E. Almaas, R.V. Kulkarni, and D. Stroud (2002), Characterizing the structure of small-world networks, *Phys. Rev. Lett.* **88**, 098101; cond-mat/0109227.

L.A.N. Amaral, A. Scala, M. Barthélémy, and H.E. Stanley (2000), Classes of small-world networks, *Proc. Natl Acad. Sci. USA* **97**, 11149; cond-mat/0001458.

L.W. Ancel, M.E.J. Newman, M. Martin, and S. Schrag (2001), Applying network theory to epidemics: Control measures for outbreaks of *Mycoplasma pneumoniae*, Working Papers of Santa Fe Institute, 01-12-078, http://www.santafe.edu/sfi/publications/Abstracts/01-12-078abs.html.

T. Araujo, R.V. Mendes, and J. Seixas (2002), A dynamical characterization of the small world phase, cond-mat/0204573.

L. de Arcangelis and H.J. Herrmann (2002), Self-organized criticality on small world networks, *Physica* A **308**, 545; cond-mat/0110239.

A. Arenas, A. Díaz-Guilera, and R. Guimerà (2001), Communication in networks with hierarchical branching, *Phys. Rev. Lett.* **86**, 3196; cond-mat/0009395.

A. Asvanund, K. Clay, R. Krishnan, and M. Smith (2001), Bigger may not be better: An empirical analysis of optimal membership rules in peer-to-peer networks, cs.CY/0109106.

F. Bagnoli and M. Bezzi (2001), Small world effects in evolution, *Phys. Rev.* E **64**, 021914; cond-mat/0007458.

P. Bak (1997), *How Nature Works* (Copernicus, New York).

J.R. Banavar, A. Maritan, and A. Rinaldo (1999), Size and form in efficient transportation networks, *Nature* **99**, 130.

A.-L. Barabási (2002), *Linked: The New Science of Networks* (Perseus Pr, New York).

A.-L. Barabási and R. Albert (1999), Emergence of scaling in random networks, *Science* **286**, 509.

A.-L. Barabási, R. Albert, and H. Jeong (1999), Mean-field theory for scale-free random networks, *Physica* A **272**, 173.

A.-L. Barabási, R. Albert, H. Jeong, and G. Bianconi (2000), Power-law distribution of the World Wide Web: *response*, *Science* **287**, 2115a.

A.-L. Barabási, V.W. Freeh, H. Jeong, and J.B. Brockman (2001), Parasitic computing, *Nature* **412**, 894.

A.-L. Barabási, H. Jeong, Z. Néda, E. Ravasz, A. Schubert, and T. Vicsek (2002), Evolution of the social network of scientific collaborations, *Physica* A **311**, 590; cond-mat/0104162.

A.-L. Barabási, E. Ravasz, and T. Vicsek (2001), Deterministic scale-free networks, *Physica* A **299**, 559; cond-mat/0107419.

A.-L. Barabási and H.E. Stanley (1995), *Fractal Concepts of Surface Growth* (Cambridge University Press, Cambridge).

M. Barahona and L.M. Pecora (2002), Synchronization in small-world systems, *Phys. Rev. Lett.* **89**, 054101; nlin.CD/0112023.

P. Baran (1964), *Introduction to Distributed Communications Networks*, RM-3420-PR, August 1964, http://www.rand.org/publications/RM/baran.list.html.

A.D. Barbour and G. Reinert (2001), Small worlds, *Random Struct. Algorithms* **19**, 54; cond-mat/0006001.

N. Barkai and S. Leibler (1997), Robustness in simple biochemical networks, *Nature* **387**, 913.

A. Barrat (1999), Comment on "Small-world networks: Evidence for a crossover picture", cond-mat/9903323.

A. Barrat and M. Weigt (2000), On the properties of small-world network models, *Eur. Phys. J.* B **13**, 547.

M. Barthélémy and L.A.N. Amaral (1999), Small-world networks: Evidence for a crossover picture, *Phys. Rev. Lett.* **82**, 3180; erratum: *Phys. Rev. Lett.* **82**, 5180.

U. Bastolla, M. Porto, H.E. Roman, and M. Vendruscolo (2001), Connectivity of neutral networks and structural conservation in protein evolution, cond-mat/0107112.

M. Bauer and D. Bernard (2002), A simple asymmetric evolving random network, cond-mat/0203232.

M. Bauer and O. Golinelli (2001), Core percolation in random graphs: a critical phenomena analysis, *Eur. Phys. J.* B **24**, 39; cond-mat/0102011.

R.J. Baxter (1982), *Exactly Solved Models in Statistical Mechanics* (Academic Press, London).

A. Becskei and L. Serrano (2000), Engineering stability in gene networks by autoregulation, *Nature* **405**, 590.

A. Bekessy, P. Bekessy, and J. Komlos (1972), Asymptotic enumeration of regular matrices, *Stud. Sci. Math. Hung.* **7**, 343.

R. Belew (2000), *Finding Out About: A Cognitive Perspective on Search Engine Technology and the WWW* (Cambridge University Press, Cambridge).

E.A. Bender and E.R. Canfield (1978), The asymptotic number of labelled graphs with given degree sequences, *J. Combinatorial Theor.* A **24**, 296.

V.L. Berezinskii (1970), Destruction of long range order in one-dimensional and two-dimensional systems having a continuous symmetry group, *Sov. Phys. JETP* **32**, 493.

J. Berg and M. Lässig (2002), Correlated random networks, cond-mat/0205589.

A.T. Bernardes, D. Stauffer, and J. Kertész (2002), Election results and the Sznajd model on Barabási network, *Eur. Phys. J.* B **25**, 123; cond-mat/0111147.

U.S. Bhalla and R. Iyengar (1999), Emergent properties of networks of biological signaling pathways, *Science* **283**, 381.

P. Bialas, L. Bogacz, Z. Burda, and D. Johnston (2000), Finitesize scaling of the balls in boxes model, *Nucl. Phys.* B **575**, 599; hep-lat/9910047.

P. Bialas, Z. Burda, and D. Johnston (1997), Condensation in the Backgammon model, *Nucl. Phys.* B **493**, 505; cond-mat/9609264.

P. Bialas, Z. Burda, and D. Johnston (1999), Phase diagram of the mean field model of simplicial gravity, *Nucl. Phys.* B **542**, 413; gr-qc/9808011.

G. Bianconi (2002a), Mean-field solution of the Ising model on a Barabási–Albert network, cond-mat/0204455.

G. Bianconi (2002b), Growing Cayley trees described by Fermi distribution, cond-mat/0204506.

G. Bianconi and A.-L. Barabási (2001a), Competition and multiscaling in evolving networks, *Europhys. Lett.* **54**, 439; cond-mat/0011224.

G. Bianconi and A.-L. Barabási (2001b), Bose–Einstein condensation in complex networks, *Phys. Rev. Lett.* **86**, 5632; cond-mat/0011029.

S. Bilke and C. Peterson (2001), Topological properties of citation and metabolic networks, *Phys. Rev.* E **64**, 036106; cond-mat/0103361.

M. Boguñá and R. Pastor-Satorras (2002), Epidemic spreading in correlated complex networks, cond-mat/0205621.

B. Bollobás (1980), A probabilistic proof of an asymptotic formula for the number of labelled random graphs, *Eur. J. Combinatorics* **1**, 311.

B. Bollobás (1985), *Random Graphs* (Academic Press, London).

B. Bollobás (1998), *Modern Graph Theory* (Springer, New York).

B. Bollobás and O. Riordan (2001), The diameter of a scale-free random graph, preprint.

B. Bollobás, O. Riordan, J. Spencer, and G. Tusnády (2001), The degree sequence of a scale-free random graph process, *Random Struct. Algorithms* **18**, 279.

S. Bornholdt and H. Ebel (2001), World-Wide Web scaling exponent from Simon's 1955 model, *Phys. Rev.* E **64**, 035104; cond-mat/0008465.

S. Bornholdt and T. Rohlf (2000), Topological evolution of dynamical networks: Global criticality from local dynamical rules, *Phys. Rev. Lett.* **84**, 6114, cond-mat/0003215.

S. Bornholdt and H.G. Schuster (Eds.) (2002), *Handbook of Graphs and Networks*, (Wiley-VCH, Berlin).

I. Bose (2002), Biological networks, cond-mat/0202192.

J.P. Bouchaud and M. Mézard (2000), Wealth condensation in a simple model of economy, *Physica* A **282**, 536.

J.P. Bouchaud and M. Potters (2000), *Theory of Financial Risks: From Statistical Physics to Risk Management* (Cambridge University Press, Cambridge).

M. Brede and U. Behn (2001), Architecture of idiotypic networks: Percolation and scaling behaviour, *Phys. Rev.* E **64**, 011908.

S. Brin and L. Page (1998), The anatomy of a large-scale hypertextual web search engine, Proceedings of the Seventh International World Wide Web Conference, pp. 107–117.

A. Broder, R. Kumar, F. Maghoul, P. Raghavan, S. Rajagopalan, R. Stata, A. Tomkins, and J. Wiener (2000), Graph structure in the web, *Comput. Netw.* **33**, 309.

M. Buchanan (2002), *Nexus: Small Worlds and the Groundbreaking Science of Networks* (W.W. Norton & Co, New York).

A. Bunde and S. Havlin (1994), *Fractals in Science* (Springer, Berlin).

Z. Burda, J.D. Correia, and A. Krzywicki (2001), Statistical ensemble of scale-free random graphs, *Phys. Rev.* E **64**, 046118; cond-mat/0104155.

Z. Burda, D. Johnston, J. Jurkiewicz, M. Kaminski, M.A. Nowak, G. Papp, and I. Zahed (2002), Wealth condensation in Pareto macro-economies, *Phys. Rev.* E **65**, 026102; cond-mat/0101068.

D. Butler (2000), Souped-up search engines, *Nature* **405**, 112.

G. Caldarelli, R. Marchetti, and L. Pietronero (2000), The fractal properties of Internet, *Europhys. Lett.* **52**, 386; cond-mat/0009178.

D.S. Callaway, J.E. Hopcroft, J.M. Kleinberg, M.E.J. Newman, and S.H. Strogatz (2001), Are randomly grown graphs really random?, *Phys. Rev.* E **64**, 041902; cond-mat/0104546.

D.S. Callaway, M.E.J. Newman, S.H. Strogatz, and D.J. Watts (2000), Network robustness and fragility: Percolation on random graphs, *Phys. Rev. Lett.* **85**, 5468; cond-mat/0007300.

F. Calvo, J.P.K. Doye, and D.J. Wales (2002), Collapse of Lennard-Jones homopolymers: Size effects and energy landscapes, *J. Chem. Phys.* **116**, 2642.

A. Calvó-Armengol and Y. Zenou (2001), Job matching, social network and word-of-mouth communication, nep/0111003.

J. Camacho, R. Guimerà, and L.A.N. Amaral (2001), Analytical solution of a model for complex food webs, *Phys. Rev.* E **65**, 030901; cond-mat/0102127.

J. Camacho, R. Guimerà, and L.A.N. Amaral (2002), Robust patterns in food web structure, *Phys. Rev. Lett.* **88**, 228102; cond-mat/0102127.

A. Capocci, G. Caldarelli, R. Marchetti, and L. Pietronero (2001), Growing dynamics of Internet providers, *Phys. Rev.* E **64**, 035105; cond-mat/0106084.

J.M. Carlson and J. Doyle (1999), Highly optimized tolerance: A mechanism for power laws in designed systems, *Phys. Rev.* E **60**, 1412.

J.M. Carlson and J. Doyle (2000), Highly optimized tolerance: Robustness and design in complex systems, *Phys. Rev. Lett.* **84**, 2529.

S. Chakrabarti, B. Dom, D. Gibson, J. Kleinberg, S.R. Kumar, P. Raghavan, S. Rajagopalan and A. Tomkins (1999), Hypersearching the Web, *Sci. Am.* June, 54.

K. Christensen, R. Donangelo, B. Koiller, and K. Sneppen (2000), Evolution of random networks, *Phys. Rev. Lett.* **81**, 2380.

F.R.K. Chung (1997), *Spectral Graph Theory* (American Mathematical Society, Providence, RI).

K.C. Claffy (1999), Internet measurement and data analysis: Topology, workload, performance and routing statistics, Proceedings of NAE workshop, http://wwwdev.caida.org/outreach/papers/Nae/.

K. Claffy, T.E. Monk, and D. McRobb (1999), Internet tomography, Nature web matters, 7 January,
http://www.helix.nature.com/webmatters/tomog/tomog.html.

J.E. Cohen, F. Briand, and C.M. Newman (1990), *Community food webs: Data and theory* (Springer, Berlin).

R. Cohen, D. ben-Avraham, and S. Havlin (2002a), Percolation Critical Exponents in Scale-Free Networks, cond-mat/0202259.

R. Cohen, D. ben-Avraham, and S. Havlin (2002b), Efficient immunization of populations and computers, cond-mat/0207387.

R. Cohen, K. Erez, D. ben-Avraham, and S. Havlin (2000), Resilience of the Internet to random breakdowns, *Phys. Rev. Lett.* **85**, 4625; cond-mat/0007048.

R. Cohen, K. Erez, D. ben-Avraham, and S. Havlin (2001a), Breakdown of the Internet under intentional attack, *Phys. Rev. Lett.* **86**, 3682; cond-mat/0010251.

R. Cohen, K. Erez, D. ben-Avraham, and S. Havlin (2001b), Reply on the comment of S.N. Dorogovtsev and J.F.F. Mendes, *Phys. Rev. Lett.* **87**, 219802.

R. Cohen and S. Havlin (2002), Ultra small world in scale-free networks, cond-mat/0205476.

F. Comellas and M. Sampels (2000), Deterministic small-world communication networks, *Inf. Process. Lett.* **76**, 83.

F. Comellas and M. Sampels (2001), Deterministic small-world networks, cond-mat/0111194.

M. Copelli, R.M. Zorzenon dos Santos, and J.S. Sa Martins (2001), Emergence of Hierarchy on a Network of Complementary Agents, cond-mat/0110350.

D. Cvetcović, M. Doob, and H. Sachs (1979), *Spectra of Graphs* (Cambridge University Press, Cambridge).

J. Dall and M. Christensen (2002), Random geometric graphs, Phys. Rev. E **66**, 016121; cond-mat/0203026.

J. Davidsen, H. Ebel, and S. Bornholdt (2001), Emergence of a small world from local interactions: Modeling acquaintance networks, *Phys. Rev. Lett.* **88**, 128701; cond-mat/0108302.

J. Dean and M.R. Henzinger (1999), Finding related pages in the World Wide Web, in Proceedings of the WWW8 Conference, Toronto, http://www.unizh.ch/home/mazzo/reports/www8conf/2148/html.

Z. Dezső and A.-L. Barabási (2002), Halting viruses in scale-free networks, *Phys. Rev.* E **65**, 055103; cond-mat/0107420.

M. Dodge and R. Kitchin (2000), *Mapping Cyberspace*, (Routledge, London).

M. Dodge and R. Kitchin (2001), *Atlas of Cyberspace*, (Addison-Wesley, Harlow).

S.N. Dorogovtsev, A.V. Goltsev, and J.F.F. Mendes (2002a), Pseudofractal scale-free web, *Phys. Rev.* E **65**, 066122; cond-mat/0112143.

S.N. Dorogovtsev, A.V. Goltsev, and J.F.F. Mendes (2002b), Ising model on networks with an arbitrary distribution of connections, *Phys. Rev.* E **66**, 016104; cond-mat/0203227.

S.N. Dorogovtsev and J.F.F. Mendes (2000a), Exactly solvable small-world network, *Europhys. Lett.* **50**, 1; cond-mat/9907445.

S.N. Dorogovtsev and J.F.F. Mendes (2000b), Evolution of networks with aging of sites, *Phys. Rev.* E **62**, 1842; cond-mat/0001419.

S.N. Dorogovtsev and J.F.F. Mendes (2000c), Scaling behaviour of developing and decaying networks, *Europhys. Lett.* **52**, 33; cond-mat/0005050.

S.N. Dorogovtsev and J.F.F. Mendes (2001a), Effect of the accelerated growth of communications networks on their structure, *Phys. Rev.* E **63**, 025101; cond-mat/0009065.

S.N. Dorogovtsev and J.F.F. Mendes (2001b), Scaling properties of scale-free evolving networks: Continuum approach, *Phys. Rev.* E **63**, 056125; cond-mat/0012009.

S.N. Dorogovtsev and J.F.F. Mendes (2001c), Comment on "Breakdown of the Internet under Intentional Attack", *Phys. Rev. Lett.* **87**, 219801; cond-mat/0109083.

S.N. Dorogovtsev and J.F.F. Mendes (2001d), Language as an evolving Word Web, *Proc. R. Soc. London* B **268**, 2603; cond-mat/0105093.

S.N. Dorogovtsev and J.F.F. Mendes (2002a), Evolution of networks, *Adv. Phys.* **51**, 1079; cond-mat/0106144.

S.N. Dorogovtsev and J.F.F. Mendes (2002b), Accelerated growth of networks, *Handbook of Graphs and Networks: From the Genome to the Internet*, ed. S. Bornholdt and H.G. Schuster (Wiley-VCH, Berlin), pp. 320–343; cond-mat/0204102.

S.N. Dorogovtsev, J.F.F. Mendes, and A.N. Samukhin (2000a), Structure of growing networks with preferential linking, *Phys. Rev. Lett.* **85**, 4633.

S.N. Dorogovtsev, J.F.F. Mendes, and A.N. Samukhin (2000b), WWW and Internet models from 1955 till our days and the "popularity is attractive" principle, cond-mat/0009090.

S.N. Dorogovtsev, J.F.F. Mendes, and A.N. Samukhin (2000d), Growing networks with heritable connectivity of nodes, cond-mat/0011077.

S.N. Dorogovtsev, J.F.F. Mendes, and A.N. Samukhin (2001a), Giant strongly connected component of directed networks, *Phys. Rev.* E **64**, 025101; cond-mat/0103629.

S.N. Dorogovtsev, J.F.F. Mendes, and A.N. Samukhin (2001b), Anomalous percolation properties of growing networks, *Phys. Rev.* E **64**, 066110; cond-mat/0106141.

S.N. Dorogovtsev, J.F.F. Mendes, and A.N. Samukhin (2001c), Generic scale of 'scale-free' networks, *Phys. Rev.* E **63**, 062101; cond-mat/0011115.

S.N. Dorogovtsev, J.F.F. Mendes, and A.N. Samukhin (2002a), Multifractal properties of growing networks, *Europhys. Lett.* **57**, 334; cond-mat/0106142.

S.N. Dorogovtsev, J.F.F. Mendes, and A.N. Samukhin (2002b), Principles of statistical mechanics of random networks, cond-mat/0204111.

S.N. Dorogovtsev, J.F.F. Mendes, and A.N. Samukhin (2002c), Modern architecture of random graphs: Constructions and correlations, cond-mat/0206467.

J.P.K. Doye (2002), The network topology of a potential energy landscape: A static scale-free network, *Phys. Rev. Lett.* **88**, 238701; cond-mat/0201430.

J. Doyle and J.M. Carlson (2000), Power laws, highly optimized tolerance, and generalized source coding, *Phys. Rev. Lett.* **84**, 5656.

J.A. Dunne, R.J. Williams, and N.D. Martinez (2002a), Small networks but not small worlds: Unique aspects of food web structure, *Proc. Natl Acad. Sci. USA* **99**, October; Working Papers of Santa Fe Institute, 02-03-010, http://www.santafe.edu/sfi/publications/wpabstract/200203010.

J.A. Dunne, R.J. Williams, and N.D. Martinez (2002b), Network topology and species loss in food webs: Robustness increases with connectance, Working Papers of Santa Fe Institute, 02-03-013, http://www.santafe.edu/sfi/publications/wpabstract/200203013.

H. Ebel, L.-I. Mielsch, and S. Bornholdt (2002), Scale-free topology of e-mail networks, *Phys. Rev.* E **66**, October; cond-mat/0201476.

J.-P. Eckmann and E. Moses (2001), Curvature of co-links uncovers hidden thematic layers in the World Wide Web, cond-mat/0110338.

L. Egghe and R. Rousseau (1990), *Introduction to Informetrics: Quantitative Methods in Library, Documentation and Information Science* (Elsevier, Amsterdam).

M.B. Elowitz and S. Leibler (2000), A synthetic oscillatory network of transcriptional regulators, *Nature* **403**, 335.

D. Eppstein and J. Wang (2002), A steady state model for graph power laws, cs.DM/0204001.

P. Erdős and A. Rényi (1959), On random graphs, *Publ. Math. Debrecen* **6**, 290.

P. Erdős and A. Rényi (1960), On the evolution of random graphs, *Publ. Math. Inst. Hung. Acad. Sci.* **5**, 17.

G. Ergun (2002), Human sexual contact network as a bipartite graph, *Physica* A **308**, 483; cond-mat/0111323.

G. Ergun and G.J. Rodgers (2002), Growing random networks with fitness, *Physica* A **303**, 261; cond-mat/0103423.

M. Faloutsos, P. Faloutsos, and C. Faloutsos (1999), On power-law relationships of the Internet topology, *Comput. Commun. Rev.* **29**, 251.

I.J. Farkas, I. Derényi, A.-L. Barabási, and T. Vicsek (2001), Spectra of "Real-World" graphs: Beyond the semi-circle law, *Phys. Rev.* E **64**, 026504; cond-mat/0102335.

I.J. Farkas, H. Jeong, T. Vicsek, A.-L. Barabási, and Z.N. Oltvai (2002), The topology of the transcription regulatory network in the yeast, *S. cerevisiae*, cond-mat/0205181.

R. Ferrer i Cancho, C. Janssen, and R. Solé (2001), The topology of technology graphs: Small world patterns in electronic circuits, *Phys. Rev. E* **64**, 046119.

R. Ferrer i Cancho and R.V. Solé (2001a), The small-world of human language, Proc. R. Soc. B **268**, 2261.

R. Ferrer i Cancho and R.V. Solé (2001b), Optimization in complex networks, cond-mat/0111222.

R. Ferrer i Cancho and R. Solé (2001c), Two regimes in the frequency of words and the origins of complex lexicons: Zipf's law revised, *J. Quant. Linguist.* **8**, 165.

P. Flajolet, D.E. Knuth, and B. Pittel (1989), The first cycles in an evolving graph, *Discrete Math.* **75**, 167.

P.J. Flory (1941), Molecular size distribution in three dimensional polymers: I. Gelation; II. Trifunctional branching units; III. Tetrafunctional branching units, *J. Am. Chem. Soc.* **63**, 3083; 3091; 3096.

P.J. Flory (1971), *Principles of Chemistry* (Cornell University Press, Ithaca, NY).

P.J. Flory (1976), Statistical thermodynamics of random networks, *Proc. R. Soc. London* A **351**, 351.

C.M. Fortuin and P.W. Kasteleyn (1972), On the random-cluster model. I. Introduction and relation to other models, *Physica* **57**, 536.

J.J. Fox and C.C. Hill (2001), From topology to dynamics in biochemical networks, *Chaos* **11**, 809.

S.M. Fraser and C.M. Reidys (1997), Evolution of random catalytic networks, Proceedings of the 4th European Conference on Artificial Life, pp. 92–100.

L.C. Freeman (1977), A set of measures of centrality based on betweenness, *Sociometry* **40**, 35.

H. Fuks and A.T. Lawniczak (1999), Performance of data networks with random links, *Math. Comput. Simulation* **51**, 103; adap-org/9909006.

E. Garfield (1972), Citation analysis as a tool in journal evaluation, *Science* **178**, 471.

E. Garfield (1979), *Citation Indexing: Its Theory and Application in Science* (Wiley, New York).

R. Gibrat (1931), *Les inégalités économiques* (Sirey, Paris).

D. Gibson, J.M. Kleinberg, and P. Raghavan (1998), Inferring Web communities from link topology, Proceedings of the 9th ACM Conference on Hypertext and Hypermedia, 225–234.

N. Gilbert (1997), A simulation of the structure of academic science, *Sociol. Res. Online* **2**, 2.

M. Girvan and M.E.J. Newman (2002), Community structure in social and biological networks, *Proc. Natl Acad. Sci. USA* **99**, 8271; cond-mat/0112110.

M. Gitterman (2000), Small-world phenomena in physics: The Ising model, *J. Phys.* A **33**, 8373.

L. Glass and C. Hill (1998), Ordered and disordered dynamics in random networks, *Europhys. Lett.* **41**, 599.

P.M. Gleiss, P.F. Stadler, A. Wagner, and D.A. Fell (2000), Small cycles in small worlds, cond-mat/0009124.

K.-I. Goh, B. Kahng, and D. Kim (2001a), Spectra and eigenvectors of scale-free networks, *Phys. Rev.* E **64**, 051903; cond-mat/0103337.

K.-I. Goh, B. Kahng, and D. Kim (2001b), Universal behavior of load distribution in scale-free networks, *Phys. Rev. Lett.* **87**, 278701; cond-mat/0106565.

K.-I. Goh, B. Kahng, and D. Kim (2002), Fluctuation-driven dynamics of the Internet topology, *Phys. Rev. Lett.* **88**, 108701; cond-mat/0108031.

K.-I. Goh, E.S. Oh, H. Jeong, B. Kahng, and D. Kim (2002), Classification of scale free networks, *Proc. Natl Acad. Sci. USA* **99**, October; cond-mat/0205232.

A.V. Goltsev, S.N. Dorogovtsev, and J.F.F. Mendes (2002), Critical phenomena in networks, cond-mat/0204596.

R. Govindan and H. Tangmunarunkit (2000), Heuristics for Internet map discovery, Proceedings of the 2000 IEEE INFOCOM Conference, Tel Aviv, Israel, March, pp. 1371–1380, http://citeseer.nj.nec.com/govindan00heuristics.html.

P. Grassberger (1983), On the critical behavior of the general epidemic process and dynamical percolation, *Math. Biosci.* **63**, 157.

X. Guardiola, A. Díaz-Guilera, C. J. Perez, A. Arenas, and M. Llas (2002), Modelling diffusion of innovations in a social network, cond-mat/0204141.

V. Gudkov and J.E. Johnson (2001), New approach for network monitoring and intrusion detection, cs.CR/0110019.

R. Guimerà, A. Arenas, and A. Díaz-Guilera (2001), Communication and optimal hierarchical networks, Physica A **299**, 247; cond-mat/0103112.

N. J. Gunther (2002), Hypernets – Good (G)news for Gnutella, cs.PF/0202019.

J. Han and W. Li (2002), How structure affects power-law behavior, cond-mat/0205259.

T.E. Harris (1989), *The Theory of Branching Processes* (Dover, New York).

J.H. Holland (2001), Exploring the evolution of complexity in signaling networks, Working Papers of Santa Fe Institute, 01-10-062, http://www.santafe.edu/sfi/publications/Abstracts/01-10-062abs.html.

P. Holme, M. Huss, and H. Jeong (2002), Subnetwork hierarchies of biochemical pathways, cond-mat/0206292.

P. Holme and B.J. Kim (2001), Growing scale-free networks with tunable clustering, *Phys. Rev.* E **65**, 026135; cond-mat/0110452.

P. Holme and B.J. Kim (2002), Vertex overload breakdown in evolving networks, *Phys. Rev.* E **65**, 066109; cond-mat/0204120.

P. Holme, B.J. Kim, C.N. Yoon, and S.K. Han (2002), Attack vulnerability of complex networks, *Phys. Rev.* E **65**, 056109; cond-mat/0202410.

H. Hong, M.Y. Choi, and B.J. Kim (2002a), Synchronization on small-world networks, *Phys. Rev.* E **65**, 026139; cond-mat/0110359.

H. Hong, M.Y. Choi, and B.J. Kim (2002b), Phase ordering on small-world networks with nearest-neighbor edges, *Phys. Rev.* E **65**, 047104; cond-mat/0203177.

H. Hong, B.J. Kim, and M.Y. Choi (2002), Comment on 'Ising model on a small world network', *Phys. Rev.* E **66**, 018101; cond-mat/0204357.

M. Hörnquist (2001), Scale-free networks are not robust under neutral evolution, *Europhys. Lett.* **56**, 461.

B. A. Huberman (2001), *The Laws of the Web*, (MIT Press, Cambridge, MA).

B.A. Huberman and L.A. Adamic (1999), Growth dynamics of the World-Wide Web, *Nature* **401**, 131.

B.A. Huberman, P.L.T. Pirolli, J.E. Pitkow, and R.J. Lukose (1998), Strong regularities in World Wide Web surfing, *Science* **280**, 95.

C. Huiterna (2000), *Routing in the Internet*, 2nd ed. (Prentice Hall, Upper Saddle River, NJ).

M. Ipsen and A.S. Mikhailov (2001), Evolutionary reconstruction of networks, nlin.AO/0111023.

T. Ito, T. Chiba, R. Ozawa, M. Yoshida, M. Hattori, and Y. Sakaki (2001), A comprehensive two-hybrid analysis to explore the yeast protein interactome, *Proc. Natl Acad. Sci. USA* **98**, 4569.

T. Ito, K. Tashiro, S. Muta, R. Ozawa, T. Chiba, M. Nishizawa, K. Yamamoto, S. Kuhara, and Y. Sakaki (2000), Toward a protein–protein interaction map of the budding yeast: A comprehensive system to examine two-hybrid interactions in all possible combinations between the yeast proteins, *Proc. Natl Acad. Sci. USA* **97**, 1143.

S. Jain and S. Krishna (1998), Autocatalytic sets and the growth of complexity in an evolutionary model, *Phys. Rev. Lett.* **81**, 5684; adap-org/9809003.

S. Jain and S. Krishna (2001), A model for the emergence of cooperation, interdependence and structure in evolving networks, *Proc. Natl Acad. Sci. USA* **98**, 543; nlin.AO/0005039.

S. Jain and S. Krishna (2002), Crashes, recoveries, and 'Core-Shifts' in a model of evolving networks, from real data to community assembly models, *Phys. Rev.* E **65**, 026103.

S. Jalan and R.E. Amritkar (2002), Self-organized and driven phase synchronization in coupled map scale free networks, nlin.AO/0201051.

W. Janke, D.A. Johnston, and M. Stathakopoulos (2002), Kertész on fat graphs?, cond-mat/0201496.

S. Janson, D.E. Knuth, T. Luczak, and B. Pittel (1993), The birth of the giant component, *Random Struct. Algorithms* **4**, 233.

S. Janson, T. Luczak, and A. Rucinski (2000), *Random Graphs* (Wiley, New York).

F. Jasch and A. Blumen (2001), Target problem on small-world networks, *Phys. Rev.* E **63**, 041108.

H. Jeong, A.-L. Barabási, B. Tombor, and Z.N. Oltvai (2001), The global organization of cellular networks, Proceedings of Workshop on Computation of Biochemical Pathways and Genetic Networks, Heidelberg.

H. Jeong, S.P. Mason, A.-L. Barabási, and Z.N. Oltvai (2001), Lethality and centrality in protein networks, *Nature* **411**, 41; cond-mat/0105306.

H. Jeong, Z. Néda, and A.-L. Barabási (2001), Measuring preferential attachment for evolving networks, cond-mat/0104131.

H. Jeong, B. Tombor, R. Albert, Z.N. Oltvai, and A-L. Barabási (2000), The large-scale organization of metabolic networks, *Nature* **407**, 651.

S. Jespersen and A. Blumen (2000), Small-world networks: Links with long-tailed distributions, *Phys. Rev.* E **62**, 6270. cond-mat/0009082.

S. Jespersen, I.M. Sokolov, and A. Blumen (2000a), Small-world rose networks as models of cross-linked polymers, *J. Chem. Phys.* **113**, 7652; cond-mat/0004392.

S. Jespersen, I.M. Sokolov, and A. Blumen (2000b), Relaxation properties of small-world networks, *Phys. Rev.* E **62**, 4405; cond-mat/0004214.

E.M. Jin, M. Girvan, and M.E.J. Newman (2001), Structure of growing social networks, *Phys. Rev.* E **64**, 046132.

J. Jost and M.P. Joy (2002), Evolving networks with distance preferences, cond-mat/0202343.

S. Jung, S. Kim, and B. Kahng (2002), A geometric fractal growth model for scale free networks, Phys. Rev. E **65**, 056101; cond-mat/0112361.

B. Kahng, Y. Park, and H. Jeong (2001), Robustness of the in-degree exponent for the World Wide Web, cond-mat/0112358.

A.R. Kansal and S. Torquato (2001), Globally and locally minimal weight spanning tree networks, *Physica* A **301**, 601; cond-mat/0112149.

V. Karimipour and A. Ramzanpour (2002), Correlation effects in a simple small-world network, *Phys. Rev.* E **65**, 036122; cond-mat/0012313.

P.W. Kasteleyn and C.M. Fortuin (1969), Phase transitions in lattice systems with random local properties, *Phys. Soc. Jpn (Suppl.)* **26**, 11.

R. Kasturirangan (1999), Multiple scales in small-world graphs, cond-mat/9904055.

S.A. Kauffman (1969), Metabolic stability and epigenesis in randomly constructed genetic nets, *J. Theor. Biol.* **22**, 437.

S.A. Kauffman (1993), *The Origins of Order: Self-organization and Selection in Evolution* (Oxford University Press, New York, Oxford).

S.A. Kauffman (1995), *At Home in the Universe: The Search for the Laws of Self-Organization and Complexity* (Oxford University Press, Oxford).

S.A. Kauffman (2000), *Investigations* (Oxford University Press, Oxford).

B.J. Kim, H. Hong, P. Holme, G.S. Jeon, P. Minnhagen, and M.Y. Choi (2001), XY model in small-world networks, *Phys. Rev.* E **64**, 056135; cond-mat/0108392.

B.J. Kim, C.N. Yoon, S.K. Han, and H. Jeong (2001), Path finding strategies in scale-free networks, *Phys. Rev.* E **65**, 027103; cond-mat/0111232.

H.-J. Kim, Y. Lee, I.-M. Kim, and B. Kahng (2001), Scale-free networks in financial correlations, cond-mat/0107449.

J. Kim, P.L. Krapivsky, B. Kahng, and S. Redner (2002), Evolving protein interaction networks, cond-mat/0203167.

O. Kinouchi, A.S. Martinez, G.F. Lima, G.M. Lourenco, and S. Risau-Gusman (2001), Deterministic walks in random networks: An application to thesaurus graphs, cond-mat/0110217.

J.M. Kleinberg (1998), Authoritative sources in a hyperlinked environment, Proceedings of the 9th Annual ACM-SIAM Symposium on Discrete Algorithms, San Francisco, California, pp. 668–677.

J.M. Kleinberg (1999a), Hubs, authorities, and communities, *ACM Comput. Surv.* **31**, No. 4, article No. 5.

J. Kleinberg (1999b), The small-world phenomenon: An algorithmic perspective. Cornell University Computer Science Department Technical Report 99-1776, http://www.cs.cornell.edu/home/kleinber/swn.ps.

J.M. Kleinberg (2000), Navigation in a small world, *Nature* **406**, 845.

J. Kleinberg, R. Kumar, P. Raphavan, S. Rajagopalan, and Tomkins (1999), The Web as a graph: measurements, models, and methods, Proceedings of the 5th International Conference on Combinatorics and Computing.

K. Klemm and V.M. Eguíluz (2002a), Highly clustered scale-free networks, *Phys. Rev.* E **65**, 036123; cond-mat/0107606.

K. Klemm and V.M. Eguíluz (2002b), Growing scale-free networks with small-world behavior, *Phys. Rev.* E **65**, 057102; cond-mat/0107607.

D.E. Knuth (1977), *The art of computer programming* vol. 1 *Fundamental algorithms* (Addison-Wesley, Reading, MA).

J.M. Kosterlitz and D.J. Thouless (1973), Ordering, metastability and phase transitions in two-dimensional systems, *J. Phys.* C **6**, 1181.

M. Kotrla, F. Slanina, and J. Steiner (2002), Dynamic scaling and universality in evolution of fluctuating random networks, cond-mat/0204476.

P.L. Krapivsky and S. Redner (2001), Organization of growing random networks, *Phys. Rev.* E **63**, 066123; cond-mat/0011094.

P.L. Krapivsky, S. Redner, and F. Leyvraz (2000), Connectivity of growing random network, *Phys. Rev. Lett.* **85**, 4629; cond-mat/0005139.

P.L. Krapivsky, G.J. Rodgers, and S. Redner (2001), Degree distributions of growing networks, *Phys. Rev. Lett.* **86**, 5401; cond-mat/0012181.

V. Krebs (2002), Mapping networks of terrorist cells, *Connections* **24**, 43; net/0203001.

A. Krzywicki (2001), Defining statistical ensembles of random graphs, cond-mat/0110574.

R.V. Kulkarni, E. Almaas, and D. Stroud (1999), Evolutionary dynamics in the Bak-Sneppen model on small-world networks, cond-mat/9905066.

R.V. Kulkarni, E. Almaas, and D. Stroud (2000), Exact results and scaling properties of small-world networks, *Phys. Rev. E* **61**, 4268; cond-mat/9908216.

L. Kullmann and J. Kertész (2001a), Preferential growth: Exact solution of the time-dependent distributions, *Phys. Rev.* E **63**, 051112; cond-mat/0012410.

L. Kullmann and J. Kertész (2001b), Preferential growth: Solution and application to modelling stock market, *Physica* A **299**, 121; cond-mat/0105473.

L. Kullmann, J. Kertész, and K. Kaski (2001), Time dependent cross correlations between different stock returns: A directed network of influence, *Phys. Rev.* E **64**, 057105; cond-mat/0203256.

R. Kumar, P. Raghavan, S. Rajagopalan, D. Sivakumar, A. Tomkins, and E. Upfal (2000a), The Web as a graph, Proceedings of the 19th ACM Symposium on Principles of Database Systems, pp. 1–10.

R. Kumar, P. Raghavan, S. Rajagopalan, D. Sivakumar, A. Tomkins, and E. Upfal (2000b), Stochastic models for the web graph, Proceedings of the 41th IEEE Symposium on Foundations of Computer Science.

R. Kumar, P. Raghavan, S. Rajagopalan, and A. Tomkins (1999a), Extracting large-scale knowledge bases from the web, Proceedings of the 25th VLDB Conference, Edinburgh, pp. 639–650.

R. Kumar, P. Raphavan, S. Rajagopalan, and A. Tomkins (1999b), Trawling the Web for emerging cyber-communities,
http://www.almaden.ibm.com/cs/k53/trawling.ps.

M. Kuperman and G. Abramson (2001a), Complex structures in generalized small worlds, *Phys. Rev.* E **64**, 047103.

M. Kuperman and G. Abramson (2001b), Small world effect in an epidemiological model, *Phys. Rev. Lett.* **86**, 2909.

M. Kuperman and D. Zanette (2002), Stochastic resonance in a model of opinion formation on small-world networks, *Eur. Phys. J.* B **26**, 387; cond-mat/0111289.

L.F. Lago-Fernández, R. Huerta, F. Corbacho, and J.A. Sigüenza (2000), Fast response and temporal coherent oscillations in small-world networks, *Phys. Rev. Lett.* **84**, 2758; cond-mat/9909379.

J. Laherrère and D. Sornette (1998), Stretched exponential distributions in nature and economy: Fat tails with characteristic scales, *Eur. Phys. J.* B **2**, 525; cond-mat/9801293.

J. Lahtinen, J. Kertész, and K. Kaski (2001), Scaling of random spreading in small world networks, *Phys. Rev.* E **64**, 057105; cond-mat/0108199.

J. Lahtinen, J. Kertész, and K. Kaski (2002), Random spreading phenomena in annealed small world networks, *Physica* A **311**, 571; cond-mat/0110365.

D. Lancaster (2002), Cluster growth in two growing network models, *J. Phys.* A **35**, 1179; cond-mat/0110111.

M. Lässig, U. Bastolla, S.C. Manrubia, and A. Valleriani (2001), The shape of ecological networks, *Phys. Rev. Lett.* **86**, 4418; nlin.AO/0101026.

V. Latora and M. Marchiori (2001), Efficient behaviour of small-world networks, *Phys. Rev. Lett.* **87**, 198701; cond-mat/0101396.

V. Latora and M. Marchiori (2002a), Is the Boston subway a small-world network?, cond-mat/0202299.

V. Latora and M. Marchiori (2002b), Economic small-world behavior in weighted networks, cond-mat/0204089.

S. Lawrence (2000), Context in Web search, *IEEE Data Eng. Bull.* **23**, 25.

S. Lawrence and C.L. Giles (1998a), Searching the World Wide Web, *Science* **280**, 98.

S. Lawrence and C.L. Giles (1998b), Context and page analysis for improved Web search, *IEEE Internet Comput.* **2**, 38.

S. Lawrence and C.L. Giles (1999a), Accessibility of information on the web, *Nature* **400**, 107.

S. Lawrence and C.L. Giles (1999b), Searching the Web: General and scientific information access, *IEEE Commun.* **37**, 116.

M. Leone, A. Vázquez, A. Vespignani, and R. Zecchina (2002), Ferromagnetic ordering in graphs with arbitrary degree distribution, cond-mat/0203416.

M. Levene, T. Fenner, G. Loizou, and R. Wheeldon (2002), A stochastic model for the evolution of the Web, *Comput. Netw.* **39**, 277; cond-mat/0110016.

J.-S. Lih and J.-L. Chern (2001), Power-law scaling for communication networks with transmission errors, cond-mat/0109009.

F. Liljeros, C.R. Edling, L.A.N. Amaral, H.E. Stanley, and Y. Åberg (2001), The web of human sexual contacts, *Nature* **411**, 907; cond-mat/0106507.

S. Lise and M. Paczuski (2002), A nonconservative earthquake model of self-organized criticality on a random graph, *Phys. Rev. Lett.* **88**, 228301; cond-mat/0204491.

A.L. Lloyd and R.M. May (2001), How viruses spread among computers and people, *Science* **292**, 1316.

A.J. Lotka (1926), The frequency distribution of scientific productivity, *J. Washington Acad. Sci.* **16**, 317.

T. Lux and M. Marchesi (1999), Scaling and criticality in a stochastic multi-agent model of a financial market, *Nature* **397**, 498.

C. Lynch (1997), Searching the Internet, *Sci. Am.*, June, 52.

B.B. Mandelbrot (1983), *The Fractal Geometry of Nature* (Freeman, New York).

S.S. Manna and P. Sen (2002), Modulated scale-free network in the Euclidean space, cond-mat/0203216.

R.N. Mantegna and H.E. Stanley (1999), *An Introduction to Econophysics: Correlations and Complexity in Finance*, (Cambridge University Press, Cambridge).

M. Marchiori and V. Latora (2000), Harmony in the small-world, *Physica* A **285**, 539; cond-mat/0008357.

A. Maritan, F. Colaiori, A. Flammini, M. Cieplak, and J. Banavar (1996), Universality classes of optimal channel networks, *Science* **272**, 984.

S. Maslov and K. Sneppen (2002a), Specificity and stability in topology of protein networks, *Science* **296**, 910; cond-mat/0205380.

S. Maslov and K. Sneppen (2002b), Pattern detection in complex networks: Correlation profile of the Internet, cond-mat/0205379.

S. Maslov and Y.-C. Zhang (2001), Towards information theory of knowledge networks, *Phys. Rev. Lett.* **87**, 248701; cond-mat/0104121.

N. Mathias and V. Gopal (2001), Small-worlds: How and why, *Phys. Rev.* E **63**, 021117; cond-mat/0002076.

S.M. Maurer and B.A. Huberman (2000), Competitive dynamics of Web sites, nlin.CD/0003041.

S.M. Maurer and B.A. Huberman (2001), Restart strategies and Internet congestion, *J. Econ. Dyn. Control* **25** 641; nlin.CD/9905036.

R.M. May and A.L. Lloyd (2001), Infection dynamics on scale-free networks, *Phys. Rev.* E **64**, 066112.

M.A. de Menezes, C.F. Moukarzel, and T.J.P. Penna (2000), First-order transition in small-world networks, *Europhys. Lett.* **50**, 574; cond-mat/9903426.

S. Milgram (1967), The small world problem, *Psychol. today* **2**, 60.

M. Molloy and B. Reed (1995), A critical point for random graphs with a given degree sequence, *Random Struct. Algorithms* **6**, 161.

M. Molloy and B. Reed (1998), The size of the giant component of a random graph with a given degree sequence, *Combinatorics, Probab. Comput.* **7**, 295.

R. Monasson (1999), Diffusion, localization, and dispersion relations on small-world lattices, *Eur. Phys. J.* B **12**, 555; cond-mat/9903347.

J.M. Montoya and R.V. Solé (2001), Topological properties of food webs: From

real data to community assembly models, Working Papers of Santa Fe Institute, 01-11-069,
http://www.santafe.edu/sfi/publications/Abstracts/01-11-069abs.html.

J.M. Montoya and R.V. Solé (2002), Small world patterns in food webs, *J. Teor. Biol.* **214**, 405; cond-mat/0011195.

C. Moore and M.E.J. Newman (2000a), Epidemics and percolation in small-world networks, *Phys. Rev.* E **61**, 5678; cond-mat/9911492.

C. Moore and M.E.J. Newman (2000b), Exact solution of site and bond percolation on small-world networks, *Phys. Rev.* E **62**, 7059.

A.A. Moreira, J.S. de Andrade Jr., and L.A.N. Amaral (2002), Extremum statistics in scale-free network models, cond-mat/0205411.

Y. Moreno, J.B. Gómez, and A.F. Pacheco (2002), Instability of scale-free networks under node-breaking avalanches, *Europhys. Lett.* **58**, 630; cond-mat/0106136.

Y. Moreno, R. Pastor-Satorras, and A. Vespignani (2002), Epidemic outbreaks in complex heterogeneous networks, *Eur. Phys. J.* **26**, 521; cond-mat/0107267.

Y. Moreno and A. Vázquez (2002), The Bak-Sneppen model on scale-free networks, *Europhys. Lett.* **57**, 765; cond-mat/0108494.

S. Mossa, M. Barthélémy, H.E. Stanley, and L.A.N. Amaral (2002), Truncation of power law behaviour in "scale-free" network models due to information filtering, *Phys. Rev. Lett.* **88**, 138701; cond-mat/0201421.

C.F. Moukarzel (2000), Spreading and shortest paths in systems with sparse long-range connections, *Phys. Rev.* E **60**, 6263.

C.F. Moukarzel and M.A. de Menezes (1999), Infinite characteristic length in small-world systems, cond-mat/9905131.

M.E.J. Newman (2000a), Small worlds: The structure of social networks, cond-mat/0001118.

M.E.J. Newman (2000b), Models of the small world, *J. Stat. Phys.* **101**, 819.

M.E.J. Newman (2001a), The structure of scientific collaboration networks, *Proc. Natl Acad. Sci. USA* **98**, 404; cond-mat/0007214.

M.E.J. Newman (2001b), Clustering and preferential attachment in growing networks, *Phys. Rev.* E **64**, 025102 cond-mat/0104209.

M.E.J. Newman (2001c), Ego-centered networks and the ripple effect, cond-mat/0111070.

M.E.J. Newman (2001d), Exact solutions of epidemic models on networks, Working Papers of Santa Fe Institute, 01-12-073,
http://www.santafe.edu/sfi/publications/Abstracts/01-12-073abs.html; cond-mat/0201433.

M.E.J. Newman (2001e), Who is the best connected scientist? A study of scientific coauthorship networks, Scientific collaboration networks. I. Network construction and fundamental results, *Phys. Rev.* E **64**, 016131; II. Shortest paths, weighted networks, and centrality, *Phys. Rev.* E **64**, 016132; cond-mat/0010296.

M.E.J. Newman (2002a), Random graphs as models of networks, cond-mat/0202208.

M.E.J. Newman (2002b), The spread of epidemic disease on networks, *Phys. Rev.* E **66**, 016128; cond-mat/0205009.

M.E.J. Newman (2002c), Assortative mixing in networks, cond-mat/0205405.

M.E.J. Newman, S. Forrest, and J. Balthrop (2002), Email networks and the spread of computer viruses, *Phys. Rev.* E **66**, 035101.

M.E.J. Newman, I. Jensen, and R.M. Ziff (2002), Percolation and epidemics in a two-dimensional small world, *Phys. Rev.* E **65**, 021904; cond-mat/0108542.

M.E.J. Newman, C. Moore, and D.J. Watts (2000), Mean-field solution of small-world networks, *Phys. Rev. Lett.* **84**, 3201.

M.E.J. Newman, S.H. Strogatz, and D.J. Watts (2001), Random graphs with arbitrary degree distribution and their applications, *Phys. Rev.* E **64**, 026118; cond-mat/0007235.

M.E.J. Newman and D.J. Watts (1999a), Renormalization group analysis of the small-world network model, *Phys. Lett.* A **263**, 341.

M.E.J. Newman and D.J. Watts (1999b), Scaling and percolation in the small-world network model, *Phys. Rev.* E **60**, 7332.

M. Ozana (2001), Incipient spanning cluster on small-world networks, *Europhys. Lett.* **55**, 762.

S.A. Pandit and R.E. Amritkar (1999), Characterization and control of small-world networks, *Phys. Rev.* E **60**, R1119; chao-dyn/9901017, cond-mat/0004163.

S.A. Pandit and R.E. Amritkar (2001), Random spread on the family of small-world networks, Phys. Rev. E **63**, 041104; cond-mat/0004163.

J.-J. Pansiot and D. Grad (1998), On routes and multicast trees in the Internet, *Comput. Commun. Rev.* **28**, 41.

V. Pareto (1897), *Cours d'Economie Politique* (Macmillan, London).

R. Pastor-Satorras, E. Smith, and R.V. Solé (2002), Evolving protein interaction networks through gene duplication, Working Papers of Santa Fe Institute, 02-02-008,
http://www.santafe.edu/sfi/publications/Abstracts/02-02-008abs.html.

R. Pastor-Satorras, A. Vázquez, and A. Vespignani (2001), Dynamical and correlation properties of the Internet, *Phys. Rev. Lett.* **87**, 258701; cond-mat/0105161.

R. Pastor-Satorras and A. Vespignani (2000), Epidemic spreading in scale-free networks, *Phys. Rev. Lett.* **86**, 3200; cond-mat/0010317.

R. Pastor-Satorras and A. Vespignani (2001), Epidemic dynamics and endemic states in complex networks, *Phys. Rev.* E **63**, 066117; cond-mat/0102028.

R. Pastor-Satorras and A. Vespignani (2002a), Epidemic dynamics in finite size scale-free networks, *Phys. Rev.* E **65**, 035108; cond-mat/0202298.

R. Pastor-Satorras and A. Vespignani (2002b), Optimal immunisation of complex networks, *Phys. Rev.* E 65, 036104; cond-mat/0107066.

R. Pastor-Satorras and A. Vespignani (2002c), Epidemics and immunization in scale-free networks, cond-mat/0205260.

D.M. Pennock, G.W. Flake, S. Lawrence, E.J. Glover, and C.L. Giles (2002), Winners don't take all: Characterising the competition for links on the web, *Proc. Natl Acad. Sci. USA* **99**, 5207.

J. Podani, Z.N. Oltvai, H. Jeong, B. Tombor, A.-L. Barabási, and E. Szathmáry (2001), Comparable system-level organization of Archaeva and Eukaryotes, Nature Publishing Group, nature genetics, advance online publication, http://genetics.nature.com, 1-3.

D.J. de S. Price (1965), Networks of scientific papers, *Science* **149**, 510.

A.R. Puniyani and R.M. Lukose (2001), Growing random networks under constraints, cond-mat/0107391.

A.R. Puniyani, R.M. Lukose, and B.A. Huberman (2001), Intentional walks on scale-free small worlds, cond-mat/0107212.

C. Quince, P.G. Higgs, and A.J. McKane (2001), Food web structure and the evolution of ecological communities, nlin.AO/0105057.

D. Rafiei and A.O. Mendelzon (2000), What is this page known for? Computing Web page reputations, Proceedings of the WWW9 Conference, Amsterdam.

A. Ramzanpour and V. Karimipour (2002), Simple models of small world networks with directed links, cond-mat/0205244.

S. Redner (1998), How popular is your paper? An empirical study of citation distribution, *Eur. Phys. J.* B **4**, 131.

G.J. Rodgers and K. Darby-Dowman (2001), Properties of a growing random directed network, *Eur. Phys. J.* B **23**, 267.

T. Roehl and S. Bornholdt (2001), Self-organized critical neural networks, cond-mat/0109256.

T. Rohlf and S. Bornholdt (2002), Criticality in random threshold networks: Annealed approximation and beyond, *Physica* A **310**, 245; cond-mat/0201079.

M.L. Sachtjen, B.A. Carreras, and V.E. Lynch (2000), Disturbances in a power transmission system, *Phys. Rev.* E **61**, 4877.

A.D. Sánchez, J.M. López, and M.A. Rodríguez (2002), Non-equilibrium phase transitions in directed small-world networks, *Phys. Rev. Lett.* **88**, 048701; cond-mat/0110500.

A. Scala, L.A.N. Amaral, and M. Barthélémy (2001), Small-world networks and the conformation space of a lattice polymer chain, *Europhys. Lett.* **55**, 594; cond-mat/0004380.

N. Schwartz, R. Cohen, D. ben-Avraham, A.-L. Barabási, and S. Havlin (2002), Percolation in directed scale-free networks, Phys. Rev. E **66**, 015104; cond-mat/0204523.

J. Scott (1979), *Social Network Analysis: A Handbook* (Sage Publications, London).

P. Sen and B.K. Chakrabarti (2001), Small-world phenomena and the statistics of linear polymer networks, *J. Phys.* A **34**, 7749; cond-mat/0105346.

Z. Silagadze (1997), Citations and the Zipf-Mandelbrot's law, *Complex Syst.* **11**, 487; physics/9901035.

H.A. Simon (1955), On a class of skew distribution functions, *Biometrica* **42**, 425.

H.A. Simon (1957), *Models of Man* (Wiley, New York).

W. Shockley (1957), On the statistics of individual variations of productivity in research laboratories, *Proc. IRE* **45**, 279; 1409.

F. Slanina and M. Kotrla (1999), Extremal dynamics model on evolving networks, *Phys. Rev. Lett.* **83**, 5587; cond-mat/9901275.

F. Slanina and M. Kotrla (2000), Random networks created by biological evolution, *Phys. Rev.* E **62**, 6170; cond-mat/0004407.

R.V. Solé and J.M. Montoya (2001), Complexity and fragility in ecological networks, *Proc. R. Soc. London* B **268**, 2039; cond-mat/0011196.

R.V. Solé, R. Pastor-Satorras, E.D. Smith, and T. Kepler (2001), A model of large-scale proteome evolution, Working Papers of Santa Fe Institute, 01-08-041, http://www.santafe.edu/sfi/publications/Abstracts/01-08-041abs.html.

S. Solomon and M. Levy (1996), Spontaneous scaling emergence in generic stochastic systems, Int. J. Phys. C **7**, 745.

S. Solomon and S. Maslov (2000), Pareto laws in financial autocatalytic/multiplicative stochastic systems, http://www.unifr.ch/econophysics/.

S. Solomon and P. Richmond (2001), Stability of Pareto-Zipf law in non-stationary economics, *Economics with heterogeneous interacting agents*, ed. A. Kirman and J.B. Zimmermann, Lecture Notes in Economics and Mathematical Systems (Springer, Berlin), p. 141; cond-mat/0012479.

D. Sornette and R. Cont (1997), Convergent multiplicative processes repelled from zero: Power laws and truncated power laws, *J. Phys.* I (France) **7**, 431.

W. Souma, Y. Fujiwara, and H. Aoyama (2001), Small-world effects in wealth distribution, cond-mat/0108482.

D. Stauffer and A. Aharony (1991), *Introduction to Percolation Theory* (Taylor & Francis, London).

M. Steyvers and J.B. Tenenbaum (2001), The large-scale structure of semantic networks: Statistical analyses and a model for semantic growth, cond-mat/0110012.

W.H. Stockmayer (1943/1944), Theory of molecular size distribution and gel formation in branched chain polymers, *J. Chem. Phys.* **11**, 45 (1943); **12**, 125 (1944).

S.H. Strogatz (2001), Exploring complex networks, *Nature* **410**, 268.

P. Svenson (2001), From N'eel to NPC: Colouring small worlds, cs.CC/0107015.

P. Svenson and D.A. Johnston (2002), Damage spreading in small world Ising models, *Phys. Rev.* E **65**, 036105; cond-mat/0107555.

G. Szabó, M. Alava, and J. Kertész (2002), Shortest paths and load scaling in scale-free trees, cond-mat/0203278.

B. Tadić (2001a), Dynamics of directed graphs: The World Wide Web, *Physica* A **286**, 509; cond-mat/0011442.

B. Tadić (2001b), Access time of an adaptive random walk on the World Wide Web, cond-mat/0104029.

B. Tadić (2001c), Adaptive random walks on the class of Web graph, *Eur. Phys. J.* B **23**, 221; cond-mat/0110033.

B. Tadić (2001d), Temporal fractal structures: Origin of power-laws in the World-Wide Web, cond-mat/0112047.

C. Tsallis and M.P. de Albuquerque (2000), Are citations of scientific papers a case of nonextensivity?, *Eur. Phys. J.* B **13**, 777; cond-mat/9903433.

P. Uetz, L. Giot, G. Cagney, T.A. Mansfield, R.S. Judson, V. Narayan, D. Lockshon, M. Srinivasan, P. Pochart, A. Qureshi-Emili, Y. Li, B. Godwin, D. Conover, T. Kalbfleisch, G. Vijayadamodar, M. Yang, M. Johnston, S. Fields, and J.M. Rothberg (2000), A comprehensive analysis of protein–protein interactions in *Saccharomyces cerevisiae*, *Nature* **403**, 623.

S. Valverde, R. Ferrer i Cancho, and R.V. Solé (2002), Scale-free networks from optimal design, cond-mat/0204344.

S. Valverde and R.V. Solé (2002), Self-organized critical traffic in parallel computer networks, *Physica* A **312**, 636.

N. Vandewalle and M. Ausloos (1997), Construction and properties of fractal trees with tunable dimension: The interplay of geometry and physics, *Phys. Rev.* E **55**, 94.

A. Vázquez (2001a), Disordered networks generated by recursive searches, *Europhys Lett.* **54**, 430; cond-mat/0006132.

A. Vázquez (2001b), Statistics of citation networks, cond-mat/0105031.

A. Vázquez, A. Flammini, A. Maritan, and A. Vespignani (2001), Modeling of protein interaction networks, cond-mat/0108043.

A. Vázquez, R. Pastor-Satorras, and A. Vespignani (2002), Large-scale topological and dynamical properties of Internet, *Phys. Rev.* E **65**, 066130; cond-mat/0112400.

D. Volchenkov and Ph. Blanchard (2002), An algorithm generating scale free graphs, cond-mat/0204126.

A. Wagner (2001a), The yeast protein interaction network evolves rapidly and contains few redundant duplicate genes, *Mol. Biol. Evol.* **18**, 1283.

A. Wagner (2001b), How to reconstruct a large genetic network from n gene perturbations in fewer than n^2 easy steps, *Bioinformatics* **17**, 1183.

A. Wagner (2001c), Reconstructing pathways in large genetic networks from genetic perturbations, Working Papers of Santa Fe Institute, 01-09-050, http://www.santafe.edu/sfi/publications/Abstracts/01-09-050abs.html.

A. Wagner (2002), Estimating coarse gene network structure from large-scale gene perturbation data, *Genome Research* **12**, 309.

A. Wagner and D.A. Fell (2001), The small world inside large metabolic networks, *Proc. R. Soc. London* B, 268, 1803.

X.F. Wang and G. Chen (2002), Synchronization in scale-free dynamical networks: Robustness and fragility, *IEEE T. Circuits-I* **49**, 54; cond-mat/0105014.

S. Wasserman and K. Faust (1994), *Social Network Analysis* (Cambridge University Press, Cambridge).

D.J. Watts (1999), *Small Worlds* (Princeton University Press, Princeton, NJ).

D.J. Watts, P.S. Dodds, and M.E.J. Newman (2002), Identity and search in social networks, *Science* **296**, 1302; cond-mat/0205383.

D.J. Watts and S.H. Strogatz (1998), Collective dynamics of small-world networks, *Nature* **393**, 440.

M. Weigt and A.K. Hartmann (2000), The number of guards needed by a museum: A phase transition in vertex covering of random graphs, *Phys. Rev. Lett.* **84**, 6118; cond-mat/0001137.

M. Weigt and A.K. Hartmann (2001a), Statistical mechanics perspective on the phase transition in vertex covering finite-connectivity random graphs, *Theor. Comput. Sci.* **256**, 199; cond-mat/0006316.

M. Weigt and A.K. Hartmann (2001b), Typical solution time for a vertex-covering algorithm on finite-connectivity random graphs, *Phys. Rev. Lett.* **86**, 1658; cond-mat/0009417.

D.R. White and M.E.J. Newman (2001), Fast approximation algorithms for finding node-independent paths in networks, Working Papers of Santa Fe Institute, 01-07-035,
http://www.santafe.edu/sfi/publications/Abstracts/01-07-035abs.html.

R.J. Williams and N.D. Martinez (2000), Simple rules yield complex food webs, *Nature* **404**, 180.

R.J. Williams, N.D. Martinez, E.L. Berlow, J.A. Dunne, and A.-L. Barabási (2001), Two degrees of separation in complex food webs, Working Papers of Santa Fe Institute, 01-07-036,
http://www.santafe.edu/sfi/publications/Abstracts/01-07-036abs.html.

N.C. Wormald (1981a), The asymptotic connectivity of labelled regular graphs, *J. Combinatorial Theor.* B **31**, 156.

N.C. Wormald (1981b), The asymptotic distribution of short cycles in random regular graphs, *J. Combinatorial Theor.* B **31**, 168.

R. Xulvi-Brunet and I.M. Sokolov (2002), Evolving networks with disadvantaged long-range connections, cond-mat/0205136.

S.-H. Yook, H. Jeong, and A.-L. Barabási (2001), Modeling the Internet's large-scale topology, cond-mat/0107417.

S.-H. Yook, H. Jeong, A.-L. Barabási, and Y. Tu (2001), Weighted evolving networks, *Phys. Rev. Lett.* **86**, 5835; cond-mat/0101309.

D.H. Zanette (2001a), Critical behavior of propagation on small-world networks, *Phys. Rev.* E **64**, 050901; cond-mat/0105596.

D.H. Zanette (2001b), Criticality of rumor propagation on small-world networks, cond-mat/0109049.

D.H. Zanette (2002), Dynamics of rumor propagation on small-world networks, *Phys. Rev.* E **65** 041908; cond-mat/0110324.

D.H. Zanette and M. Kuperman (2002), Effects of immunization in small-world epidemics, *Physica* A **309**, 445; cond-mat/0109273.

D.H. Zanette and S.C. Manrubia (2001), Vertical transmission of culture and the distribution of family names, *Physica* A **295**, 1; nlin.AO/0009046.

N. Zekri and J.-P. Clerc (2001), Statistical and dynamical study of disease propagation in a small world network, *Phys. Rev.* E **64**, 056116; cond-mat/0107562.

D. Zheng and G. Ergun (2001), Coupled growing networks, cond-mat/0112052.

G.K. Zipf (1949), *Human Behaviour and the Principle of Least Effort* (Addison-Wesley, Cambridge, MA).

INDEX